Recycled Concrete

Woodhead Publishing Series in Civil and Structural Engineering

Recycled Concrete

Technologies and Performance

Edited by

Vivian W.Y. Tam

Mahfooz Soomro

Ana Catarina Jorge Evangelista

ELSEVIER

WP
WOODHEAD
PUBLISHING
An imprint of Elsevier

Woodhead Publishing is an imprint of Elsevier
50 Hampshire Street, 5th Floor, Cambridge, MA 02139, United States
The Boulevard, Langford Lane, Kidlington, OX5 1GB, United Kingdom

ISBN: 978-0-323-85210-4

For information on all Woodhead Publishing publications visit our website at https://www.elsevier.com/books-and-journals

Publisher: Matthew Deans
Acquisitions Editor: Gwen Jones
Editorial Project Manager: Rafael Guilherme Trombaco
Production Project Manager: Fizza Fathima
Cover Designer: Miles Hitchen

Typeset by TNQ Technologies

Contents

List of contributors

Lara Carral Autonomous University of Aguascalientes, Aguascalientes, Ags., Mexico

Carlos Chastre CERIS, Department of Civil Engineering, Faculdade de Ciências e Tecnologia, Universidade Nova de Lisboa, Caparica, Portugal

Wei Chen School of Civil Engineering and Architecture, Ningbo Technical University, Ningbo, Zhejiang Province, People's Republic of China

Jorge de Brito CERIS, Department of Civil Engineering, Architecture and Georesources, Instituto Superior Técnico, Universidade de Lisboa, Lisbon, Portugal

Ana Catarina Jorge Evangelista Engineering Institute of Technology, Perth, WA, Australia

Luís Evangelista CERIS, Department of Civil Engineering, Instituto Superior de Engenharia de Lisboa, Rua Conselheiro Emídio Navarro, Lisbon, Portugal

Karoline Figueiredo Programa de Engenharia Ambiental (PEA), Universidade Federal do Rio de Janeiro (UFRJ), Rio de Janeiro, Brazil

Fernando L. Gayarre University of Oviedo, Gijón/Xixón, Asturias, Spain

Assed Haddad Programa de Engenharia Ambiental (PEA), Universidade Federal do Rio de Janeiro (UFRJ), Rio de Janeiro, Brazil

Ruoyu Jin School of Built Environment and Architecture, London South Bank University, London, United Kingdom

Luiz Sarti Junior Universidade Federal de São Carlos (UFSCar), São Carlos, São Paulo, Brazil

Katarzyna Kalinowska Faculty of Civil and Environmental Engineering, Bialystock University of Technology, Poland

Ni Nyoman Kencanawati Civil Engineering, University of Mataram, Mataram, Indonesia

Nariman J. Khalil Department of Civil & Environmental Engineering, University of Balamand, Koura, Lebanon

Oleksandr Kovalchuk V.D.Glukhovsky Scientific Research Institute for Binders and Materials, Kyiv National University of Construction and Architecture, Kyiv, Ukraine

Pavel V. Krivenko V.D.Glukhovsky Scientific Research Institute for Binders and Materials, Kyiv National University of Construction and Architecture, Kyiv, Ukraine

Carlos López-Colina University of Oviedo, Gijón/Xixón, Asturias, Spain

Rebeca Martínez University of León, León, Spain

Takao Namihira Institute of Industrial Nano Materials, Kumamoto University, Japan

João Pacheco CERIS, C5Lab - Sustainable Construction Materials Association, Edifício Central Park, Linda a Velha, Portugal

Sheyla Serra Universidade Federal de São Carlos (UFSCar), São Carlos, São Paulo, Brazil

Miguel A. Serrano University of Oviedo, Gijón/Xixón, Asturias, Spain

Mitsuhiro Shigeishi Department of Civil and Environmental Engineering and Architecture, Kumamoto University, Japan

Carlos Soares Universidade Federal Fluminense (UFF), Niterói, Rio de Janeiro, Brazil

Mahfooz Soomro School of Engineering, Design and Built Environment, Western Sydney University, Penrith, NSW, Australia

Jesús M. Suárez University of Oviedo, Gijón/Xixón, Asturias, Spain

Vivian W.Y. Tam School of Engineering, Design and Built Environment, Western Sydney University, Penrith, NSW, Australia

Victoria Zozulynets V.D.Glukhovsky Scientific Research Institute for Binders and Materials, Kyiv National University of Construction and Architecture, Kyiv, Ukraine

Acknowledgment

This study was financed in part by the Coordenação de Aperfeiçoamento de Pessoal de Nível Superior—Brasil (CAPES)—Finance Code 001.

Introduction

1

Ana Catarina Jorge Evangelista [1], *Vivian W.Y. Tam* [2] *and Mahfooz Soomro* [2]
[1]Engineering Institute of Technology, Perth, WA, Australia; [2]School of Engineering, Design and Built Environment, Western Sydney University, Penrith, NSW, Australia

1.1 Construction and demolition waste – overview

Worldwide, the construction sector is one of the principal consumers of natural resources and presents a significant carbon footprint related to the production and transportation of construction materials, e.g., cement, and aggregates. Due to COVID-19, the construction industry has been negatively impacted considering the lockdown and restrictions by various governments, most of the projects were halted since most of the construction could not get the raw materials in time. Naturally, the aggregate market is affected on a short-term basis, but the market is recovering especially in Asia–Pacific after the relaxation of lockdown in many countries and the global outlook to 2030 indicates that the aggregates production will be driven by three key factors, namely, increase in global population (reaching 8.5 billion by 2030), continued economic development and increase in global urbanisation (reaching 60% by 2030). Accordingly, it is estimated that global aggregates production will increase to some 60 billion tonnes by 2030. According to O'Brien (2019), the strongest growth will be in Asia, particularly in India, with lower but positive growth in Africa and Latin America.

It is important to mention that the land for landfill becomes scarce and the world demand for concrete aggregate has reached 60 billion tons, hence, sustainable alternatives such as, using C&D waste is gaining importance due to legislation and availability. Regulations and legislation by the governments in various countries around the world have structured and constituted a market for building materials and products derived from the C&D waste streams (Tam et al., 2018).

As per the definition of EPA USA ("Construction and Demolition Debris: Material-Specific Data | US EPA,"), the Construction and Demolition (C&D) materials consist of the debris generated during the construction, renovation, and demolition of buildings, roads, and bridges, often contains bulky, heavy materials such as Concrete, and Wood (from buildings), Asphalt, Gypsum, Metals, Bricks, Glass and Plastics. Usually, C&D waste is divided into five main fractions i.e., metal, concrete and mineral, wood, miscellaneous and unsorted mixed fractions. More precisely, it may contain the following; (Tam et al., 2018):

- Concrete
- Bricks, tiles and ceramics
- Wood
- Glass

Recycled Concrete. https://doi.org/10.1016/B978-0-323-85210-4.00002-3

- Plastic
- Bituminous mixtures and tars
- Metals (ferrous and nonferrous)
- Soils (contaminated) and stones
- Insulation materials (including asbestos)
- Gypsum based materials (including plasterboard)
- Waste electronic and electrical equipment
- Chemicals (including solvents)
- Packaging materials
- Hazardous substances

Lu et al. (2017) observed that in developed economies, the construction industry generally accounts for around 25%−30% of the total solid waste disposal and it has damaged the image of construction industry as one of the main contributors to environmental impacts around the globe. Furthermore, emerging economies such as Brazil, China, India, Russia, and South Africa are facing the challenges of economic growth and the need to avoid depletion of natural resources to sustain the construction activities. The literature (Erebor et al., 2019; Supino et al., 2016; Zeng et al., 2018) recommends several measures of waste management to reduce the disposal in landfills, however, many countries lack efficient C & D waste management. The circular economy, which includes the C&D waste recycling industry, is relatively new to emerging economies (Bao & Lu, 2020).

More recently, the concept of circular economy has been applied in developed economies, e.g., in Europe, and some countries have managed to reduce to zero or close to zero the C & D waste landfills (Bao & Lu, 2020). Many EU countries have succeeded in establishing markets for recovered C&D materials. This indicates that the European construction sector is highly circular.

In Europe, the construction and demolition waste generated reached 700 million tonnes per year. Eurostat estimated that a total of 857.2 million tons of C & D waste was generated in the EU-27 Member States in 2010. In 2012, the largest declared quantities of waste were generated in the countries that have the largest population, such as France and Germany (Iacoboaea et al., 2019). However, to comply with Waste Framework Directive (Directive (2008/98/EC), 2008), the European countries put significant effort to achieve the 70% recovery target by 2020, with most countries already exceeding the target in 2016 (European Enviroment Agency, 2020).

It is estimated that China produces about 1.5 billion tons of C&D waste every year and the United States produces 569 million tons/year. In some developed economies, the C&D waste reuse and the recycling rate has reached 70%−95% while in China it is just 5% (Bao & Lu, 2020). In India and Brazil, due to inadequate recycling facilities, illegal dumping of C&D waste is a common practice in the absence of regulation and a more effective landfill levy (Tam et al., 2018).

In Australia, the management of C&D waste is not legislated by the Central Government but the environmental issues, including all waste streams, are primarily the responsibility of Australian state and territory governments. From 2006 to 2007, in Australia 43.78 million tons of waste were generated, 38% of which was from the C&D waste stream (Tam et al., 2018). However, in most Australian jurisdictions, a landfill levy is applied to discourage waste disposal and encourage waste recovery. It is promising to embrace the reuse and recycling of materials by imposing a landfill levy on the disposal of waste (Shooshtarian et al., 2020).

1.2 Sustainability and the benefits of LCA – recycled concrete

The concept of sustainability first came in 1987 in the document entitled "Our Common Future through the Brundtland Report: Humanity can make development sustainable – to ensure that it meets the needs of the present without compromising the ability of future generations to meet their own needs". This document was prepared by the World Commission on Environment and Development and created by the United Nations (UN). The term Sustainable Development has emerged in this document and refers to development that meets present needs without compromising the ability of future generations to meet their own needs (Brundtland, 1987).

In 2015, the United Nations established 17 sustainable development goals for 2030, which were duly adopted worldwide. Two of these directly affect buildings and infrastructure. They recognise the social impact of developed world actions on the under-developed world's progressing welfare and prosperity and encourage more responsible and dispersed development in infrastructure and industrialisation (UN Commission for Sustainable Development (2015)- Sustainable Development Goals: Goal 9). Ethical procurement and sourcing of materials is an important part of social sustainability to ensure renewal and equitable conditions of production and reward. There is a need to track production and identify unfair social conditions (Fewings & Henjewele, 2019).

Environmental issues are gaining importance and are increasingly being discussed worldwide. With this, architecture and construction needed to adapt to this new reality. Several countries have developed systems for evaluating sustainable buildings. Economic and social issues are crucial for sustainable development and also deserve special treatment. Worldwide, the academics and researchers have shown their concerns around the topic of Sustainability and the Construction industry.

In order to promote the sustainable development of the construction engineering discipline, it is important to simultaneously account for all sustainability pillars in a fully harmonious way. In addition, the concern to minimise the impacts must be related to the whole project life cycle, from the extraction and processing of the raw materials to the final disposal of the project (or its elements). This gave rise to life cycle analysis approaches, focusing on environmental, economic, and social impacts throughout the various cycles involved in a project.

Since the late 1960s, researchers have been developing a basis for the current methods of life cycle analysis (Albertí et al., 2017). Initially, the impacts of interest were energy consumption, solid waste production, and air and water pollutants. The concept of Environmental Life Cycle Assessment (E-LCA) began to be implemented, commonly referred to as Life Cycle Assessment (LCA).

Life cycle assessment is one of the most used methodologies to compute the damage impact categories of building materials. This method analyses the different life-cycle phases of products from the extraction of raw materials and manufacturing, packaging and transporting to the site, constructing and installing and operation, until demolition and recycling (Figueiredo et al., 2021).

The LCA methodology is based on ISO 14040 and 14044 guidelines covering: Goal and Scope aiming to identify the functional equivalent, system boundary, and the set of building materials, Life Cycle Inventory (LCI)'s main challenge is the difficulty of collecting reliable and applicable data Life Cycle Impact Assessment (LCIA), and Interpretation. The quantities of materials, energy consumption, and input data are gathered using the sustainability indicators. The final step is Interpretation, which allows for the evaluation of the results achieved from LCI and LCIA (UNEP, SETAC, & LC-Initiative, 2009).

Over the recent year, researchers have been using LCA to evaluate the environmental impacts of Portland cement, concrete, and aggregate and compare these against recycled concrete and recycled aggregate. Devaki and Shanmugapriya (2022) found that almost 70% of LCA studies were done in European countries; this shows the deficiencies in research in this area in other countries, especially in developing countries, for example in India.

Zhang et al. (2019) pointed out that LCA is an important tool to evaluate the environmental impact of recycled aggregate concrete (RAC) to provide support in decision making when selecting environment-friendly products, but special attention is needed considering five main aspects:

- A clear indication of the mixture design methods of RAC should be given at first, and then, the comparison of the environmental impact of RAC can be conducted based on the same mixture design method;
- The selection of functional unit has a significant influence on the LCA results of RAC. A benchmark method to define the functional unit is urgently needed to make different LCA studies comparable;
- A method that combines the effect of quality, mass, and market price on the allocation procedure should be further investigated in future research to make the assessment result of RAC more scientific and reasonable;
- Ignoring CO_2 uptake in the life cycle of RAC will lead to an overestimated CO_2 emission.
- The relationship between the total environmental impact of concrete and the transport distance of recycled aggregate (RA) has been established. Based on this relationship, a critical distance (Dcri) for recycled aggregate (RA) (the environmental impact of RAC is less than that of NAC) can be obtained.

Considering sustainability of concrete, it is important to highlight the efforts by the cement industry to reduce greenhouse gas emissions and embodied energy. According to ACI (American Concrete Institute Committee E–701, 2013), there are increased uses of alternative cementitious materials, reduction in embodied energy, beneficial reuse/recycle of materials and reduction in clinker factor. In summary "discussing the sustainability of concrete, it is important to recognize that concrete is more sustainable than cement. It is an engineered system that can be designed or optimized for any number of properties. In that design process, a number of decisions must occur that can have a profound effect on sustainability."

1.3 Outline of the book

Cement is one of the most important building materials in the world. It is used mainly as a binding and strength-giving medium for the production of conventional and recycled concrete. Since the title of the book deals with various aspects of recycled concrete, hence cement has been discussed in detail in Chapter 2. This chapter deals with cement manufacture, cement properties, different types of cement, blended cements, composite cements (mostly produced in European countries), manufactured in United States of America, Europe and Australia and the life cycle assessment of different types of cement.

After detailed evaluation of the entire life cycle of cement production, which evaluates impacts of producing cement from raw materials (their extraction, crushing, grinding) to processing of cement, packaging and shipping without taking into consideration the end use and disposal of cement. The cradle to gate GHG emissions evaluated for CEM-1 show total GHG effect of 0.817 kg/kg of cement produced, while blended cements provide higher environmental savings of approximately 5% (utilisation of CKD for sequestration) in cement.

The use of pozzolan, industry by-product and agro-waste containing siliceous, alumino-siliceous and calcium alumino-siliceous compounds, as partial substitution for cement or in concrete mixture proportioning is not only an effective strategy for cost effectiveness, but it improves the durability performance of concrete and most importantly, reduce the CO_2 emissions associated with concrete industry. Chapter 3 discusses the production, physical and chemical properties of industrial and agro waste, their use in cement and concrete industry and the improvement in the durability of concrete structures and concrete products.

Chapter 4 describes the chemical and physical properties of some granular materials, which due to their nonhazardous and mainly inert nature can be used in the manufacture of concrete. The initial sections present the characteristics of the recycled aggregate obtained from construction and demolition waste. Considering the current trend of circular economy in the construction sector, there are incentives to enhance the recycling practices, and it is a common practice to reduce the amount of urban solid waste. On the other hand, despite the COVID-19 restrictions on the construction activities, there is a predicted annual growth rate of 3.3% showing the need for recycled materials to avoid depletion of a natural resource.

Various programs were created to circulate the economy, encourage the reuse of building components and materials after selective demolition, and stimulate the sale of these products. In this sense, an evaluation of the chemical and physical properties of some granular materials, which due to their nonhazardous and mainly inert nature can be used in the manufacture of concrete, are presented. In order to enhance the application of recycled materials, it is important to identify the different types of waste and production according to their specific country and the region.

Recycled aggregate, produced only by crushing waste concrete into coarse or fine aggregate, do not have the same quality as natural aggregate, because of the high water absorption and low density of recycled aggregate. This is due to the presence of old residual mortar on the surface of the recycled aggregate. Chapter 5, presents the treatments applied to remove the adhered mortar on recycled concrete aggregate (RCA) covering the technologies needed to improve the RCA quality to conform to the specifications regulated by various countries around the world and to make the aggregate useful for utilisation in the nonstructural, precast as well as structural concrete.

The huge popularity of concrete in today's world leads to the search for effective methods allowing for the processing of concrete rubble into coarse aggregates and separated cement mortar that will allow the preservation of valuable natural resources.

Chapter 6 deals with the suitability of recycled aggregate concrete (RAC) for structural applications, with emphasis on coarse recycled aggregates produced from concrete waste. The feasibility of RAC has been consistently reported worldwide and technical bodies are becoming increasingly receptive to the use of recycled aggregates for structural concrete. The application of RAC has been considered worldwide and technical bodies are more open to the use of recycled aggregates in structural concrete.

However, the developers must accept that the structural behavior of RAC should conform to the code of practice, and currently, there is still a gap and this prevents the proper guidelines for its structural design in order to enhance its usage. The main findings show that the structural design of RAC may be unsafe when code provisions developed for natural aggregate concrete are applied, namely when design concerns the bond length of anchorages or lap splices or shear resistance. This chapter deals with the structural behavior of concrete made with coarse recycled aggregates produced from concrete waste. This is relevant because, despite the abundant research arguing in favor of RAC as a structural material, clear design guidelines are lacking and comprehensive research on the understanding of the influence of recycled aggregates on structural behavior is needed for the development of such guidelines.

Chapter 7 presents the applications of recycled aggregate (RA) for non-structural applications, which are a common practice of reusing and recycling construction and demolition (C&D) waste. One of the most common uses of recycled aggregate is its nonstructural applications. RA can be used in the mix design of new concrete. Nonstructural concrete is the type of concrete with generally lower strength (i.e., lower than 25 MPa), such as permeable concrete, sidewalks, and other elements in buildings. In this chapter several case studies are presented aiming to explore the current trend in existing research and field practices in adopting RA for different types of nonstructural elements in building and infrastructure construction. It can be pointed out that more emerging practices for using RA in various nonstructural applications can be adopted by following the different circular strategies and by extending the traditional down cycling approach.

In fact, it is worth for a more comprehensive investigation from technical, economic, environmental, and other dimensions to evaluate the feasibility of using RA in various nonstructural applications. Various factors for adopting RA into new applications are listed in this chapter, including legislation, market, social, technical, financial, and educational. It is therefore important to evaluate the use of RA by considering both, the technical research and practical implementation.

RAC has different properties and exhibits long-term durability problems. Chapter 8 discusses the RAC durability properties and its behavior under various environmental conditions. Worldwide several studies have favoured the use of recycled aggregate as a potential solution to mitigate environmental impacts, and its characteristics can have a significant influence on the durability of recycled concrete. Therefore, it is essential to understand the mechanism by which such factors influence the service life of concrete structures and how recycled concretes differs from conventional concrete made with natural aggregates.

Current studies demonstrated remarkable improvement in the properties of treated RCA when compared to untreated RCA. However, it is noted that a limited research has been done to investigate the effect of different types of chemicals and solutions for coating and permeation in RCA. Also, there are promising technologies for RCA treatment using chemical permeation combined with compressed CO_2 conditioning. More research needs to be conducted on the long-term mechanical properties of RAC and its durability behavior, to assess the RAC performance from both material and structural aspects.

Chapter 9 deals with an alternative to solve red mud disposal issues to the environment. The increase in the production of aluminum metal has been increasing this issue and it is likely to continue for the coming years considering its peak demand for several industrial applications in the transport, construction, packaging, electrical and other sectors. However, the production process also results in the production of significant amounts of a highly alkaline, hazardous and toxic waste called bauxite residue or red mud. Therefore, the problems related to the toxicity of red mud and its disposal has become an increasingly important matter, attracting continued research into remediation methods by recycling it into useful products or by devising much safer storage methods. The main goal of the research work presented in this chapter is to optimize the formation process of alkali activated cements and concrete in order to allow the incorporation of a large portion of red mud.

The red mud is by-product waste, which can be used as precursor active component in alkali-activated cements and concrete. However, it contains alkalis, hazardous chemicals and heavy metals, limiting its use in the production of Ordinary Portland cements. Use of red mud in alkali-activated materials makes it possible to obtain binder cement with red mud content up to 60% by mass, meeting the requirements of cement Grades 32.5 and 42.5 at the same time. Also, red mud could be used as fine aggregate in conventional concretes, where red mud content could be 15%−40% by mass (350−900 kg/m^3) for concrete classes C15−C40. For use in roller compacted road base, red mud content could reach 90% by mass of the material, meeting the specified requirements for road base construction.

Concrete has been increasingly investigated regarding environmental and sustainability assessments. The amount of CO_2 and other greenhouse gases generated in the manufacture of concrete materials can be minimized if the concrete industry's manufacturing processes are more efficient. However, it is not sufficient to evaluate the environmental impacts of the production process. In this context, the Life Cycle Assessment (LCA) methodology seems to be a suitable tool to assess the potential environmental impacts of structural and nonstructural concrete. Chapter 10 presents exploratory studies in the literature that try to substitute ordinary Portland cement with other less damaged materials for the environment. Additionally, the use of recycled aggregate to replace natural aggregates in concrete has proven to be a great solution not only for the environment but also for saving on construction costs. Some recent studies of concrete LCA present a comparison between natural aggregates (NA) and recycled aggregate (RA), proving the advantages obtained in favor of using RA replacing NA.

Through improved and consistent data collection and monitoring methods, it will be possible to accurately measure the environmental impacts caused during the concrete life cycle. In this context, the LCA application is an excellent way to improve the management of the environmental impacts and identify improvements through the concrete value chain.

Chapter 11 presents the main issues related to Asset Management of Recycled Concrete (RC) applications. The terms related to asset management are conceptualized by presenting the most widespread asset management systems, standards, and decision-support tools. Concrete is one of the most used materials in the modern world, and losses inevitably occur during its use, whether in the stages of manufacturing, transport, use, and selective demolition. These steps generate waste, which can be reused or recycled to compose new products. The concrete waste can be processed to RCA when the residues meet specific characteristics and situations. RCA and derivative products can be classified as "assets" of an organisation. Asset management is conceptualised by presenting the most widespread asset management systems, standards, and decision-support tools.

The asset management of recycled concrete applications is a challenging topic due to the diversity of conceptual approaches related to asset management and the characteristics and specificities of recycled concrete. Therefore, the strategy considered by the authors as the most appropriate to contribute to the theme, was to make an expanded approach to the aspects associated with asset management of recycled concrete applications.

References

Albertí, J., Balaguera, A., Brodhag, C., & Fullana-i-Palmer, P. (2017). Towards life cycle sustainability assessment of cities. A review of background knowledge. *Science of the Total Environment, 609*, 1049–1063. https://doi.org/10.1016/j.scitotenv.2017.07.179

American Concrete Institute Committee E-701. (2013). American concrete Institute education bulletin E3-01 cementitious materials for concrete. *Concrete*, 1–25. Retrieved from http://www.concrete.org/general/fE3-01.pdf.

Bao, Z., & Lu, W. (2020). Developing efficient circularity for construction and demolition waste management in fast emerging economies: Lessons learned from Shenzhen, China. *Science of the Total Environment, 724*, 138264. https://doi.org/10.1016/j.scitotenv.2020.138264

Brundtland, G. (1987). UN Brundtland commission Report. *Our Common Future*. https://sustainabledevelopment.un.org/content/documents/5987our-common-future.pdf.

Construction and Demolition Debris: Material-Specific Data | US EPA. (n.d.). Retrieved January 4, 2022, from https://www.epa.gov/facts-and-figures-about-materials-waste-and-recycling/construction-and-demolition-debris-material.

Devaki, H., & Shanmugapriya, S. (2022). LCA on construction and demolition waste management approaches: A review. *Materials today: Proceedings*. https://doi.org/10.1016/j.matpr.2022.03.286

Directive (2008/98/EC). (2008). Directive 2008/98/EC of the European parliament. In *Official Journal of the European union*. Retrieved from http://eur-lex.europa.eu/LexUriServ/LexUriServ.do?uri=CELEX:32008L0098:EN:NOT.

Erebor, E., Ibem, E. O., & Adewale, B. A. (2019). Current research trends on sustainable construction project delivery. *International Journal of Mechanical Engineering and Technology, 10*(1), 1973–1986.

European Enviroment Agency. (2020). *Construction and demolition waste : Challenges and opportunities in a circular economy. Briefing No. 14/2019, (January), 8.* Retrieved from https://www.eea.europa.eu/publications/construction-and-demolition-waste-challenges/at_download/file.

Fewings, P., & Henjewele, C. (2019). Construction project management. In *Construction Project Management*. https://doi.org/10.1201/9781351122030

Figueiredo, K., Pierott, R., Hammad, A. W. A., & Haddad, A. (2021). Sustainable material choice for construction projects: A life cycle sustainability assessment framework based on BIM and Fuzzy-AHP. *Building and Environment, 196*, 107805. https://doi.org/10.1016/j.buildenv.2021.107805

Iacoboaea, C., Aldea, M., & Petrescu, F. (2019). Construction and demolition waste-a challenge for the European Union? *Theoretical and Empirical Researches in Urban Management, 14*(1), 30–52.

Lu, W., Webster, C., Peng, Y., Chen, X., & Zhang, X. (2017). Estimating and calibrating the amount of building-related construction and demolition waste in urban China. *International Journal of Construction Management, 17*(1), 13–24. https://doi.org/10.1080/15623599.2016.1166548

O'Brien, Jim (2019). Global aggregate growth examined. *Aggregateresearch.com*, 1–5. pp. 1–5.

Shooshtarian, S., Maqsood, T., Khalfan, M., Yang, R. J., & Wong, P. (2020). Landfill levy imposition on construction and demolition waste: Australian stakeholders' perceptions. *Sustainability (Switzerland), 12*(11). https://doi.org/10.3390/su12114496

Supino, S., Malandrino, O., Testa, M., & Sica, D. (2016). Sustainability in the EU cement industry: The Italian and German experiences. *Journal of Cleaner Production, 112*, 430–442. https://doi.org/10.1016/j.jclepro.2015.09.022

Tam, V. W. Y., Soomro, M., & Evangelista, A. C. J. (2018). A review of recycled aggregate in concrete applications (2000–2017). *Construction and Building Materials, 172*, 272–292. https://doi.org/10.1016/j.conbuildmat.2018.03.240

UNEP, S. E. T. A. C., & LC-Initiative. (2009). Guidelines for social life cycle assessment of products. In *Management, 15*. Retrieved from http://www.unep.fr/shared/publications/pdf/DTIx1164xPA-guidelines_sLCA.pdf.

Zeng, N., Liu, Y., Mao, C., & König, M. (2018). Investigating the relationship between construction supply chain integration and sustainable use of material: Evidence from China. *Sustainability (Switzerland), 10*(10). https://doi.org/10.3390/su10103581

Zhang, Y., Luo, W., Wang, J., Wang, Y., Xu, Y., & Xiao, J. (2019). A review of life cycle assessment of recycled aggregate concrete. *Construction and Building Materials, 209*, 115–125. https://doi.org/10.1016/j.conbuildmat.2019.03.078

Production of cement and its environmental impact

2

Mahfooz Soomro [1], *Vivian W.Y. Tam* [1] *and Ana Catarina Jorge Evangelista* [2]
[1]School of Engineering, Design and Built Environment, Western Sydney University, Penrith, NSW, Australia; [2]Engineering Institute of Technology, Perth, WA, Australia

2.1 Introduction

Global cement production in 2018 was approximately 3.99 billion tons, with the cement production volume contraction 2.7% compared to 2017 due to slower growth in China, whilst production in India has been growing significantly. China still produces 54.5% of the world's cement, Cembureau members representing 6.4%, EU28 representing 4.4%, CIS representing 2.6%, Americas representing 6.8%, Asia (excluding China) representing 23.9% of the world's production (Cembureau, 2019). Fig. 2.1 shows the world cement production by countries and the main regions in 2018.

Hydraulic cements consist essentially of Portland cement and its several modifications. To understand the properties of Portland cement, it would be helpful to acquire some knowhow on its manufacturing process, its chemical and mineralogical composition and the reactivity of its constituent compounds such as calcium silicates and calcium aluminates. The properties of concrete containing Portland cement develop as a result of chemical reactions between the Portland cement compounds and water, because the hydration reactions are accompanied by the formation of cement hydration products i.e., calcium silicate hydrate (gel), calcium hydroxide (portlandite) and calcium aluminate hydrate.

In this chapter, hydration reactions of silicate compounds (C_3S or Alite) and (C_2S or Belite) with their influence on strength development, the aluminate compounds with their influence on setting behavior of cement are discussed. The relationship between the chemistry of reactions and physical aspects of setting and hardening of Portland cement are explained. Classification of Portland cement types and cement specifications are reviewed as well.

2.2 Cement manufacture

Portland cement is essentially a calcium silicate cement, which is produced by firing in a kiln; a well-homogenized and finely ground mixture of limestone or chalk (calcium carbonate) and an appropriate quantity of clay or shale (containing alumina (Al_2O_3), iron oxide (Fe_2O_3), and alkalis) to partial fusion, at temperatures of approximately 1450°C. The composition of raw mix is fine-tuned by the addition of iron oxide and/or sand.

Recycled Concrete. https://doi.org/10.1016/B978-0-323-85210-4.00010-2

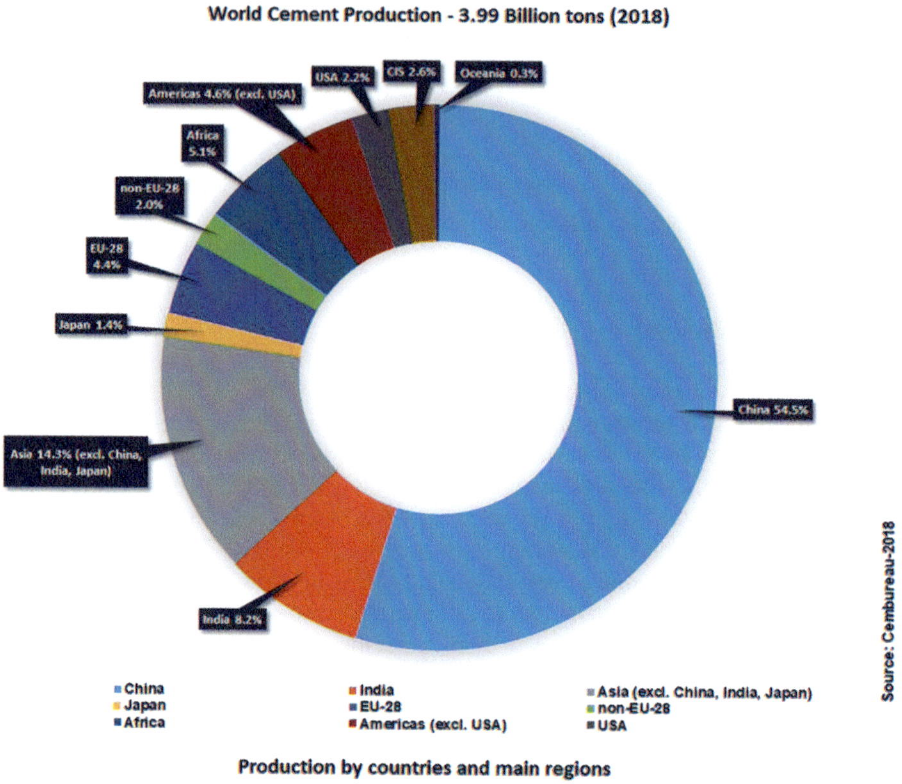

Figure 2.1 World cement production by countries and main regions (Cembureau, 2019).

The first calcium silicate cements were produced by the Greeks and Romans, who discovered that volcanic ash, if ground finely and mixed with lime and water, produced a hardened mortar, which was resistant to weathering. The reaction known as the pozzolanic reaction, was the basis of the contribution made to strength and concrete performance by pozzolanic materials such as fly ash, blast furnace (metallurgical) slag, microsilica, metakaolin and bio-organic ash (rice husk ash, bagasse ash) in modern concrete.

The cements produced in the first half of the nineteenth century did not have the same constituent composition as modern Portland cements because the lower temperatures achieved in the shaft kilns was not high enough for the formation of main constituent mineral, tri-calcium silicate (Alite) of modern cements. The only silicate present in the shaft kiln cement was a less reactive di-calcium silicate (Belite).

It was with the introduction of the rotary kiln at the end of the nineteenth century that enabled a homogeneous product (clinker) to be manufactured at consistently high temperatures ($\sim 1450°C$) ensuring the formation of higher proportions of tri-calcium silicates (Alite) from di-calcium silicates (Belite) producing improved strength development and durability by ensuring the formation of higher proportions of tri-calcium silicate from di-calcium silicate thereby reducing the proportion of di-calcium silicate.

The control of clinker composition has vastly improved to precise control using online conveyor X-ray fluorescence analyzing techniques on the production line as well as laboratory X-ray fluorescence equipment compared to the volume proportions, which were arrived at by trial and error in the late nineteenth century. The continuous improvements in manufacturing methods and quality control combined with market competitive pressures have resulted in a substantial increase in the 28-day strengths of Portland cement, since the late nineteenth century (Blezard, 2004).

2.2.1 Wet process

In the early era of rotary kilns, the burning of slurry of preground raw mix, which was homogenized in slurry basins with 30%–35% water (wet process) then fed in very long rotary kilns (~200 m). This process produced good quality clinker, but the process was highly energy intensive and carried a heavy fuel penalty, since the energy was also consumed to evaporate around 35% moisture content from the slurry together with the partial fusion of raw materials.

The fuel consumption was high (1350–1500 kcal/kg clinker), but the electrical energy consumption (in kWh/t) was less compared to dry process. To save on fuel costs, the wet process is rarely used these days and the shift is toward dry process.

To reduce the energy consumption in the burning process (pyro-processing), new processes were introduced, which used lower amounts of water for instance, semiwet and semidry processes.

2.2.2 Semiwet process

In the semiwet process, the kiln was fed with raw meal in the form of partially moisturized cakes containing 15%–20% moisture after partial dewatering of slurry by filtration (filter press). Heat consumption in this process was 1000–1200 kcal/kg clinker.

This process had the advantage of partial fuel saving; however, this process was not widely accepted in the industry due to additional energy consumption and high maintenance cost of slurry filtration unit.

2.2.3 Semidry process

This process was especially developed to counter the main drawback of the wet process i.e., high fuel consumption. In semidry process, the raw materials ground in dry condition were homogenized and then nodulized in a pan nodulizer of rotary dish type by spraying controlled quantity of water (10%–12%) on the pan nodulizer. The nodules thus produced were fed to the rotary kiln.

The semidry process can only be applied to raw materials having proper plasticity for producing nodules of adequate strength. It was generally adopted where alkali content in raw materials and fuel was on the higher side and raw material properties impeded the preparation of raw mix in the dry condition. The raw meal nodules were fed to the short rotary kiln with the help of hot traveling grates, which further reduced the water content of nodules before entering the rotary kiln.

2.2.4 Dry process

In the second half of the twentieth century significant advances were made to further reduce the energy consumption in cement manufacturing process, which have culminated in the development of precalciner dry process kilns. In this type of kiln, dry prehomogenized raw mix is fed through the precalciners to decarbonate the limestone present in the raw mix before entering the rotary kiln thereby effectively reducing the energy consuming stage. Heat consumption for wet process was higher (1350–1500 kcal/kg clinker) whereas for the dry process it may vary from 720 to 1100 kcal/kg clinker depending upon the number of preheaters, precalciners and the type of kiln being used.

Although the wet process requires approximately 18%–26% more energy than the dry process, in some cases it becomes mandatory to use the wet process for ecological and/or technological reasons. In all processes, rigid control of the composition of the raw meal for kiln feed is attained through continuous analysis of raw materials by X-ray Fluorescence, at various stages of the operation to attain the required composition.

The precalcination of the kiln feed brings many advantages, the most important of which is high kiln output from a relatively short and small-diameter rotary kiln. Almost all new kilns installed since 1980 have been of preheater, precalciner type.

The raw materials are ground to a required fineness, typically in the range of 10%–30% retained on a 90-micron sieve, which will enable satisfactory mixing and homogenization of the raw meal. The homogenized raw meal is introduced into the top of the suspension preheater tower and passes downwards through a series of suspension preheater cyclones to the precalciner vessel. The raw meal passing downwards is suspended in the hot gas stream and heat exchange is rapid. In the precalciner vessel the raw meal is flash heated to ~900°C and although the material residence time in the vessel is only a few seconds, approximately 90% of the limestone in the raw meal is decarbonated before entering the rotary kiln.

In the rotary kiln the precalcined raw meal is heated to ~1450°C and partially melted while, due to the tumbling action of the kiln, it is converted into granular material known as clinker. As the kiln feed moves forward through the final few meters in the kiln, its temperature drops rapidly, and it emerges from the rotary kiln at ~1200°C as clinker, dark gray colored nodules about 6–50 mm in diameter. The clinker drops in the grate cooler where it is cooled to ~90 to ~60°C before going to storage yard, which, after complete cooling is interground with gypsum (calcium sulfate) to the Blaine (surface area) of 330 m^2/kg to 360 m^2/kg, to produce cement. The gypsum (calcium sulfate) is added to control the initial hydration reactions of the cement and prevent it from rapid or flash setting (Fig. 2.2).

The high-temperature reactions for the formation of cement clinker have been described as follows (Bentur, 2003):

- Decomposition of the clay minerals (~500–800°C)
- Decomposition of the calcite (~700–900°C)
- Reactions of the calcite (or lime formed from it), SiO_2, and the decomposed clays to form $2CaO \cdot SiO_2$ (~1000–1300°C)

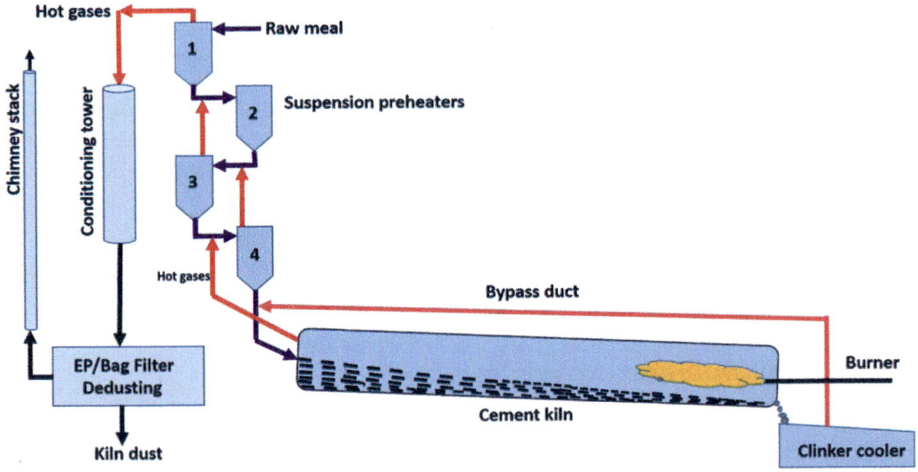

Figure 2.2 Schematic of cement manufacture (dry process) (Soomro & Hung, 2013).

- Clinkering reactions at about 1300–1450°C to form $3CaO \cdot SiO_2$, a melt of aluminate and ferrite is formed to act as a flux to facilitate the formation of $3CaO \cdot SiO_2$ by the reaction between CaO and $2CaO \cdot SiO_2$
- Cooling back to ambient temperature, during which time the melt crystallizes to form the ferrite and aluminate phases.

The chemical reactions taking place in the cement kiln may be summarized as follows:

$$\text{Limestone} \rightarrow CaO + CO_2$$
$$\text{Clay} \rightarrow SiO_2 + Al_2O_3 + Fe_2O_3$$

$$\boxed{\text{Clay + Limestone}} \begin{cases} \rightarrow 3CaO \cdot SiO_2 \text{ (tri-calcium silicate, } \textbf{Alite}) \\ \rightarrow 2CaO \cdot SiO_2 \text{ (di-calcium silicate, } \textbf{Belite}) \\ \rightarrow 3CaO \cdot Al_2O_3 \text{ (tri-calcium aluminate, } \textbf{Aluminate}) \\ \rightarrow 4CaO \cdot Al_2O_3 \cdot Fe_2O_3 \text{ (tetra-calcium alumino-ferrite, } \textbf{Ferrite}) \end{cases}$$

The air that cools the clinker in the grate cooler is used to improve the thermal efficiency of the process, by feeding the grate-cooler preheated combustion air to heat the raw meal in the suspension preheaters.

The production capacity of modern preclaciner kilns has increased steadily over the years and today the largest precalciner kilns in operation produce around 10,000 tons of clinker per day.

2.3 Cement chemical composition

The chemical composition of Portland cement is analyzed routinely in cement testing laboratories established in the cement manufacturing plants with X-ray fluorescence and reported in terms of the oxides of the various elements that are present in the cement. Cement chemists customarily use shorthand notation for cement oxides shown in Table 2.1.

To indicate chemical formulae of the relevant species in cement, the traditional abbreviated nomenclature for the oxides of various elements and the mineral phase composition is shown in Table 2.2.

2.3.1 *X-ray fluorescence analysis (XRF)*

Within the cement industry, precise control chemical and mineralogical composition is one of the principal means of maintaining the quality of the raw materials and final products and their optimum performance in cements. Moreover, pushed by environmental issues and rising customer demands in terms of quality and new functionalities like blended cements, the market is imposing new challenges on cement producers, demanding the development and the increasing use of more sustainable materials.

Table 2.1 Portland cement oxides and their shorthand notations (Nawy, 2008).

Oxide	Shorthand notation	Common name	Typical (weight%) in Portland cement
CaO	C	Lime	63
SiO_2	S	Silica	22
Al_2O_3	A	Alumina	6
Fe_2O_3	F	Ferric oxide	2.5
MgO	M	Magnesia	2.5
Na_2O	N	Alkali	0.4
K_2O	K		0.6
SO_3	S	Sulfur trioxide	2.0
TiO_2	Ti	Titanium oxide	Minor

Table 2.2 Chemical formula and cement nomenclature for major OPC phases (Hewlett & Liska (2019).

Abbreviation	Chemical formula	Oxide composition	Main mineral
C_3S (Alite)	Ca_3SiO_5	$3CaO \cdot SiO_2$	Tricalcium silicate
C_2S (Belite)	Ca_2SiO_4	$2CaO \cdot SiO_2$	Dicalcium silicate
C_3A (Aluminate)	$Ca_3Al_2O_3$	$3CaO \cdot Al_2O_3$	Tricalcium aluminate
C_4AF (Ferrite)	$Ca_4Al_2Fe_2O_{10}$	$4CaO \cdot Al_2O_3 \cdot Fe_2O_3$	Tetra-calcium alumino-ferrite

Cement manufacturers therefore must know and examine the exact composition of each of the raw materials more accurately, as well as their effects, not only on the final product, but also on the production process.

X-ray fluorescence (XRF) provides detailed information for proportioning raw materials, controlling the main variables in the production process and plays essential part in maintaining the quality and the consistency of cement produced. It also plays significant role in the environmental control of waste recovery, alternative fuels, analysis of trace elements and chemical characterization of liquids such as additives and slurries.

XRF is now universally employed in cement manufacturing plants and cement testing laboratories as an essential tool for the analysis of raw meals for cement manufacture, clinker, and cement (Wirth & Barth, 2020). Numerous XRF instruments are available, and they can be broadly classified in to *two* types (a) wave-disperse X-ray fluorescence (WDXRF) and (b) energy-disperse X-ray fluoresce (EDXRF). Both types of these machines use either pressed powder in the form of pellets or fused disks, as samples and analyze the elements of interest either simultaneously or sequentially.

The XRF analytical technique is far superior to older wet-chemical analysis or by atomic absorption spectrometer (AAS) because it is rapid and does not require extensive training or experience on the part of the analyst; hence it has become technique of choice of most of the cement chemists. The relative ease and low cost of sample preparation, and the stability and ease of use of X-ray spectrometers make this as one of the most widely used methods for analysis of major and trace elements in clinker, cement, as well as raw materials for cement manufacture.

2.4 Clinker phase composition

Some essential properties of Portland cement for instance, heat of hydration, setting behavior, hardening and its soundness mostly depend on the content of clinker phases, while the chemical composition of clinker phases, their size and form as well as the structure of the matrix depend on the quality of the raw meal, burning temperature and cooling rate of the clinker. These factors affect the distribution of mineral phases in clinker, and hence on the performance of the Portland cement. The mineralogical composition of clinker and its microstructure influence the quality of cement, for instance, there is a relation between its Alite content and its strength development, whereas the content of Aluminate has influence on setting and hardening behavior of the cement. The determination of quantitative composition of the phases and the microstructure of the clinker are thus important indicators for assessing the characteristics of the Portland cement.

2.4.1 Determination of clinker phases

Since, the phase content of cement clinker greatly influences its properties, therefore study of its mineralogy becomes important. Various direct and indirect methods have been used for the determination of the phase content of clinker. In the cement industry, it is common practice to compute the compound composition of Portland cement from its oxide analysis by:

2.4.1.1 Indirect methods

- Bogue Method/Modified Bogue Method

2.4.1.2 Direct methods

- XRF (Chemical composition)
- Gravimetric (wet) analysis (Chemical composition)
- Rietveld refinement Method
- Optical Microscopy
- Scanning Electron Microscopy (SEM) coupled with EDXA

2.4.1.2.1 Clinker phase calculations

The four phases of clinker can be calculated using a set of equations, which were first proposed by Bogue in 1929. The calculations of the particular clinker phases from Bogue equations are carried out solely on the basis of the oxide composition, and thus do not take account of possible process influences. The Bogue equations assume that the chemical reactions of the formation of clinker compounds have proceeded to completion, and that the presence of impurities such as MgO and alkalis can be ignored. Both assumptions are not valid; therefore, in some cases the computed compound compositions, give low values for Alite and Aluminate, too high for Belite and relatively correct for Ferrite (Taylor, 1989). Taylor et al., suggested modified Bogue formulae, which takes into account the presence of solid solutions of minor components in clinker phases.

It is useful to note that even small differences in the oxide analyses of two cements can result in large differences in their compound composition. However, clinker phase composition gives a good indication about cement properties in terms of strength development, heat of hydration and sulfate resistance.

(a) Modified Bogue Method

The method involves the following assumptions:

- all the Fe_2O_3 is combined as C_4AF.
- the remaining Al_2O_3 (i.e., after deducting that combined in C_4AF) is combined as C_3A.

The CaO combined in the calculated levels of C_3A and C_4AF and any free lime are deducted from the total CaO and the level of SiO_2 determines the proportions of C_3S and C_2S. The procedure can be expressed mathematically (in mass %) as follows:

CASE A: ($A/F \geq 0.64$) (For normal Portland clinker)

$$C_3S = 4.071(\text{total CaO} - \text{free lime}) - 7.600SiO_2 - 6.718Al_2O_3 - 1.430Fe_2O_3 - 2.852SO_3$$

$$C_2S = 2.867SiO_2 - 0.7544C3S$$

$C_3A = 2.65Al_2O_3 - 1.692Fe_2O_3$
$C_4AF = 3.043Fe_2O_3$

The equations are applicable to Portland cements with an alumina to ferric (A/F) ratio 0.64 or higher in the raw meal; should the ratio be less than 0.64 another set of equations apply, which are mentioned below as per the ASTM C 150.

CASE B: (A/F $\leq$ 0.64) (For cements rich in iron oxide, no C_3A)

$C3S = 4.071(\text{total CaO} - \text{free lime}) - 7.60SiO_2 - 4.479Al_2O_3 - 2.859Fe_2O_3 - 2.852SO_3$

$C2S = 2.867SiO_2 - 0.7544C_3S$

$C3A = 0$

$C4AF = 2.100\ Al_2O_3 + 1.702\ Fe_2O_3$

The disadvantages of this indirect method are the possibility of error and the lack of knowledge about the clinker matrix. Thus, relatively small deviations in the chemical composition can lead to significant variations in the Bogue values. On the other hand, even if the chemical composition of two clinkers is identical, they may differ considerably in their actual phase composition.

The composition of the four principal chemical compounds in the clinker and their normal range of levels in current UK and European Portland cement clinkers are summarized in Table 2.3.

Clinker phase composition has decisive influence on several properties of cement paste, thus also on concrete. The most important are the following:

- strength development depends principally on Alite content
- sulfate resistance depends principally on Aluminate content and
- rheological paste properties depend chiefly on Aluminate content.

The characteristics of these compounds when cement is hydrated are shown in Table 2.4. The two calcium silicates (Alite and Belite) are primarily responsible for the strength that the cement will develop upon hydration. The Aluminate can lead to durability problems when the concrete is in contact with soils or groundwater containing sulfates. By making relatively small changes in the proportions of raw

Table 2.3 Ranges of principal phases in European clinkers (Newman & Choo, 2003).

Mineral	Chemical formula	Typical level (Mass %)	Typical range (Mass %)
C_3S (Alite)	$3CaO \cdot SiO_2$	57	45–65
C_2S (Belite)	$2CaO \cdot SiO_2$	16	10–30
C_3A (Aluminate)	$3CaO \cdot Al_2O_3$	9	5–12
C_4AF (Ferrite)	$4CaO \cdot Al_2O_3 \cdot Fe_2O_3$	10	6–12

Table 2.4 Characteristics of cement phases on hydration (Nawy, 2008).

Compound	Reaction rate	Heat liberated	Contribution to strength
C_3S	Moderate	High	High
C_2S	Slow	Low	Low initially, higher later
$C_3A + CSH_2$	Fast	Very high	Low
$C_4AF + CSH_2$	Moderate	Moderate	Low

CSH_2 = gypsum

materials, one can bring about relatively large changes in the proportions of the Portland cement phases thereby reducing the negative effects of cement on the durability performance of the concrete.

2.5 Properties of clinker phases

The Portland cement clinker consists of up to 95% of the four crystalline chemical phases; Alite (C_3S), Belite (C_2S), aluminate phase (C_3A) and ferrite phase (C_4AF) in a close interpenetrating association. In addition, the clinker contains voids and usually some free (uncombined) lime i.e., (free CaO) and sometimes periclase (free MgO).

Clinker phase composition has decisive influence on several properties of cement paste, thus also on concrete. The most important are the following:

- strength, which depends principally on Alite content
- sulfate resistance, which depends principally on aluminate phase content,
- rheological paste properties, which depends chiefly on aluminate phase content.

2.5.1 *Alite (tri-calcium silicate)*

Alite is the main constituent of clinker containing up to 50 to 70% of its total mass. Quantitatively and with regards to the properties of the cement (especially its strength development), Alite possesses the most hydraulic characteristics of the clinker phases and determines the strength development by creating calcium silicate hydrates (C–S–H). It makes the clinker easier to grind, hydrates rapidly generating high heat, develops an early hardness and strength and increases resistance to freezing and thawing. However, raising of Alite content in cement beyond the specified limits, increases the heat of hydration as well as solubility of cement in water, which may contribute to shrinkage cracks in mortars or concrete.

Because of the inclusions of some quantity of foreign ions e.g., approximately 2% of MgO, also Al_2O_3, Fe_2O_3, TiO_2 and others, originating from normal and alternative raw materials as well as fuels and alternative fuels, it is not a pure tri-calcium silicate compound in stoichiometric sense. Because it contains foreign ions, it was named Alite by Bogue & Tornbeum (Bogue, 1955). The inclusion of foreign ions usually has a stabilizing effect on the crystal matrix and usually increases its strength. The temperature at which they form is approximately 1250°C. If cooling occurs too slowly, Alite may decompose in to Belite and CaO, especially if it contains Fe^{2+} as a result of burning under reducing conditions (Fig. 2.3).

2.5.2 Belite (di-calcium silicate)

Chemically pure di-calcium silicate (C_2S) is not found in cement clinker either, hence called Belite; it likewise contains incorporated foreign oxides. It occurs mainly in solid form at the clinkering temperature and usually comprises 5%–30% of the clinker by mass. Up to 6% of the mass of Belite is composed of foreign constituents such as alkali metals, iron, sulfur or magnesium. It hydrates and hardens slowly and takes long time to add to the strength and after 1 year its contribution to the strength and hardness is proportionately almost equal to C_3S. It imparts resistance to chemical attack. Raising of C_2S content renders clinker harder to grind, reduces early strength development, decreases resistance to freezing and thawing at early ages and decreases heat of hydration. At early ages, less than a month, C_2S has little influence on the strength development and hardness.

The distribution of Belite in clinker is often a function of the fineness and homogeneity of the raw meal. Belite forms round crystals that are usually between 20 and 40 microns in size and under the microscope appear with intersecting lamellae or sometimes parallel lamellae (Fig. 1.2). These lamellae found in Belite are an indication of an upper burn temperature of maximum 1420°C. These varying crystal structures reflect the creation of different forms of Belite during the sintering and cooling processes. Technical clinker primarily contains β-Belite modification, also present as α-Belite if cooled quickly. β-Belite modification is the form in which this compound phase

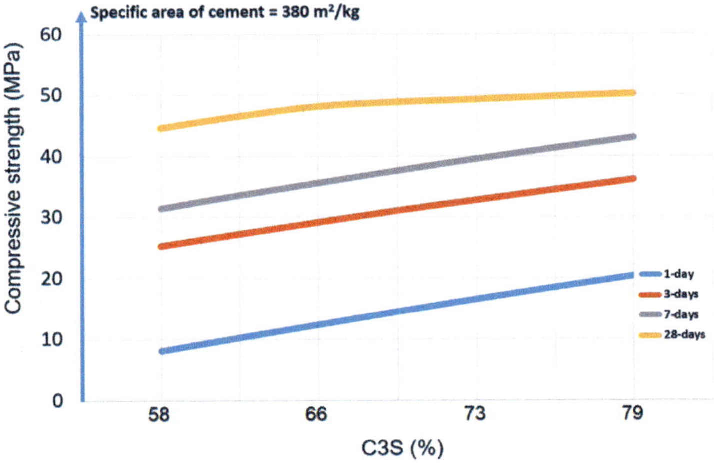

Figure 2.3 Effect of Alite content on strength of cement (Blaine = 380 m^2/kg) (Kurdowski, 2014).

is predominantly present in the clinker, it may at room temperature change to γ-modification, which is the more stable form, but virtually lacking in hydraulic properties. This change is accompanied by a volume increase of about 10%, which is considered to be the cause of the so-called "falling" of the clinker, a rapid disintegration, which can be inverted by rapid cooling. With present day technology of cement manufacture, the risk of clinker failing has been eliminated.

2.5.3 Aluminate phase (tri-calcium aluminate)

The aluminate phase is about 5%−11% of cement (about 8.0% in European cements). The aluminate phase is not in its pure form of C_3A but contains foreign constituents. In this phase the incorporation of alkalis (Na_2O and K_2O) each in amounts exceeding 5%, is possible. The aluminate phase possesses a high degree of reactivity, which is further increased by the incorporation of alkalis and is responsible for flash set of finely grounded clinker. In order to retard the reaction of aluminate phase at the start of hydration, all cement types must contain some added sulfate (in the form of gypsum at the time of grinding cement), which works as cement set retarder. Aluminate phase is responsible for the initial set, high heat of hydration and has greater tendency to volume changes causing cracking. Raising the aluminate content reduces the setting time, weakens the resistance to sulfate attack and lowers the ultimate strength; however, its contribution to hydraulic properties are slight. Aluminate phase present in large quantities is prone to and not resistant to sulfate attack, and also reduces the resistance of hardened paste or concrete against sulfates and other chemical attacks, which is why the amount of C_3A content in sulfate resisting cements have been limited to 3%.

2.5.4 Ferrite phase (tetra-calcium alumino-ferrite)

It is the ferrite phase that contributes specifically to giving cement its color: pure $C_2(AF)$ is brown, while $C_2(AF)$ containing MgO is of dark gray/green color. This phase is very slow reacting and of little importance to the physical properties of cement.

Composition of ferrite phase in clinker depends in great degree on the A/F ratio and also on the burning conditions. The compositions range from C_6AF_2 through C_4AF to C_6A_2F. Many authors found that in Portland cement clinkers of common composition ferrite phase has the composition close to brownmillerite i.e., C_4AF (Vernet & Demoulin, 1980).

The ferrite phase does not possess a constant chemical composition but in fact it is a member of a solid solution series extending theoretically from C_2A to C_2F as follows:

$$C_2A \rightarrow C_6A_2F \rightarrow C_4AF \rightarrow C_6AF_2 \rightarrow C_2F$$

Depending on the availability of iron and aluminum, the members of solid solution series will be situated nearer the iron rich or nearer the aluminum rich end of the series. However, quite often the composition of this phase in cement clinker corresponds more or less to C_4AF. Similar to the calcium silicates, C_4AF contain significant amounts of magnesium, sodium, potassium and silica in their crystal structure.

The ferrite phase is about 8%–14% (normally about 9%) of cement. It is responsible for flash set but generates less heat. It has poorest cementing value. Raising the C_4AF content reduces the strength slightly (Table 2.5).

Table 2.5 Properties of phases in cement clinker (Kohlhaas & Labahn, 1983).

Pure phases in clinker	Tri-calcium silicate (Alite)	Di-calcium silicate (Belite)	Tri-calcium aluminate (Aluminate)	Calcium alumino-ferrite (Ferrite)
Notation	C_3S	C_2S	C_3A	C_2AF
Composition	$3CaO \cdot SiO_2$	$2CaO \cdot SiO_2$	$3CaO \cdot Al_2O_3$	$2CaO$ $(Al_2O_3 \cdot Fe_2O_3)$
Color of pure phase	White	White	White	Dark brown due to incorporation of MgO/or dark-grey-green
Percent content in cement clinker Maximum Average Minimum	80 60 40	30 15 0	15 11 7	15 8 4
Incorporated foreign ions	Mg, Al, Fe	Alkalis, Al, Fe, fluoride	Alkalis, Fe, Mg	Si, Mg
Technical properties	Rapid hydration, high initial and good final strength, moderate heat of hydration, main strength contributor to Portland cement.	Slow hydration, good final strength, low heat of hydration.	Rapid hydration, exothermic high heat of hydration, which promotes early strength, shrinks appreciably on hydration, reacts with sulfates and undergoes volume expansion.	Slow and moderate hydration, hardly any strength development, moderate heat of hydration, MgO incorporation gives cement its color.

2.6 Cement hydration products

The term "hydration of cement" covers the whole complex process of the reaction of cement with water. The individual cement grains vary in size and chemical composition and as a consequence, the hydration reactions that take place between fine grained Portland cement (comprising calcium silicate and aluminate phases) and water are highly complex. The hydration products vary in their chemical composition and the microstructural characteristics vary with time, temperature of surrounding environment, temperature of cement, water and aggregates in mortars and concrete. As the clinker minerals go through these chemical reactions with water, the resulting products are referred to as hydrates or hydrated compounds.

After mixing with water, the reaction takes place between the active components of cement (C_3A, C_3S, C_2S and C_4AF) and water. The hydration products start depositing on the outer periphery of the nucleus of hydrated cement particles. This reaction proceeds slowly for 2–5 h and is called *induction* or *dormant period.* As the hydration proceeds, the deposit of hydration products on the original cement grain makes the diffusion of water to unhydrated nucleus more and more difficult, consequently reducing the rate of hydration with time. At any stage of hydration, the cement paste consists of gel (a fine-grained product of hydration having large surface area collectively), the unreacted cement, calcium hydroxide, water, and some minor compounds. The hydrated products form a plastic mass, which is workable for a certain period of time and gradually over time, the plastic mass becomes stiff and less workable, the material sets and its strength increases over a long period of time. The factors responsible for the physical properties of concrete are the hydration products of cement and the resultant microstructure of the hydrated cement. Hydration occurs both in air and under water and the hydration products are virtually insoluble in water.

Since the chemical reactions of Portland cement with water are responsible for the setting and hardening of mortar and concrete, the underlying chemistry must be described to understand and appreciate these processes. Knowledge of the hydrated substances formed is important because the cement itself is not a cementing material rather its hydration products have the cementing action. Knowledge of the amount of heat released is important because the heat is sometimes a help and sometimes a hindrance. Knowledge of reaction speed is important because it determines the time of setting and hardening. The initial reaction must be slow enough to enable the concrete to be poured into place. On the other hand, after the concrete has been placed rapid hardening is often desirable (Brunauer & Copeland, 1964).

The reactions of the compounds and their products are as follows:

$$C_3A + 3CaSH_2 + 26H_2O \rightarrow C_3A \cdot 3CS \cdot H_{32} \text{ (Ettringite, AFt)} \tag{2.1}$$

[$\Delta H =$ approximately -452 kJ/mol, fast reaction]
Lack of sulfate

$$C_3A + CaSH_2 + 10H_2O \rightarrow C_3A \cdot CS \cdot H_{12} \text{ (Monosulfate, AFm)} \tag{2.2}$$

[ΔH = approximately −245 kJ/mol, fast reaction]
Reaction of Ettringite with C_3A

$$2C_3A + C_3A \cdot 3CS \cdot H_{32} + 4H_2O \rightarrow 3C_3A \cdot CS \cdot H_{12} \text{ (Monosulfate, AFm)} \quad (2.3)$$

$$2C_3S + 6H_2O \rightarrow (C_3 \cdot S_2 \cdot H_3) + 3CH \text{ (Calcium silicate hydrate + calcium hydroxide)} \quad (2.4)$$

[ΔH = approximately −115 kJ/mol, moderately fast reaction]

$$2\ C_2S + 5H_2O \rightarrow (C_3 \cdot S_2 \cdot H_4) + CH \text{ (Calcium silicate hydrate + calcium hydroxide)} \quad (2.5)$$

[ΔH = approximately - 45 kJ/mol, slow reaction]

Chemical reactions between Portland cements, water and admixtures are usually well controlled and follow a definite sequence (Seligmann & Greening, 1969). Initial partial dissolution of alkali sulfates, calcium sulfates, calcium aluminates and silicates rapidly increase concentrations of ions in solution to super saturation levels (except for alkali-hydroxides) and produce calcium aluminate, sulfo-aluminate, and silicate hydrates on the surfaces of the cement and other particles. Within minutes, these reactions are slowed by coatings of hydrates on the reaction surfaces that produce a "dormant" period of slow reaction as Ettringite continues to form and calcium superiority persists. Eventually, calcium hydroxide begins to crystallize from solution, reducing the calcium concentration and accelerating the dissolution and reaction of the silicates.

Setting results from the growth of acicular crystallites of calcium silicates hydrates that bridge the water filled void spaces. Reaction of aluminates to form Ettringite continues until the readily soluble calcium sulfates are depleted. The concentration of sulfate in the solution then decreases, causing dissolution of some Ettringite and a period of acceleration of hydration of the aluminate and ferrite phases.

The hydrated crystals of C−S−H gel are extremely small, fibrous, platy or tubular in shape varying from less than 2−10 nm (nanometers) or more. The C−S−H phase makes up 50%−60% of the volume of solids in a completely hydrated Portland cement paste and is, therefore, the most important in determining the properties of the paste. The proposed surface area for C−S−H is of the order of 100−700 m^2/g, the solid-to-solid distance being about 18 Å and the $Ca(OH)_2$ liberated during the silicate phase crystallizes in the available free space.

The calcium hydroxide crystals (Portlandite) consist of 20−25% volume of the solids in the hydrated paste. These have lower surface area and their strength contributing potential is limited. The gel must be saturated with water if hydration is to continue. The calcium hydroxide crystals formed in the process dissolve in water providing hydroxyl ions (OH−) and increase the pH to 13, which plays its part in the protection of reinforcement in concrete. As hydration proceeds, the crystals become more heavily interlocked thereby increasing the strength, however, the main cementing action is provided by the C−S−H gel, which occupies two-thirds of the total mass of hydrate.

During the reaction prior to setting, when the paste is plastic, if the surface becomes partly dried, capillary tension in the water can cause plastic shrinkage and cracks. Chemical shrinkage is the volume change that results from formation of hydrates that have less solid volume than the volume of water and the reacted solids. While the paste is plastic, the entire volume of paste undergoes chemical shrinkage. After the setting, the external dimensions remain essentially fixed, and additional water must be imbibed to keep the pores saturated with water. If sufficient water is not imbibed, the paste becomes self-desiccated (Klieger & Lamond, 1994).

If the early reactions of C_3A and C_4AF are not well controlled by sufficiently rapid dissolution of calcium sulfates, thin platy crystals of calcium alumino-ferrite monosulfate hydrate (AFm) form in the solution instead of dense coatings of Ettringite, the (Aft) or tri-sulfate phase, resulting in premature stiffening, or in extreme cases "flash set" (Taylor, 1997). If on the other hand, the aluminate reactions are too slow, to consume all of the rapidly dissolving calcium sulfates, crystallization of fine needles of gypsum from solution will cause early stiffening "false set".

If there is slight imbalance in reaction rate control, rapid slump loss of concrete may result from adsorption of water-reducing admixtures from solution by high surface reaction products.

2.7 Types of cements and blended cements

Properties of concrete in fresh and hardened states depend mostly on the properties of cement used and different concrete applications require cements with different properties. Some applications require a rapid gain in strength to speed up the construction process. Other applications may require a low heat of hydration to control volume change and the associated shrinkage cracking. In some cases, concrete is exposed to sulfates that can deteriorate concrete made from ordinary Portland cement. Therefore, it is important to understand how various types of cement influence and affect the fresh and hardened concrete properties. An attempt is made in this chapter to provide general background and description of various General Purpose and Special Purpose cement types.

Different Portland cement types, their grades and varieties are based on the following production steps:

- Manipulation of cement phases
- Control of Blaine fineness of cement
- Substitution of supplementary cementitious materials
- Addition of special chemicals

2.7.1 Ordinary Portland cements

Ordinary Portland cement consists of the phases C_3S (Alite), C_2S (Belite), C_3A (Aluminate), C_4AF (Ferrite), free CaO, Alkali sulfates (R_2SO_4) and $CaSO_4 \cdot nH_2O$. The first four are the main and important phases constituting 85%–90% while the other constituents are present in small proportions totaling 10%–15%. By manipulating the relative proportions of the four principal phases and concurrently adjusting the fineness of the product, OPC can be graded in to various strength classes defined by 28-day compressive strengths of cement mortars, conforming to national standard specifications. The following Special Purpose cement varieties are produced by adjusting the proportions of the main phases together with the fineness of the product.

- Rapid hardening cement
- Low heat cement
- Sulfate resisting cement

2.7.2 Blended cements (composite cements)

Blended cements are produced by mixing different supplementary cementitious materials (SCMs) such as flyash, metal slag, silica fume and limestone powder individually or a combination of two or more SCMs at the final grinding stage of the product.

Cements in which a portion of Portland cement is replaced by mineral admixtures/ SCMs also known as such as flyash (also known as pulverized fuel ash or PFA) metakaolin, blast furnace slag (BFS), burnt clay (new type cement blending material), volcanic ash, diatomaceous earth (skeletons of algae – diatoms) and limestone constitute the term blended cements or composite cements.

In blended cements a portion of Portland cement clinker or Portland cement is replaced by industrial by-products such as granulated blast furnace slag (gbfs) aptly named Portland blast furnace slag cement (Type IS), or fly ash (artificial pozzolan) and certain types of volcanic ashes or diatomaceous earth (natural pozzolan), named Portland-pozzolan cement (Type IP), Portland limestone cement (Type 1L) as defined in ASTM C595/C595M, 2020, (AASHTO, 2018). Burnt clay is a new cement blending material. When hydrated, the gbfs, fly ash, pozzolan react with the cement hydration products, producing additional hydrates, which make the concrete denser thus, making a positive contribution to concrete strength development and its durability (Massazza, 1998).

The SCMs react with the hydration products of Portland cement and produce additional hydrates, which lessen the heat of hydration, improve workability, improve the microstructure of concrete, decrease permeability thus concrete strength development is increased. Chemical and microstructural improvements make the concrete matrix denser, which improves the resistance to environmental attacks such as sulfate attack and chloride ingress (sea water attack) while carbonation depths remain unaffected. The expansion induced by alkali-silica reactivity can be avoided by using cements containing pozzolan, while the concrete resistance to freezing and thawing remains unaffected, since freeze-thaw attack primarily depends on the content of entrained air (Massazza, 1998, 2003, Moranville-Regourd, 1998). The addition of SCMs in cement makes it economically cheaper and also environmentally friendly. SCMs reduce energy consumption in cement production, hence reduce CO_2 emissions associated with the production of concrete with the added benefit that the industrial waste streams are recycled and turned into value-added streams.

SCMs can be divided into three main categories:

1. Calcined pozzolan
2. Natural pozzolan
3. Inert (nonreactive) materials or Fillers

The SCMs are defined in detail in Chapter 3 together with their performance in concrete. However, comparative strength development of blended cements containing 20% flyash, slag, natural pozzolan, limestone powder compared with Portland cement is shown in Fig. 2.4.

2.7.3 Australian cements

In Australia, cements covered by the Standard AS 3972 have been designated as "general purpose" or "special purpose". The difference between the two is shown in Table 2.6. Seven major types of cement are produced and sold in the Australian market. However, the most commonly used are the types GP (General Purpose) and GB (General Blended) cements (Guirguis, 1993).

The Australian standards AS 3972 incorporates a number of significant changes that distinguish it from the previous two standards AS 1315 "Portland Cement" and AS 1317 "Blended Cement". As per AS 3972, "General Purpose and Blended Cements", the mineral addition have increased from 5% to 7.5% of limestone powder (containing more than 80% calcium carbonate by weight) or flyash or slag or a combination thereof. Flyash and slag are required to comply with the requirements of AS 3582 "Supplementary Cementitious Materials for use in Portland and Blended cements" (AS 3582.1), "Supplementary Cementitious Materials for direct addition in to Concrete" (AS 3582.2) (Basheer and Barbhuiya, 2009) and "Fly Ash in Australian Standards, opportunities with using Fly Ash in concrete" (ADAA, 2009).

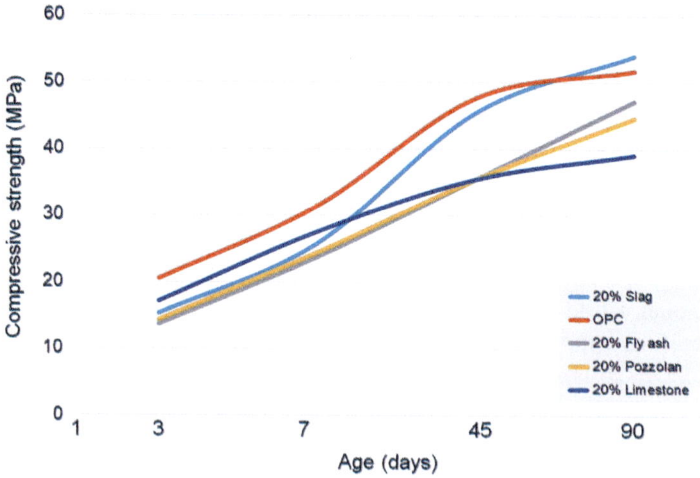

Figure 2.4 Comparative strength development of Blended cements containing 20% flyash, slag, pozzolan and limestone with OPC. (W/C = 0.6, curing 20°C) (Moir, 2003).

Table 2.6 Major types of cements produced and sold in Australia (Guirguis, 1993).

General purpose cement	Special purpose cement[c]
Type GP (General purpose)	Type HE (high early strength)
Type GB[a] (General blended)	Type LH (low heat)
Type GL[b] (General containing limestone)	Type SL (shrinkage limited)
	Type SR (sulfate resisting)

[a]Type GB – no upper limit for flyash/slag. Maximum 10% silica fume.
[b]Type GL – upper limit of 20% limestone (integrated).
[c]Special Purpose cement – no limit on composition, performance limits nominated for tests specific to each cement type (Guirguis, 1993).

In AS 3972, the cement properties have been divided into two categories, specified properties (for which limits are specified) and reportable properties (for which no limits are specified) but for which test results must be reported if required by the purchaser. Table 2.7 shows the summary of AS 3972 requirements (AS 3972, 2010).

2.7.3.1 General purpose type (GP) cement

GP cement is the most widely used type cement suitable for all conventional concrete construction. GP cement complies with the requirements specified in Australian Standard AS 3972 "General Purpose and Blended Cements". AS 3972-2010 features an increase in the allowable level of "mineral addition" from 5% to 7.5%. Typically, mineral additions such as limestone, flyash or slag are usually interground with the GP cement clinker.

The benefits of GP cement are (a) its high versatility and can be used in a wide variety of concrete applications. It can also be blended with most supplementary cementitious materials and admixtures. It is also ideal for a wide range of applications including,

- Structural concrete such as columns, beam slabs, tilt-up walls, pipes, concrete
- Domestic house construction such as concrete slabs, driveways, footpaths, renders, grouts, tile adhesive, pipes, concrete blocks, pavers etc.
- Major civil engineering projects requiring cement of high quality and consistency
- Mining applications

2.7.3.2 General Blended type (GB) cement

Blended cement is a blend of Portland cement and flyash or slag, which complies with the requirements specified in Australian Standard AS 3972 for Type GB cement. It also complies with the AS 3972 requirements for Type SR (sulfate resisting) and Type SL (shrinkage limited) cements. The use of GB cements does not guarantee sulfate resistant or low shrinkage concrete, since there are other factors, which may influence concrete performance such as cementitious content, water to cement ratio, curing, as well as aggregate type (Cement Australia, 2011).

Table 2.7 Summary of Australian Standards AS 3972 requirements (Guirguis, 1993).

	Cement types				
	General purpose		Special purpose		
	General purpose Portland (GP)	General purpose blended (GB)	High early strength (HE)	Low heat (LH)	Sulfate resisting (SR)
Chemical properties					
Major oxide composition	Reported if required, no limit is specified				
Sulfuric anhydrite (SO_3) max. (%)	3.5	3.5	3.5	3.5	3.5
MgO	Portland cement clinker shall contain less than 4.5%				
Chloride content	Reported if in excess of 0.05%				
Loss on ignition (LOI)	Reported if required, no limit is specified				
C_3A content, max. (%)	–	–	–	–	5.0
Setting time					
Minimum (mins.)	45	45	45	45	45
Maximum (hrs.)	10	10	10	10	10
Soundness					
Maximum (mm)	5	5	5	5	5
Compressive strength					
3 days, min. MPa	–	–	20	–	–
7 days, min. MPa	25	15	30	10	20
28 days, min. MPa	40	30	–	30	30
Fineness	Reported if required, no limit is specified				
Heat of hydration					
7 days, max.(J/g)	–	–	–	280	–
28 days, max.(J/g)	–	–	–	320	–

2.7.4 European cements

EN 197-1 distinguishes cements into five main groups (CEM I to CEM V) and provides classification of 27 common cements based on Portland cement clinker and incorporation of one or more other main constituents. In addition, EN 197-1 sets out performance requirements for strength and volume stability. These categories indicate the relative proportions of cement clinker and other main constituents, as follows:

- CEM I Portland cement: This comprises mainly ground clinker (>95% clinker) and up to 5% of minor additional constituents.

- CEM II Portland-composite cement: This comprises of seven cement types, which contain 65%–94% clinker, and from 6% to 35% of another single constituent.
- CEM III Blast-furnace cement: This comprises of 5%–64% clinker and 36%–95% blast-furnace slag.
- CEM IV Pozzolanic cement: This comprises of 45%–89% clinker, and 11%–55% of silica fume, or pozzolana or fly ash or a combination thereof.
- CEM V Composite cement: This comprises of 20%–64% clinker, and 18%–50% blast-furnace slag or 18%–50% pozzolana or siliceous fly ash or a combination thereof.

The EN 197-1 standard for common cements further specifies seven sulfate resisting cements, three low early-strength blast furnace cements and two sulfate resisting low early strength blast furnace cements. There are also a number of special cements that are covered by specific standards, such as super sulfate-cement (EN 15743), very low-heat cement (EN 14216) and calcium aluminate cement (EN 1464) (European Commission, 2018). The types of cements and their composition (as per EN 197-1) are summarized in Table 2.8.

The composite cements contain a proportion of Portland cement clinker substituted with industrial by-products such as flyash, slag, certain types of volcanic materials (natural pozzolans) or limestone. In addition, silica-fume/microsilica (D) and

Table 2.8 Types of cements and their composition (Basheer & Barbhuiya, 2009).

Cement	EN cement notification	Clinker content (%)	Content of other main constituents (%)
Portland cement	CEM I	95–100	<5
Portland-slag cement	CEM II/A-S	80–94	6–20
	CEM II/B-S	65–79	21–35
Portland-silica fume cement	CEM II/A-D	90–94	6–10
Portland-pozzolan cement	CEM II/A-P	80–94	6–20
	CEM II/B-P	65–79	21–35
	CEM II/A-Q	80–94	6–20
	CEM II/B-Q	65–79	21–35
Portland-pulverized fuel ash cement	CEM II/A-V	80–94	6–20
	CEM II/B-V	65–79	21–35
	CEM II/A-W	80–94	6–20
	CEM II/B-W	65–79	21–35
Portland-burnt shale cement	CEM II/A-T	80–94	6–20
	CEM II/B-T	65–79	21–35
Portland–limestone cement	CEM II/A-L	80–94	6–20
	CEM II/A-LL	80–94	6–20
	CEM II/B-L	65–79	21–35
	CEM II/B-LL	65–79	21–35

Continued

Table 2.8 Continued

Cement	EN cement notification	Clinker content (%)	Content of other main constituents (%)
Portland-composite cement	CEM II/A-M	80–94	6–20
	CEM II/B-M	65–79	21–35
Blast furnace cement	CEM III/A	35–64	36–95
	CEM III/B	20–34	66–80
	CEM III/C	5–19	81–95
Pozzolanic cement	CEM IV/A	65–89	11–35
	CEM IV/B	45–64	36–55
Composite cement	CEM V/A	40–64	36–60
	CEM V/B	20–38	62–80
	Explanation of notations used. CEM II – Main cement type indicates higher (A), medium (B) and lower (C) proportions of Portland cement clinker. S – subtype, indicates the second main constituent i.e., blast furnace slag, 42.5N – standard strength class, where N indicates normal early strength, while R indicates rapid early strength as in 42.5R. The code letters used indicate sub-type of cement; S = blast-furnace slag, D = silica fume, P = natural pozzolan, Q = natural calcined pozzolan, V- siliceous flyash (e.g., type F flyash), W = calcareous flyash (e.g., type C flyash), T = burnt shale, L = limestone, LL = limestone of high quality, M = two or more of the above constituents (composite). CEM V/A composite cement contains 18%–30% blast furnace slag. CEM V/B composite cement contains 31%–50% blast furnace slag.		

metakaolin (MK) are also used as supplementary cementitious materials in concrete for improving various physical and durability properties of concrete.

2.7.4.1 Strength classes of cement

There are three strength classes of cement based on their compressive strength determined in accordance with EN 196-1 at 28 days conforming to the requirements of EN 197-1 and summarized in Table 2.9.

2.7.5 USA cements

ASTM C150/C150M-16 specifications for Portland cement covers 10 types of cements (ASTM, 2016). Out of these 10 types, five are the main types of Portland cement (Type I through Type V) described below and their applications are tabulated in Table 2.10.

- Type I regular cement, general use.
- Type II moderate sulfate resistance, moderate heat of hydration

Table 2.9 Compressive strength and setting time requirements of three strength classes of cement (EN 197-1, 2011, pp. 1–44).

Strength class	Compressive strength at 28 days (MPa)	Initial setting time (mins)
32.5 N	≥32.5, ≤52.5	≥75
32.5 R	≥32.5, ≤52.5	≥75
42.5 N	≥42.5, ≤62.5	≥60
42.5 R	≥42.5, ≤62.5	≥60
52.5 N	≥52.5	≥45
52.5 R	≥52.5	≥45

Table 2.10 Types and applications of standard Portland cement (Dehayes & Tennis, 2006).

Type	Name	Application
I	Normal	For use in general concrete work. When the special properties of other cement types are not needed. No limits are imposed on any of the four phase compounds of cement clinker.
II	Moderate sulfate resistant.	For use when moderate sulfate resistance is required. The C_3A content of the cement is limited to max. 8%. Also, an additional max. Limit of 58% of the sum of C_3S and C_3A applies when moderate heat of hydration is desired.
III	High early strength	For use when high early strength is needed. The C_3A content of the cement is limited to max. 15%. The early high strength of type III cement is partially due to the higher specific surface of approximately 500 m^2/kg Blaine, as compared to 330–400 m^2/kg for type I cement.

Continued

Table 2.10 Continued

Type	Name	Application
IV	Low heat of hydration	For use when low heat of hydration is required. The content of C_3S and C_3A are limited to 35% and 7% respectively and requires a minimum of 40% C_2S.
V	High sulfate resistant	For use when high sulfate resistance is required. The C_3A content of the cement is limited to 5%.

Type I is usually used as a reference, which contains 50% Alite, 25% Belite, 12% Aluminate, 8% Ferrite, and 5% gypsum. Compared to type I, type III has more Alite (60%) and less Belite (15%) (Li, 2011).

- Type III high early strength
- Type IV low heat
- Type V high sulfate resistance

In addition to these five main types, air-entrainers are added to Type I, II and III cements during manufacturing to produce Type IA, IIA and IIIA cements, which provide better resistance to freeze and thaw than the nonair-entrained cements. Type II cements also contain two additional cement types, moderate heat of hydration, Type II(MH) and moderate sulfate resistant, Type II (MH)A.

Moreover, type III has higher fineness number than type I. As a result, the early strength of type III at 1 day is doubled as compared to that of type I. Meanwhile, the heat released by type III increases to 500 J/g (type I is 330 J/g). On the other hand, type IV has less Alite (25%) and more Belite (50%). Hence, the early strength of type IV at 1 day is only half of that of type I.

However, the heat released by type IV greatly decreases to 210 J/g, and thus is called-low heat Portland cement. As for type V, its sum of Aluminate and Ferrite is only 14% and much less than the 20% for type I. Since these two compounds readily react with sulfate, the lower content gives it less opportunity to be attacked by sulfate ions.

From the information provided in Tables 2.2–2.8, we can evaluate the behavior of each type of cement. The various behaviors and performances provide the basic justification in selecting cements for engineering practice as shown in (Table 2.11).

Low heat Portland cement is no longer made in the United States because the addition of mineral admixtures directly in concrete is less expensive way to control the

Table 2.11 Chemical, fineness and heat of hydration requirements for Portland cements (ASTM, 2016).

Portland cement type	Phase composition maximum (%)				Blaine fineness (m^2/kg)	Heat of hydration (7 days, J/g)
	Alite (C_3S)	Belite (C_2S)	Aluminate (C_3A)	Ferrite (C_4AF)		
I	55	19	10	7	370	330
II	51	24	6	11	370	250
III	56	19	10	7	540	500
IV	28	49	4	12	380	210
V	38	43	4	9	380	250

temperature rise. Similarly, the concrete producers prefer cements without entrained air because addition of air entraining admixtures in concrete offers a better control to produce the desired air-entrainment in concrete, hence more than 90% of the hydraulic cements produced in USA are ASTM Type-1 and Type-II Portland cements, about 3% are ASTM Type-III Portland cements and the rest are special cements such as oil-well cements (Mehta & Monteiro, 2013).

2.8 Environmental impact of cement production

2.8.1 Life Cycle Assessment of Portland cements, blended cements and recycled concrete

Over the last decades, Portland cement industry has been searching for alternatives to mitigate its environmental impacts. According to the literature (Chen et al., 2010)**, the two main problems are anthropogenic emissions and energy consumption. The global CO_2 emission from cement industry is projected to reach 2.34 Gt/year in 2050** (Khongprom et al., 2020). **The scientific community has been conducting studies to quantify the energy consumption and greenhouse gas (GHG) emissions.**

Since 1990, the life cycle assessment method has been used in the construction sector. This methodology is appropriate to analyze cement manufacturing process, in this case a cradle to gate is the most suitable as the full cycle (cradle to grave) includes a diverse range of cement application, such as pavements, slabs, columns, beams, bridges etc.). Usually, the LCA approach to evaluate the cement production is the process life cycle analysis (PLCA).

There are two aspects of cement production that result in emissions of CO_2. The first is the chemical reactions of raw meal in the kiln resulting in the production of the main component of cement, clinker, as carbonates (largely limestone, $CaCO_3$) are decomposed into CaO and CO_2 by the addition of heat. According to Andrew (2019)**, over the last 20 years, Vietnam, China and India presented an increase in the CO_2 emission from cement production of 496%, 238% and 215%, respectively. On the other hand, Spain and Germany showed a decrease of 34% and 14%, respectively.** Fig. 2.5 **presents emissions of CO_2 (color coded) of different countries in 2019.**

More recently Guo et al. (2020) reported the importance to conduct the life cycle assessment, cradle to grave, considering the carbon uptake capacity of cement materials (concrete, mortars, construction waste and cement kiln dust) to provide a more realistic analysis of the total CO_2 emission from cement industry. The results showed that 21.12 Gt CO_2 (18.12–24.54 Gt CO_2, 95% CI) had been absorbed in the cements produced from 1930 to 2019, with the 2019 annual figure mounting up to 0.90 Gt CO_2 per year (0.76–1.07 Gt CO_2, 95% CI). In summary, the cumulative uptake is equivalent to roughly 52% of the process emission, based on their analysis.

The strategy used in the literature review relies on the Scopus Elsevier Database as the primary source for retrieving publications. Search with keywords "Life Cycle

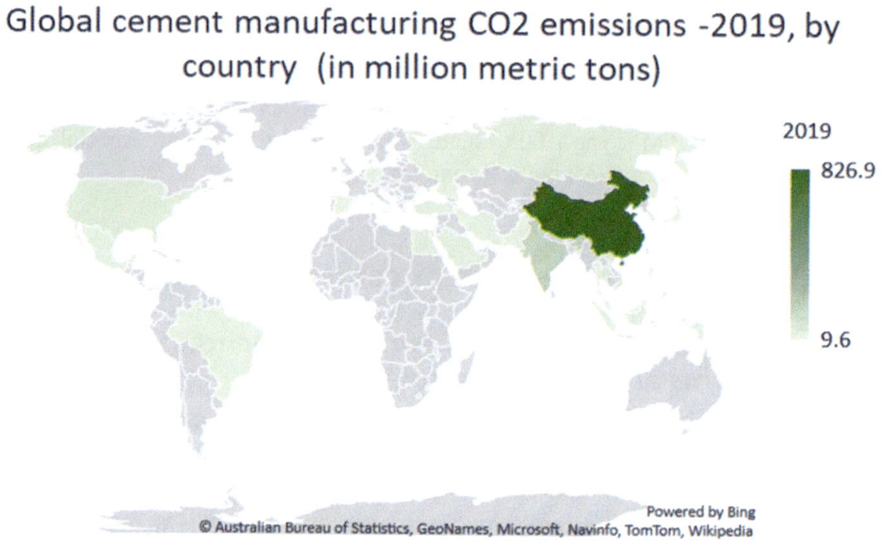

Figure 2.5 – Cement production: CO_2 emission in different counties.
Andrew, R. M. (2019). Global CO_2 emissions from cement production, 1928–2018. *Earth System Science Data*, *11*(4), 1675–1710. https://dx.doi.org/10.5194/essd-11-1675-2019.

Assessment" and "cement" in the title and abstract fields, limited to English language, retrieved 752 documents published since 1995 (see Fig. 2.6).

The use of Bibliometric Software VOS viewer that has smooth integration with the Scopus Elsevier Database was essential to guide our steps and assure that we were on

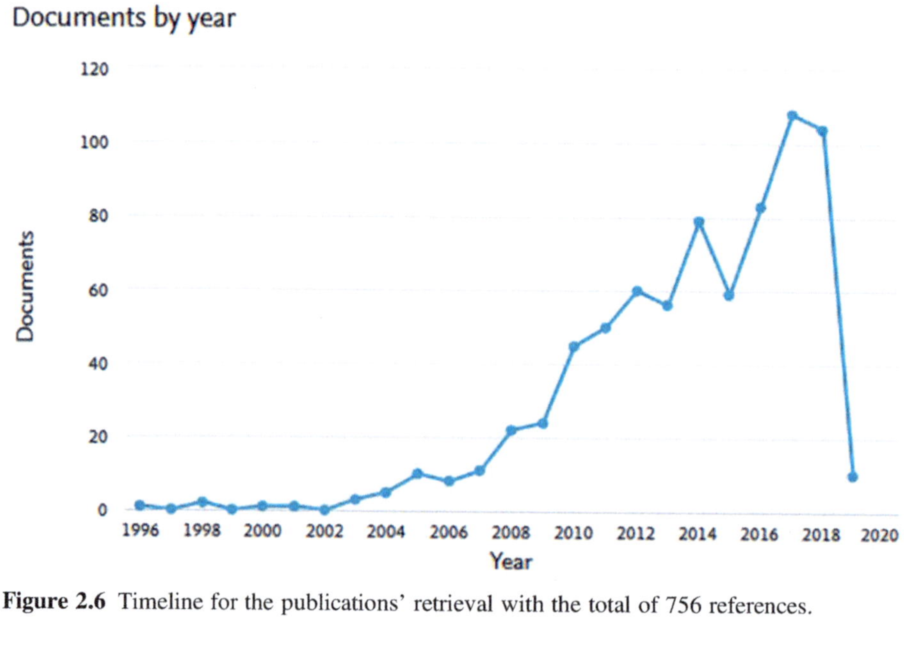

Figure 2.6 Timeline for the publications' retrieval with the total of 756 references.

the right track. Fig. 2.7 shows the keywords retrieved from 752 documents close related to LCA of Cement Portland and published in relevant sources, such as, Journal of Cleaner Production, and International Journal of Life Cycle Assessment, Cement and Concrete composites, which were in-depth analyzed and discussed in this study.

2.8.1.1 Life Cycle Assessment (LCA) applied to Portland cement production

The LCA methodology is based on ISO 14040 and 14044 guidelines covering: Goal and Scope aiming to identify the functional equivalent, system boundary and the set of building materials, Life Cycle Inventory's (LCI) main challenge is the difficulty of collecting reliable and applicable data. In Life Cycle Impact Assessment (LCIA), the quantities of materials, energy consumption, and input data are gathered using the sustainability indicators. The final step is Interpretation that allows for the evaluation of the results achieved from LCI and LCIA (UNEP, SETAC and LC-Initiative, 2009).

According to ISO 14044, Life Cycle Impact Assessment proceeds through four steps: selection of impact categories and classification (mandatory), characterization (mandatory), normalization (optional) and weighting (optional). LCA covers all the energy and resources required in all phases of the life cycle to estimate the emissions

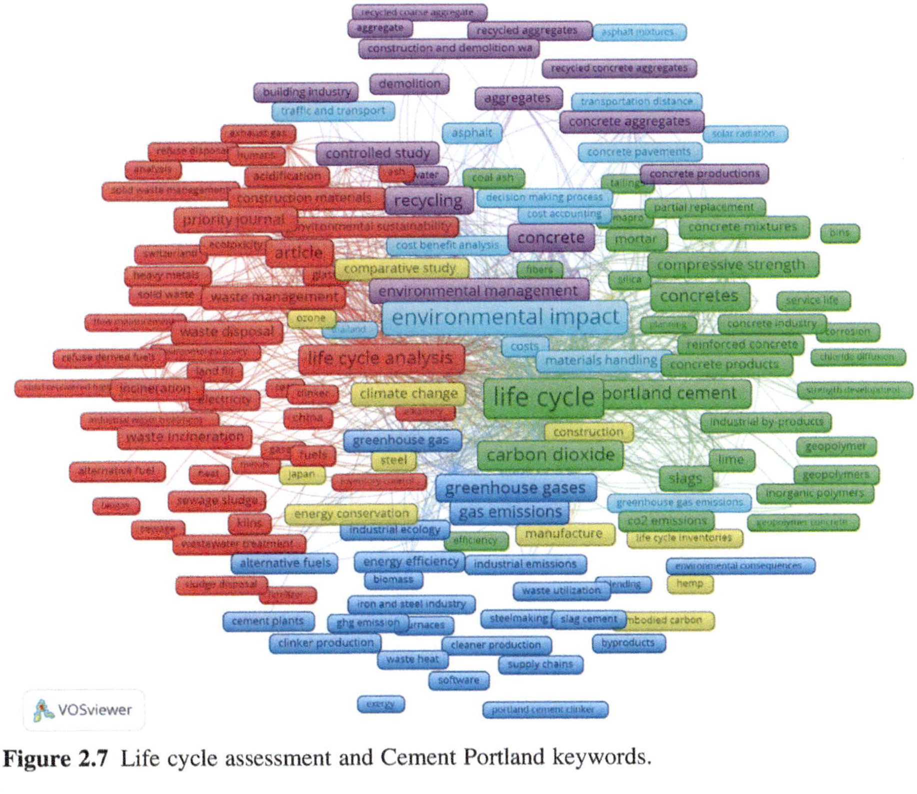

Figure 2.7 Life cycle assessment and Cement Portland keywords.

to the environment resulting from pollutants and provides opportunities for environmental management (Lee & Inaba, 2004; Mohammadi & South, 2017).

Recently, several computer tools based on LCA methodology have been developed to simplify the analysis for products and systems, such as SimaPro, GaBi, TEAM, LCE, and Open LCA. In order to provide relevant information about environmental impacts evaluation, this paper illustrates the classification of different midpoint indicators that lead to endpoint categories, as shown in Table 2.12: Environmental impacts related to human health, ecosystem quality and natural resource (European Commission - Joint Research Center—Institute for Environment and Sustainability (EC-JRC, 2010).

In principle, LCA could be used as a decision support tool by three distinct groups:

1. Product Producers: to improve the environmental performance of a production system.
2. Product Consumers: to guide purchasing decisions; and
3. Policymakers: to inform and direct long-term strategies.

Considering cement manufacturing process, CO_2 emissions have been evaluated due to the use of electrical energy for (i) Initial grinding of raw materials (ii) mixing of raw materials in the silos before preheating (iii) final grinding of clinker with gypsum & other substitutes (e.g., flyash, slag, pozzolan etc.).

2.8.2 Goal and scope

This section covers the aims for carrying out, the functional units, the system boundary, reference flows, limitation, data requirements and allocation procedures under ISO 14044 Goal and Scope.

The key functional unit (FU) used is either kg or m^3 of produced clinker, cement or concrete in cases where the target is to reduce the amount of cement in concrete.

Table 2.12 Environmental impacts related to human health, ecosystem quality and natural resource (European Commission - Joint Research Center—Institute for Environment and Sustainability EC-JRC (2010).

Midpoint indicators	Endpoint area of protection
Climate change	Human health/ecosystem quality
Ozone depletion	Human health
Ionizing radiation	Human health
Photochemical ozone formation	Human health/ecosystem quality
Particulate matter formation	Human health
Acidification	Ecosystem quality
Eutrophication	Ecosystem quality
Toxicity	Ecosystem quality
Land use	Human health/ecosystem quality
Consumptive water use	Human health/ecosystem quality
Resource depletion – fossil fuels	Natural resources
Resource depletion - minerals	Natural resources

However, Sagastume Gutiérrez et al. (2017) highlighted that is very important to consider both the durability and compressive strength when cements or concretes are assessed based on a previous research (Damineli et al., 2010). Initially, the authors, focused in to change the functional unit from a unit of concrete volume or to a unit of functional performance, such as compressive strength.

LCA applied to Portland cement production have been studied and published in the literature. Table 2.13 presents commonly used functional unit and system boundary. Additionally, literature (Seto et al., 2017) shows LCA to evaluate the benefit of cementitious materials, e.g., fly ash, silica fume and slag, as replacement of cement to reduce the environmental impacts of concrete. In this case, the functional unit is expressed in terms of 1 m^3 of concrete considering its compressive strength and type of application.

Table 2.13 Examples of functional unit and system boundary − cement Portland production.

Functional unit	System boundary	Country	Reference
1 kg of ordinary Portland cement (CEM I)	The first one (S1) is the extraction and preparation of the raw materials in the quarry. The second process (S2) is the production of clinker at the cement plant. The third process (S3) is the production of the cement, which includes the grinding of clinker and its mixing with gypsum	France	Chen et al. (2010)
1 (one) tonne of product or 20 bags of 50 kg of cement	Raw materials extraction from nature until its final disposal	Brazil	Nigri, Rocha and Filho (2010)
1000 kg of CEM I	System boundary also includes transportation of all materials and fuels from their source to the cement plant	Sweden	Mousavi (2013)
1 tonne of clinker and 1 tone of gray cement.	System boundaries include waste from refuse as secondary materials and as alternative fuels	Spain	García-Gusano et al. (2015)
bi = m/VC CS bi − binder intensity m − Mass of binder (kg) CS − 28 days compressive strength of concrete	Performance indicator as a functional unit allows avoiding the distinction between a material scale, where impacts are expressed in kg or m3, and a structural scale, where the effective volume to provide the function is used	Brazil	Damineli et al. (2010)

Continued

Table 2.13 Continued

Functional unit	System boundary	Country	Reference
CFP = m/CS*D CFP – cement Functional Performance m - Mass of binder (kg) CS – 28 days' compressive strength of concrete D – durability in years	Cement Functional performance (CFP): mass of cement needed to obtain 1 MPa of compressive strength during 1 year of durability (kg/(MPa·year)) Cements that result in concretes with the highest compressive strength and durability for the lowest mass of binder has obviously the lowest CFP.	Colombia	Sagastume Gutiérrez et al. (2017)
Packaging of 50 kg standard cement	The cradle-to-grave includes four stages: raw material, manufacture, distribution and final disposal. The raw material stage included the manufacturing of raw material for packaging. The manufacturing phase produced cement packaging as well as filling of cement. The distribution part included transportation to the warehouse, distribution center, retail terminal and then to the consumer. In the disposal stage, landfill and incineration are considered as main waste treatments.	China	Ma et al. (2019)
1 tonne of cement (PC or MgO cement)	Inventory data for the procurement of raw materials (e.g., quarrying of limestone), together with energy (electricity and heat) consumption of different procedures, was acquired from the SimaPro database. 'Cradle-to-gate' perspective consisting of environmental procedures ranging from the acquisition of raw materials to the packaging of the final product.	Singapore	Ruan and Unluer (2016)

2.8.2.1 *Life Cycle Inventory (LCI)*

The second phase of LCA is Life Cycle Inventory (LCI) where the information on raw materials, energy consumption, and emissions are collected. In summary, data collection, calculation, quantified inputs and outputs of a production system are included in the entire lifecycle. Table 2.14 shows a wide range of inputs considered to the analysis of environmental impacts presented in the literature.

2.8.2.2 *Life Cycle Impact Assessment (LCIA)*

The third phase of LCA is the Life Cycle Impact Assessment (LCIA). In the literature there are different methods and software to calculate the impacts. For example, CML, IMPACT202+, Ecoinvent. Usually, the adopted indicators include: abiotic depletion (AD), acidification (AC), eutrophication (EP), global warming potential (GWP), ozone layer depletion (OD), carbon footprint (CF) water footprint (WF), nitrogen footprint (NF) and photochemical oxidation (PO), Human toxicity (HT), Freshwater aquatic ecotox (FA), Marine aquatic ecotoxicity (MA), Greenhouse gases (GHG) (Chen et al., 2010).

Conducting LCIA (García-Gusano et al., 2015), observed that each cement subtype contribution is at the same level than productions. Portland cement Type-I contributed to 30% to each impact. Portland cement types II/A-V (fly-ashed), II/A-L (calcareous) and II/A-M(composite) also have remarkable contributions accounting for 10% to 15% each. The alternative to incorporate cementitious materials was correlated to

Table 2.14 Examples of inputs and life cycle stages.

Type of data/inputs	Life span	Reference
Raw materials/Energy/Transport/ plant installation	50 years with a production of 340,000 t/yr	Chen et al. (2010)
The clinker unit process includes the primary raw materials and the fuels required as well as the transport of the materials and the infrastructures. The cement unit process introduces the clinker and the extra constituents according to each cement subtype production in 2010.	In terms of energy, the average Spanish thermal consumption of the kiln was 3536 MJ/t of clinker, and the electricity consumption was 92 kWh/t of clinker	García-Gusano et al. (2015)
Inventory data for the procurement of raw materials (e.g., quarrying of limestone), together with energy (electricity and heat) consumption of different procedures, was acquired from the SimaPro database.	Climate change was also investigated by IPCC 2007 within a timeframe of 100 years.	Ruan and Unluer (2016)

the decarbonization of cement production process, i.e., reducing CO_2 emissions coming from limestone's calcination.

Kajaste and Hurme (2016) proposed the sum of CO_2 specific emissions ETotal (kg CO_2/t cement) is:

$$\text{ETotal} = \text{w1EClinker baseline-w2EClinkerSubstitute} + \text{w3EFossil Fuel} + \text{w4EElectricity} + \text{w5ETransport} \quad (2.6)$$

Ruan and Unluer (2016) assessed the environmental impacts of reactive MgO cement production and compared the outcomes with traditional Portland Cement production. Even considering that this type of cement is magnesite, these resources are widely available, for example, 75% of the global magnesite production is from China, North Korea, Slovakia, Turkey, Russia, Australia and India. There are potential applications, such as Refractory applications, MAP cements. This research focused on global warming potential, human heath, ecosystem quality and resources. The authors pointed main disadvantage of reactive MgO production when compared to that of PC is its effect on human health, which is attributed to the large quantity of coal used in the production of MgO. On the other hand, regarding the high carbonation capability of MgO cements within concrete mixes is considered, the net CO_2 emissions of reactive MgO cement was calculated to be up to 73% (i.e., 0.71 t/t) lower than those of Portland Cement.

2.8.2.3 Interpretation

The last phase of LCA is Interpretation. This step leads conclusions and recommendations, shows the magnitude of environmental issues, and generates appropriate decisions.

Determination of correct functional unit is crucial for the environmental impact analysis of Portland cement. According to Sagastume Gutiérrez et al. (2017), LCA of three different types of cements showed that when a Mass Functional Unit (MFU -1 kg of cement) widely used is not the most suitable one. Also, the second alternative - binder intensity, indicated that cements with the highest strength grade and the lowest clinker content have the lowest impacts. However, the CFP (combining strength and durability as primary function of cements) showed that less binder is required per unit of functional performance for the cements with higher zeolite fraction, which result in concrete mixes with higher durability. It is an indication that functionalities of cement, compressive strength and durability, give a more accurate comparison of different types of cements.

The main purpose to assess cement production environment impact is to provide recommendation to enhance the sustainability of concrete application in the engineering construction sector. It is a common practice to use blended cement, for example CEM I (100% PC content), CEM II/B−V (65% PC content, 35% FA content) and CEM III/B (30% PC content, 70% GGBS content). However, the final decision should consider not only the evaluation of environmental impact in terms of CO_2 equivalent, but incorporate cost analysis, uncertainty and sensitivity analysis using different methods (Tait & Cheung, 2016).

2.9 Summary

Life cycle assessment (LCA) is a method of evaluation of the environmental impacts of a product and/or a process. Most LCAs evaluate environmental impact over the entire life cycle (from cradle to grave) of a product or a process.

For cement, cradle to grave assessment is difficult due to the reason that cement has so many end-uses while each use has a complex life cycle. Therefore, LCA assessment of cement is limited to (from cradle to gate), which evaluates impacts of producing cement from raw materials (their extraction, crushing, grinding) to processing, packaging and shipping without taking into consideration the end use and disposal of cement. The cradle to gate GHG emissions evaluated for CEM-1 show total GHG effect of 0.817 kg/kg of cement produced, while blended cements provide greatest environmental savings of approximately 5% (utilization of CKD for sequestration) in cement.

References

AASHTO. (2018). *Standard specification for blended hydraulic cement ASTM: C 595-18, AASHTO: M 240-03. USA.*

(ADAA), A. D. A. of A. (2009). Fly ash in Australian standards, opportunities with using fly ash in concrete. *Flyash Reference Datasheet No.1.* Available at: https://www.adaa.asn.au/uploads/default/files/adaa-ref_data_sheet_1.pdf.

American Standards Testing Materials (ASTM). (2020). *ASTM C 595/C 595M standard specification for blended hydraulic cements* (pp. 1–8). ASTM.

Andrew, R. M. (2019). Global CO_2 emissions from cement production, 1928–2018. *Earth System Science Data, 11*(4), 1675–1710. https://doi.org/10.5194/essd-11-1675-2019

ASTM. (2016). *ASTM C150/C150M-standard specification for Portland cement.* https://doi.org/10.1520/C0150

Basheer, P. A. M., & Barbhuiya, S. (2009). Chapter 6: Different types of cement used in concrete. In *ICE manual of construction materials* (pp. 61–68). U.K.: Institution of Civil Engineers. https://doi.org/10.1680/mocm.35973.0061

Bentur, A. (2003). Cementitious materials - Nine Millennia and A new century: Past, present, and Future, Perspectives in Civil Engineering. *Commemorating the 150th Anniversary of the American Society of Civil Engineers, 14*(February), 243–263. https://doi.org/10.1061/(asce)0899-1561(2002)14:1(2)

Blezard, R. G. (2004). Chapter 1 - the history of calcareous cements. In P. C. B. T.-L. C. of C. Hewlett, & C. (Fourth E (Eds.), *Lea's chemistry of cement and concrete* (pp. 1–23). Oxford: Butterworth-Heinemann. https://doi.org/10.1016/B978-075066256-7/50013-8

Bogue, R. H. (1955). *The chemistry of Portland cement.* Reinhold Publishing Corporation (Reinhold reference books). Available at: https://books.google.com.au/books?id=IvZTAAAAMAAJ.

Brunauer, S., & Copeland, L. E. (1964). *Scientific American.* Scientific American. Available at: https://books.google.com.au/books?id=WTk7AAAAMAAJ.

Cembureau. (2019). *Cembureau activity report 2019.* The European Cement Association. https://doi.org/10.6009/jjrt.2020_jsrt_76.3.304

Cement Australia. (2011). Blended cement. Available at: https://www.cementaustralia.com.au/sites/default/files/2018-09/product_data_sheet_ca0628_general_blend_cement.pdf.

Chen, C., et al. (2010a). Environmental impact of cement production: Detail of the different processes and cement plant variability evaluation. *Journal of Cleaner Production, 18*(5), 478–485. https://doi.org/10.1016/j.jclepro.2009.12.014. Elsevier Ltd.

Chen, C., et al. (2010b). LCA allocation procedure used as an incitative method for waste recycling: An application to mineral additions in concrete. *Resources, Conservation and Recycling, 54*(12), 1231–1240. https://doi.org/10.1016/j.resconrec.2010.04.001. Elsevier B.V.

Damineli, B. L., et al. (2010). Measuring the eco-efficiency of cement use. *Cement and Concrete Composites, 32*(8), 555–562. https://doi.org/10.1016/j.cemconcomp.2010.07.009

Dehayes, S. M., & Tennis, P. D. (2006). Hydraulic cement-chemical properties. In J. F. Lamond, & J. H. Pielert (Eds.), *Significance of tests and properties of concrete making materials. STP 169D* (pp. 450–461). ASTM International.

European Commission - Joint Research Centre, Institute for Environment and Sustainability (EC-JRC). (2010). *Framework and requirements for life cycle impact assessment Models and indicators. First, International reference life cycle data system (ILCD) handbook.* Italy: Luxembourg: Publications Office of the European Union. https://doi.org/10.2788/38719. First.

European Commission. (2018). *Competitiveness of the European cement and lime sectors.* Available at: https://www.wifo.ac.at/jart/prj3/wifo/resources/person_dokument/person_dokument.jart?publikationsid=61003&mime_type=application/pdf.

European Standards. (2011). *EN 197-1-2011 - cement - Part 1- composition, specifications and conformity criteria for common cements* (pp. 1–44).

García-Gusano, D., et al. (2015). Life cycle assessment of the Spanish cement industry: Implementation of environmental-friendly solutions. *Clean Technologies and Environmental Policy, 17*(1), 59–73. https://doi.org/10.1007/s10098-014-0757-0. Springer Science and Business Media Deutschland GmbH.

Guirguis, S. (1993). The new cement standard in Australia-its implication and further development. *Cement, Concrete and Aggregate, 15*(2), 170–173. ASTM.

Guo, R., et al. (2020). Global CO_2 uptake of cement in 1930–2019. *Earth System Science Data Discussions*, 1–28. https://doi.org/10.5194/essd-2020-275, 2020.

Hewlett, P. C., & Liska, P. (2019). *Lea's Chemistry of Cement and Concrete (5th Edn.).* UK: Butterworth-Heinemann, Elsevier Publications, ISBN 978-0-08-100773-0.

Kajaste, R., & Hurme, M. (2016). Cement industry greenhouse gas emissions - management options and abatement cost. *Journal of Cleaner Production, 112*, 4041–4052. https://doi.org/10.1016/j.jclepro.2015.07.055. Elsevier Ltd.

Khongprom, P., Champanoi, S., & Suwanmanee, U. (2020). An input-output approach for environmental life cycle assessment of cement production. *Chemical Engineering Transactions, 81*(2017), 1345–1350. https://doi.org/10.3303/CET2081225

Klieger, P., & Lamond, J. F. (1994). Significance of tests and properties of concrete and concrete-making materials STP-169C. In American Society for Testing and Materials (Ed.), *ASTM speci.* ASTM (ASTM/STP). Available at: https://books.google.com.au/books?id=pl3Dh5sYwCUC.

Kohlhaas, B., & Labahn, O. (1983). *Cement engineers' handbook.* Bauverlag. Available at: https://books.google.com.au/books?id=KmyeQgAACAAJ.

Kurdowski, W. (2014). *Cement and concrete chemistry.* Springer Netherlands. Available at: https://books.google.com.au/books?id=XT0RngEACAAJ.

Lee, K.-M., & Inaba, A. (2004). *Life cycle assessment: Best practices of International Organization for Standardization (ISO) 14040 series* (February).

Li, Z. (2011). *Advanced concrete technology.* Wiley. Available at: https://books.google.com.au/books?id=1lthdPkTS9wC.

Ma, X., Li, C., & Li, B. (2019). Carbon emissions of China's cement packaging: Life cycle assessment. *Sustainability. MDPI AG, 11*(20), 5554. https://doi.org/10.3390/su11205554

Massazza, F. (1998). Chapter 10 - Pozzolana and pozzolanic cements. In Hewlett, P. C. B. T.-L. C. of C. and C. (Fourth E (Ed.), *Lea's chemistry of cement and concrete* (pp. 471–635). Oxford: Butterworth-Heinemann. https://doi.org/10.1016/B978-075066256-7/50022-9

Massazza, F. (2003). Pozzolana and pozzolanic cements. In *Lea's chemistry of cement and concrete* (4th ed.). Elsevier Ltd. https://doi.org/10.1016/B978-075066256-7/50022-9

Mehta, P. K., & Monteiro, P. J. M. (2013). Chapter 6 "Hydraulic cements". In P. K. Mehta, & P. J. M. Monteiro (Eds.), *Concrete: Microstructure, properties and materials* (4th ed., pp. 1–672). McGraw-Hill Education. Available at: https://books.google.com.au/books?id=c-9JAgAAQBAJ.

Mohammadi, J., & South, W. (2017). Life cycle assessment (LCA) of benchmark concrete products in Australia. *International Journal of Life cycle Assessment, 22*(10), 1588–1608. https://doi.org/10.1007/s11367-017-1266-2. Springer Verlag.

Moir, G. (2003). 1 cements. In J. Newman, & B. Seng Choo (Eds.), *Advanced concrete technology. Vol.1 Constituent materials* (pp. 1–20). Elsevier B.V.

Moranville-Regourd, M. (1998). Chapter 11 - Cements made from blastfurnace slag. In Hewlett, P. C. B. T.-L. C. of C. and C (Ed.), *Lea's chemistry of cement and concrete* (Fourth E, pp. 637–678). Oxford: Butterworth-Heinemann. https://doi.org/10.1016/B978-075066256-7/50023-0

Mousavi, M. (2013). *Life cycle assessment of Portland cement and concrete bridge.*

Nawy, E. G. (2008). *Concrete construction engineering handbook.* CRC Press. Available at: https://books.google.com.au/books?id=1OwkUrXuhjQC.

Newman, J., & Choo, B. S. (2003). Chapter.1 "Cements". In *Advanced concrete technology Vol.1: Constituents* (pp. 1–12). Elsevier Science. Available at: https://books.google.com.au/books?id=IMdF-QR_8mkC.

Nigri, E., Rocha, S., & Filho, E. (2010). Portland cement: An application of life cycle assessment. *Product: Management & Development, 8*(2), 167–172.

Ruan, S., & Unluer, C. (2016). Comparative life cycle assessment of reactive MgO and Portland cement production. *Journal of Cleaner Production, 137*, 258–273. https://doi.org/10.1016/j.jclepro.2016.07.071. Elsevier Ltd.

Sagastume Gutiérrez, A., et al. (2017). Improved selection of the functional unit in environmental impact assessment of cement. *Journal of Cleaner Production, 168*, 463–473. https://doi.org/10.1016/j.jclepro.2017.09.007

Seligmann, P., & Greening, N. R. (1969). Chemistry of cement clinker. Cement Association of Japan (Proceedings of the Fifth International Symposium on the Chemistry of Cement, Tokyo, 1968). In *Proceedings of the fifth international symposium on the chemistry of cement, Tokyo, 1968* (pp. 170–199). Available at: https://books.google.com.au/books?id=eJYtAQAAIAAJ.

Seto, K. E., Churchill, C. J., & Panesar, D. K. (2017). Influence of fly ash allocation approaches on the life cycle assessment of cement-based materials. *Journal of Cleaner Production, 157*, 65–75. https://doi.org/10.1016/j.jclepro.2017.04.093. Elsevier Ltd.

Soomro, M., & Hung, Y. T. (2013). Use of selective recycled materials in civil engineering construction. In , *2. Handbook of Environment and Waste Management: Volume 2: Land and Groundwater Pollution Control* (pp. 635–684). https://doi.org/10.1142/9789814449175_0010

Standards Australia. (2010). *AS 3972-2010 general purpose and blended cements.*

Standards Australia. (2016). *AS 3582.1 supplementary cementious materials for use with Portland and blended cements-flyash. Standards Australia.*

Tait, M. W., & Cheung, W. M. (2016). A comparative cradle-to-gate life cycle assessment of three concrete mix designs. *International Journal of Life Cycle Assessment, 21*(6), 847–860. https://doi.org/10.1007/s11367-016-1045-5. Springer Verlag.

Taylor, H. F. W. (1989). Modification of the Bogue calculation. *Advances in Cement Research, 2*(6), 73–77. https://doi.org/10.1680/adcr.1989.2.6.73

Taylor, H. F. W. (1997). *Cement Chemistry*. Thomas Telford. Available at: https://books.google.com.au/books?id=1BOETtwi7mMC.

UNEP, SETAC and LC-Initiative. (2009). *Guidelines for social life cycle assessment of products, management*.

Vernet, C., Demoulin, E., Gourdin, P., & Hawthorn, F. (1980). *7th ICCC Paris, II* pp. II–267) (Paris).

Wirth, K., & Barth, A. (2020). *X-ray fluorescence (XRF)*. Indianapolis: Indiana University, Purdue University. Available at: https://serc.carleton.edu/207633.

Industrial and agro-waste materials for use in recycled concrete

3

Mahfooz Soomro[1], *Vivian W.Y. Tam*[1] and *Ana Catarina Jorge Evangelista*[2]
[1]School of Engineering, Design and Built Environment, Western Sydney University, Penrith, NSW, Australia; [2]Engineering Institute of Technology, Perth, WA, Australia

3.1 Flyash

Flyash is a by-product of burning coal in the thermal power stations to produce electricity. To generate electricity, coal is first pulverized in grinding mills, which is then blown with air in the burning zone of the furnace. In this zone the coal combustible constituents ignite at very high temperatures producing heat with the temperatures reaching approximately 1500°C, while the noncombustible inorganic constituents melt and form very small liquid droplets. These spherical shape droplets are carried out of the burning zone by the flue gases. While leaving the chamber, they rapidly cool forming small glassy spherical particles, which are trapped and collected either in the bag houses or in the electrostatic precipitators. Flyash accounts for 75%–85% of the total coal, while the remaining 25%–15% is the ash collected at the bottom of the furnace called bottom ash. Fig. 3.1 represents schematic diagram of coal power plant.

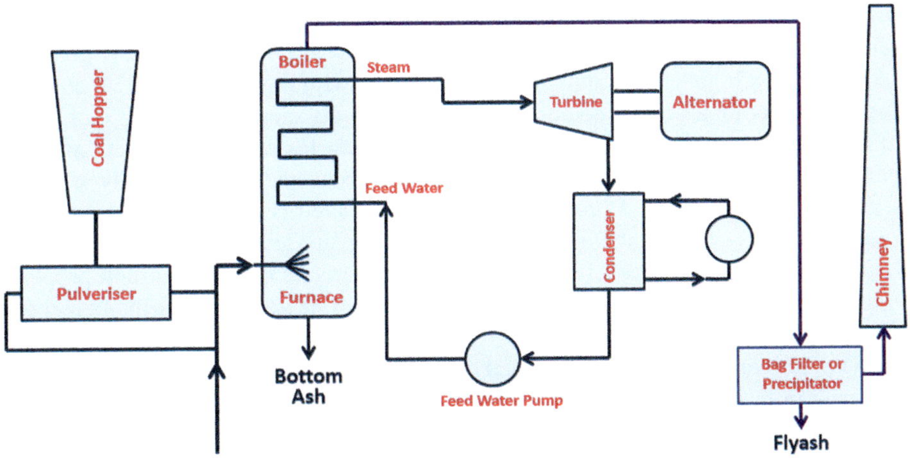

Figure 3.1 Schematic diagram of coal power plant (Das & Kumaran, 2015).

Recycled Concrete. https://doi.org/10.1016/B978-0-323-85210-4.00009-6

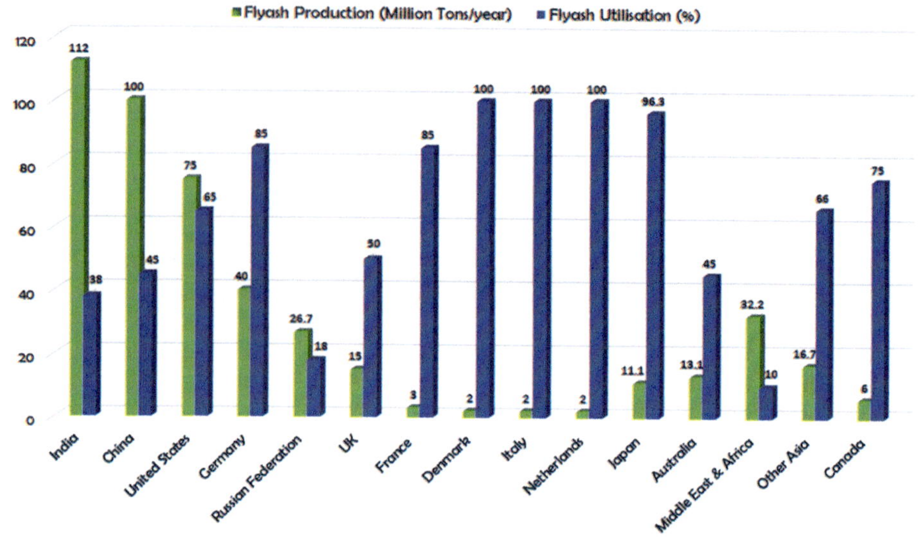

Figure 3.2 The production and utilization of flyash around the globe (Gollakota & Shu, 2019).

In 2015 the global production of flyash was approximately 750 million tons (Mt), while the global utilization of flyash was only 187.5 Mt or (1/4th) of the total production (Yao et al., 2015). The production and utilization of flyash around the globe is shown in Fig. 3.2.

3.1.1 Characteristics of flyash

3.1.1.1 Physical properties

The chemical composition of the pulverized coal and the burning conditions within the burning chamber of the power plant can affect the properties and the color of the collected flyash. Flyash colors can be light, tan to dark gray. Light to tan are generally associated with higher lime content, brownish color with higher iron content and dark gray to black color attributed to unburned carbon usually measured as Loss on Ignition (LOI) content. Usually, flyash color is very consistent for each power plant and coal source. Flyash particles are finer than Portland cement and predominantly spherical in shape either solid called precipitator or hollow called cenosheres. These particles are mostly glassy (amorphous) in nature, having density less than 1.0 gm/cm^3. Their size typically range between 10–100 μm. Fineness is one of the important property, which contributes to the pozzolanic activity of flyash. It is defined as the percent by weight of the material retained on 45 μm sieve (325 mesh). ASTM C 618 (2014) limits the maximum amount of flyash retained on 45 μm sieve to be 34%. Coarser gradation can result in less reactive ash and also could contain higher carbon content (Siddique et al., 2011). SEM micrograph of flyash at magnification (x1500) is shown in Fig. 3.3, and the summary of the physical properties of flyash are listed in Table 3.1.

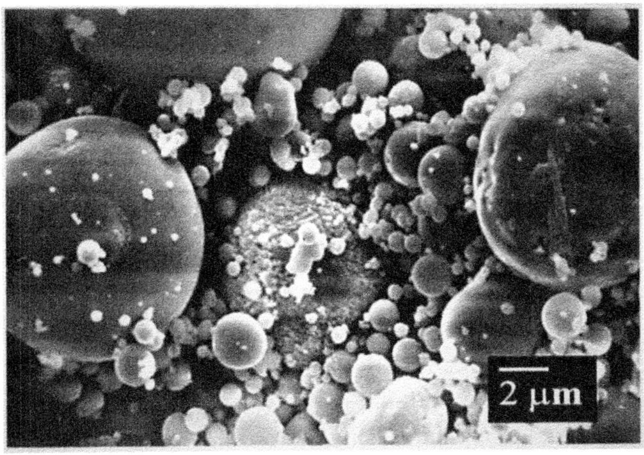

Figure 3.3 SEM micrograph of flyash (Sokolar & Nguyen, 2018).

Table 3.1 Summary of the physical properties of flyash (Gollakota & Shu, 2019).

Property	Flyash	Reference
Bulk density (compacted flyash) (kg/m^3)	1120–1500	Memphis.edu, Chap.03
Specific gravity	1.6–2.6	
Plasticity	Nonplastic	
Free swell index	Very low	
Porosity (%)	30–65	
Surface area (m^2/g)	5	
Lime reactivity (MPa)	1–8	
pH	6.0–8.0	
Extreme particle size (mm)	0.001–0.1	
Average particle size (μm)	6.92	Adjei et al. (2020)
Particle shape	Spherical (solid/hollow)	Adjei et al. (2020)
Optimum moisture content (%)	45.5	Kim et al. (2005)
California bearing ratio (%)	10	Mahvash et al. (2017)
Los Angles Abrasion (%)	28.35	Dash et al. (2016)
Permeability coefficient (cm/s)	9.5×10^{-6}	Hasan et al. (2014)

3.1.1.2 Chemical composition of flyash

The chemical composition of flyash depends on the mineral composition (inorganic part) of the coal. The summary of the chemical composition ranges of flyash produced in various countries around the world, are listed in Table 3.2.

Table 3.2 Summary of chemical composition ranges of flyash around the world.

Country	Chemical composition (%)										Reference
	SiO_2	Al_2O_3	Fe_2O_3	CaO	MgO	SO_3	K_2O	Na_2O	TiO_2	P_2O_5	LOI	
Australia	31.1−68.6	17.0−33.0	1.0−27.1	0.1−5.3	0−2.0	0−1.5	0.1−2.9	0−0.6	1.2−3.7	0−3.9	n.a	Blissett (2012), Lu and Do (1991)
Bangladesh	55.0	24.7	7.7	6.2	0.7	n.a	1.1	1.1	n.a	0.9	n.a	Cokca and Yilmaz (2004)
Bulgaria	30.1−57.4	12.5−25.4	5.1−21.2	1.5−28.9	1.1−2.9	0.4−12.7	0.8−2.8	0.4−1.9	0.6−1.0	0.1−0.4	0.8−32.8	BP Energy (2018)
Canada	35.5−62.1	12.5−23.2	3.0−44.7	1.2−13.3	0.4−3.1	0.2−7.8	0.5−3.2	0.1−7.3	0.4−1.0	0.1−1.5	0.3−9.7	Ram and Masto (2014), Ramezanianpour (2014)
China	35.6−57.2	18.8−55.0	2.3−19.3	1.1−7.0	0.7−4.8	1.0−2.9	0.8−0.9	0.6−1.3	0.2−0.7	1.1−1.5	n.a	Blissett (2012)
Denmark	53.98	22.27	11.60	3.95	1.97	0.73	Equivalent = 2.71		n.a	n.a	2.13	Nairn (2001)
France	47.0−51.0	26.0−34.0	6.9−8.8	2.3−3.3	1.5−2.2	0.1−0.6	n.a	2.3−6.4	n.a	n.a	0.5−4.5	Wesche (1991)
Germany	20.0−80.0	1.0−19.0	1.0−22.0	2.0−52.0	0.5−11.0	1.0−15.0	0−2.0	0−2.0	0.1−1.0	n.a	0−5.0	Vichaphund (2014)
Greece	21.0−35.0	10−17.9	4.5−8.4	27.3−45.0	1.5−3.8	4.0−8.6	0.4−1.0	0.2−1.0	n.a	n.a	3.0−7.0	Vichaphund (2014), BP Energy (2018)
India	50.2−59.7	14.0−32.4	2.7−16.6	0.6−9.0	0.1−2.3	n.a	0.2−4.7	0.2−1.2	0.3−2.7	n.a	0.5−7.2	Pandian (2004), Blissett (2012)
Israel	45.6−58.6	24.4−34.5	3.0−6.7	4.9−9.9	1.6−2.5	0.6−0.8	0.1	0.01	1.2−1.9	0.8−1.8	6.0	Prashanth (2001)
Italy	41.7−54	25.9−33.4	3.0−8.8	2.0−10.0	0−2.4	n.a	0−2.6	0−1	1.0−2.6	0−1.5	1.9−9.0	BP Energy (2018), Feuerborn (2009)
Japan	53.9−63.0	18.2−26.4	4.2−5.7	2.0−8.1	0.9−2.4	0.3−1.4	0.6−2.7	1.1−2.1	0.8−1.2	n.a	0.5−2.1	Kolbe (2011), Franus (2015)
Korea	50.0−55.7	24.7−28.7	3.7−7.7	2.6−6.2	0.7−1.1	0.5−1.1	1.1	n.a	n.a	0.9	4.3−4.7	Cokca (2004), Mollamahmutoğ;lu (2001), Nhan (1996)
Mexico	59.6	22.8	5.6	3.1	0.9	0.4	1.3	0.5	0.9	0.0	n.a	Mukherjee (2008)
Netherlands	45.1−59.7	24.8−28.9	3.3−9.0	0.5−6.8	0.6−3.7	0.2−1.3	0.6−2.9	0.1−1.2	0.9−1.8	0.1−1.5	2.7−8.1	BP Energy (2018), Belviso (2010)
Poland	32.2−53.3	4.0−32.2	4.5−8.9	1.2−29.9	1.2−5.9	n.a	0.2−3.3	0.2−1.5	0.6−2.2	0.1−0.9	0.5−28.0	Moreno (2005)
Russia	40.5−48.6	23.2−25.9	n.a	6.9−13.2	2.6−4.0	n.a	1.9−2.6	1.2−1.5	0.5−0.6	0.3−0.4	n.a	Chindaprasirt (2012)

South Africa	46.3–67.0	21.3–27.0	2.4–4.7	6.4–9.8	1.9–2.7	n.a	0.5–1.0	0–1.3	1.2–1.6	0.3–0.9	n.a	Mukherjee (2008), ASTM C-9 (2005), Lee (2003)
Spain	41.5–58.6	17.6–45.4	2.6–16.2	0.3–11.8	0.3–3.2	0.1–2.2	0.2–4.0	0–1.1	0.5–1.8	0.1–1.7	1.1–9.7	BP Energy (2018), Yao (2015), Koukouzas (2006)
Taiwan	48.75	23.21	4.15	3.39	1.0	n.a	1.10	0.24	n.a	n.a	10.39	Nairn (2001)
Turkey	37.9–57.0	20.5–24.3	4.1–10.6	0.2–27.9	1.0–3.2	0.6–4.8	0.4–3.5	0.1–0.6	0.6–1.5	0.2–0.3	0.4–2.7	BP Energy Outlook (2018), Wang and Wu (2006), Kolias (2005), Takada (1995)
United Kingdom	50.09	28.10	11.70	1.62	1.54	n.a	0.62	0.28	n.a	n.a	2.30	Nairn (2001)
United States of America	34.9–58.5	19.1–28.6	3.2–25.5	0.7–22.4	0.5–4.8	0.1–2.1	0.9–2.9	0.2–1.8	1.0–1.6	0.1–1.3	0.2–20.5	Binal (2016), Blissett (2012)
Minima	20.0	1.0	1.0	0.1	0.0	0.0	0.0	0.0	0.1	0.0	0.0	
Maxima	80.0	55.0	44.7	52.0	11.0	15.0	4.7	7.3	3.7	3.9	32.8	

n.a, not available.

3.1.2 Contribution of flyash elemental oxides

The bulk composition of flyash, especially silica (SiO_2), alumina (Al_2O_3), ferric oxide (Fe_2O_3), and calcium oxide (CaO), has a significant influence upon its utilization; a high CaO content makes it valuable for cement and soil stabilization, a specific ratio of SiO_2 to Al_2O_3 is important for zeolite synthesis, while a higher sulfite (SO_3) and carbon content prevents its use in concretes.

Alkalis (Na_2O and K_2O) react with silicates within aggregates, producing a gel, resulting in swelling and deterioration of concretes; however, flyash does not contribute reactive alkalis to the concrete therefore it can be added to control these reactions (Cook et al., 1992; Dhir et al., 1986). The presence of alkalis is important in the synthesis of geopolymers and zeolite where they play a charge-balancing role as part of the structure.

Magnesium oxide (MgO) can result in the formation of periclase (a crystallization reaction, which results in volume expansion of mineral in concretes), which is undesirable as it results in instability and deterioration of concretes (Dhir et al., 1986).

Sulfite (SO_3) can react with unhydrated and hydrated C_3A to form Ettringite (CaO, Al_2O_3, $3CaSO_4$, $32H_2O$), which has the capability of substituting elements into its structure, making it useful in some waste stabilization applications (Dhir et al., 1986).

Loss on ignition (LOI) is a measure of the unburned carbon content of flyash. Unburned carbon has a large surface area and has a capacity to physically adsorb compounds onto its surface (Hassett et al., 1999); hence, the carbon content in flyash is critical because of its adverse effect on air entrainment in concrete, therefore high carbon flyash or flyash with varying carbon content must be avoided whenever air entrainment in concrete is a prerequisite (Ramachandran et al., 1996).

While the quality of flyash still continues to be monitored by its LOI content, its validity as a predictive tool remains questionable (Dodson et al., 2013), it has been found that BET surface area seems to give a more realistic view of the carbon content than the LOI (Külaots et al., 2004).

Elements contained within the glass phase of the flyash are relatively stable; however, the elements that concentrate within mineral phases or remain at the surface of flyash usually participate in leaching.

3.1.3 Classification of flyash

ASTM C 618 (2019) specifications cover coal flyash and raw or calcined natural pozzolan for blending in cement or in concrete, where cementitious or pozzolanic activity is desired. ASTM C 618 has categorized flyash and natural pozzolan into three classes.

3.1.3.1 Class F

This class of flyash exhibits pozzolanic properties but rarely self-hardening properties. Class F flyash is normally produced from the burning of anthracite or bituminous coal and with CaO content of less than 6%. It is designated as low calcium ash, and is not self-hardening but generally exhibits pozzolanic properties. Class F flyash predominantly contain >70% amorphous silica, which is the determining factor for pozzolanic activity. In the presence of water, Class F flyash will react with calcium oxide (CaO) or calcium hydroxide [$Ca(OH)_2$] to form cementitious compounds similar to those generated by hydration of Portland cement. Because flyash particles are spherical in shape, substitution of flyash generally improves workability and reduces water demand when mixed with cement or concrete.

3.1.3.2 Class C

This class of flyash has pozzolanic properties as well as self-cementitious properties since most Class C flyash contain from 10% to more than 15% CaO. Class C flyash is normally produced from lignite or sub-bituminous coal. When mixed with water, Class C flyash hydrates similar as Portland cement. The degree of self-hardening depends on the CaO content. The higher the CaO content, the higher the self-cementitious value. The unburned carbon content in Class C flyash is very low with LOI value less than 1%.

3.1.3.3 Class N

Raw or calcined natural pozzolan such as some diatomaceous earths, opaline cherts, pumice, shales, tuffs and volcanic ashes calcined or uncalcined are included in this category. Calcined kaolin clay and some agro-waste ashes also fall in this Class.

The glass fraction in flyash usually varies between 70% and 89% depending on the type and coal source, degree of coal pulverization, combustion conditions in the furnace, and the rate of cooling of combustion residue. It is believed that the calcium in the glassy content of flyash affects the pozzolanic activity, rather than the free lime.

The European standard for Flyash, BS EN 450, also classifies flyash in two general classes; siliceous flyash (V), and calcareous flyash (W). Siliceous flyash must contain less than 5% (by mass) of reactive CaO and at least 25% (by mass) of reactive silica, while in calcareous flyash the content of reactive CaO must not be less than 5% by mass. Calcareous flyash, containing between 5% and 15% by mass of reactive CaO, must contain more than 25% by mass of reactive silica.

The EN 450 (2012) has been divided into two parts, EN 450-1 deals with the definitions, specifications and conformity criteria, whereas EN 450-2 covers the conformity evaluation of flyash for concrete. The flyash producer (either the power plant or marketing company) is responsible for the quality, efficiency and reliability of the product.

In 2005, the revised EN 450 included some fundamental amendments; the most important ones from the viewpoint of the producers are:

- Introduction of three categories of LOI (as listed in Table 3.3a);
- Introduction of two classes of fineness (amount retained on 45 μm sieve Table 3.3b)
- The acceptance of flyash obtained from the co-combustion of certain types of biomass and waste (mentioned in the footnote of Table 3.3a); and
- Acceptance of flyash processed e.g., by classification, selection, sieving, drying, blending, grinding, carbon reduction or by a combination of these processes in an adequate production plant as flyash, according to the standard.

The standard refers to siliceous flyash only, as calcareous flyash, obtained mostly from the combustion of lignite, cannot be used as concrete addition according to the (EN 450). This flyash is used mostly for the filling of opencast mines. To increase the use of flyash, the European Coal Combustion Products Association (ECOBA) is active in the development of European standards for flyash, especially the (EN 450) "Flyash for Concrete" and (BS EN 14227) Part 3: Flyash bound mixtures and (BS EN 14227) Part 4: Flyash for hydraulically bound mixtures.

3.1.4 Chemical and physical requirements for flyash used as mineral admixture in concrete

The performance of flyash in concrete is strongly influenced by its physical, chemical and mineralogical properties, while the chemical and mineralogical composition is dependent mainly on the composition of coal. The related properties are covered in standards including the regulations governing the utilization of flyash. Table 3.3a and 3.3b lists the international standards from USA (ASTM C 618), Europe (EN 450-1), Australia/New Zealand (AN/NZS 3582.1), Japan (JIS A 6201) to China ((GB/T 1596)) and contain chemical and physical properties of flyash. These standards are compiled to inform about options for increased utilization and avoid misunderstandings in cross border trade. Ashes can be processed in suitable production facilities through classification, selection, sieving, drying, blending and grinding or carbon reduction by heat treatment, or by a combination of these processes. Such processed fly ashes may consist of fly ashes from different sources each conforming to the required criteria of the standards.

3.1.5 Utilization of flyash

Flyash must be viewed as a commodity with a promising economic future and as a valuable raw material rather than an unwanted solid waste. The potential for flyash utilization is an excellent illustration of how an industrial by-product can be successfully reintegrated into a nation's economy. Fig. 3.4 illustrates the utilization of flyash worldwide, in various fields of engineering construction, while Fig. 3.5 gives an overview of the utilization of flyash in various fields of engineering construction in Europe.

3.1.6 Benefits of flyash in construction industry

For most concrete producers, flyash is an important ingredient in concrete mix designs. Depending on the application, the type of flyash, specification limits, geographic location and climate, flyash can be used at levels ranging from 15% to 25% (most common) to 40%–60% (when rapid setting time is not required), thereby reducing emissions by roughly the same amount, and helping to keep concrete products at an affordable price.

Flyash is a pozzolanic material. It is a finely-divided amorphous alumino-silicate with varying amounts of calcium, which when mixed with Portland cement and water, will react with the calcium hydroxide released by the hydration of Portland cement to

Table 3.3a Compilation of chemical requirements for flyash used in cement & concrete (Yao et al., 2014).

Country	USA			Europe			Australia/New Zealand			Japan	China	
Standard classifications	**ASTM C618**			**BS EN 450-1**			**AS/NZ 3582.1**			**JIS A6201**	**GB/T 1596**	
	Class		Natural Pozzolan				Spec. grade	Grade 1	Grade 2		Flyash used in cement & concrete	
	F	C	N	A	B	C	Fine	Medium	Coarse		F	C
SiO_2				<22						>45		
$SiO_2 + Al_2O_3 + Fe_2O_3$ min (%)	70	50	70	65			70	70	70	70		
$CaO_{free\ max}$ (%)		> Class F	Free CaO <2.6 Reactive CaO <11.0				10			6	1	4
MgO max (%)	5	5		4.5						5		
SO_3 max (%)	5	5	4	3				3		5	3	
Loss on Ignition (LOI) max (%)	12	6	10	7.0	<9.0	<11.0	3	4	6	10	5.0 (Class I) 8.0 (Class II) 15.0 (Class III)	
Supplementary (optional) chemical requirements												
Available alkalis, as Na_2O, max (%)	1.5	1.5		5.5								
Moisture (%)	3.0						0.5			1.0	<1.0	

Note: EN 450-1 allows the following types of co-combustion materials, which can be used as flyash.
1. Vegetable materials such as woodchips, straw, olive shells, and other vegetable fibers.
2. Green wood and cultivated bio-mass
3. Animal meal
4. Municipal sewage sludge
5. Paper sludge
6. Petroleum coke
7. Virtually ash free liquid as gaseous fuels

Table 3.3b Physical requirements for flyash (Harris et al., 2019).

Country	USA		Europe		Australia			Japan	China	
	ASTM C618		EN 450-1		AS 3582.1			JIS 6201	GB/T 1597	
Standard classifications	Class F	Class C	Cat. N	Cat. S	Spec. Grade	Grade 1	Grade 2		Class F	Class C
					Fine	Medium	Coarse			
Amount retained on 45 μm sieve	34 ± 5%		40 ± 10%	12	15	25	45		Class I – ≤ 12% Class II – ≤ 30% Class III – ≤ 45%	
Fineness, specific surface area (cm^2/g)								min 2400[1]		
Particle density (kg/m^3)	5%[1]		± 200[1]						≤2.6	
Specific gravity (min)								1.95		
Soundness (max)	0.8%[2]		10 mm[2]							≤5 mm
Setting time (max) (minutes)			120[3]							
Strength Activity Index @ 7 days, min (min %)	75[3]									
Strength Activity Index @ 28 days, min (min %)	75[4]		75[4]					60	≥70	
Strength Activity Index @ 90 days, min (min %)			85[4]							
Strength Activity Index @ 91 days, min (min %)								70		
Relative strength (MPa) min					105% (1)	75% (1)				

Water requirement, max (%)	105	95^5	—		Class I – ≤ 95 Class II – ≤ 105 Class III – ≤ 115
Water content ratio (%)	—	—	—	102	—
	1. Max deviation from average value **2.** According to ASTM C-1012, Flyash/Cement: 20/80 equal slump **3.** According to ASTM C311, Flyash/cement: 20/80 equal slump.	**1.** Max deviation from declared value **2.** Only when CaO_{free} content is >1.5%. **3.** To be tested with mix of 25% flyash: 75% cement. **4.** According to EN196, no fixed flyash/cement ratio W/C = 0.5. **5.** Only for fineness category S.	**1.** Relative strength not a measure of performance but for reactivity in a given blend (blend by trials).	The fineness shall not differ by 450 cm^2/g.	**1.** For flyash for cement use

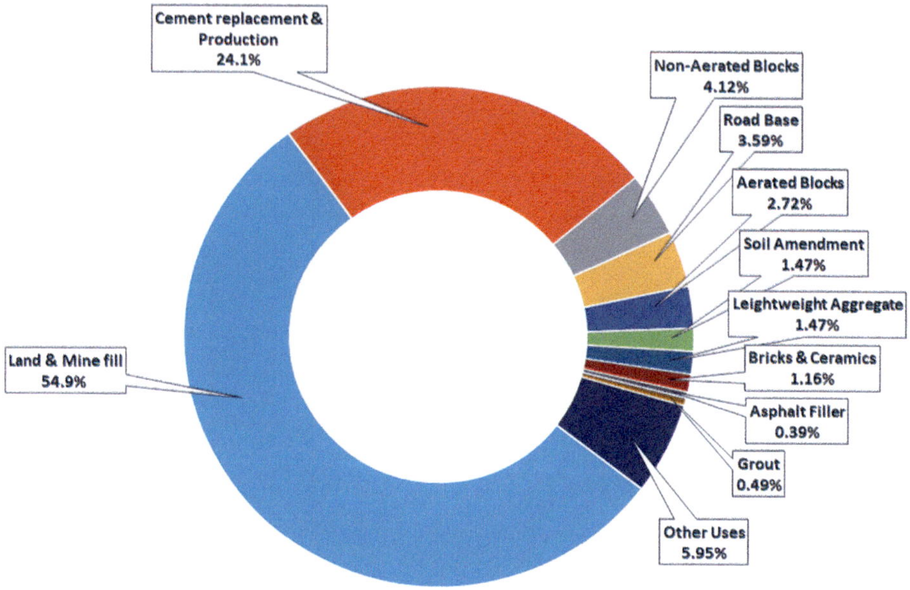

Figure 3.4 Utilization of flyash in various fields of engineering construction, worldwide (Soomro & Hung, 2014).

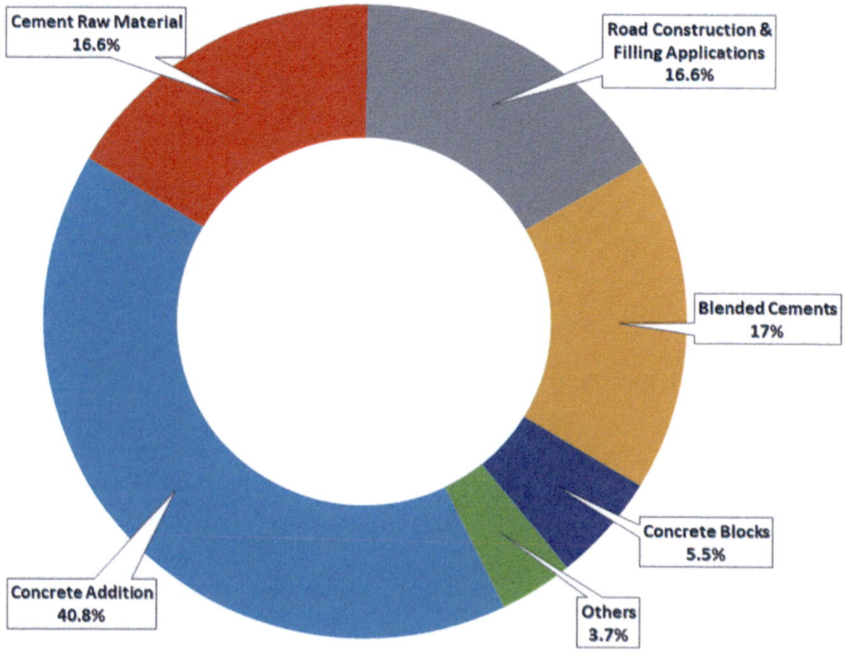

Figure 3.5 Utilization of flyash in construction industry and underground mining in Europe (EU 15) in 2016. (Total utilization = 11.4 Mt) (ECOBA, 2016, p. 1).

produce calcium-silicate hydrates (C—S—H) and calcium-aluminate hydrates (C—A—H). Some flyash with higher amounts of calcium will also display cementitious behavior by reacting with water to produce hydrates in the absence of alkali source such as calcium hydroxide. These pozzolanic reactions are beneficial to the concrete in that they increase the quantity of the cementitious binder phase (C—S—H) and, to a lesser extent, (C-A-H), reducing the permeability and improving the long term strength of the system thus both of these mechanisms enhance the durability of the concrete.

For mix design purposes, flyash itself may be considered similar like Portland cement, except that the specific gravity for flyash is different. The specific gravity of Portland cement is typically 3.15, while the specific gravity of flyash may range from 2.2 to 2.8, depending on flyash composition. Therefore, if a certain percentage of cement is replaced with flyash on a mass basis, simply multiply the initial Portland cement quantity by the percent replacement. For some flyash, particularly low calcium Class F flyash, higher replacement rates (1.2:1 up to 2:1) are required to maintain equivalent early concrete strength. The next modification to the concrete mix design involves the water content. Due to the spherical shape of flyash particles, the water demand is typically reduced up to 5% less with Class F flyash, and up to 10% less with Class C flyash (this may also be accomplished by a partially lower chemical admixture dosage).

3.1.6.1 Advantages of flyash in concrete

Today, there are at least two ways to make flyash more beneficial: a dry process that involves triboelectric static separation and a wet process based on froth flotation. These procedures generally lower the carbon content and the LOI of flyash. The cost of an additional storage bin should be easily covered by the reduction in the cost of the concrete and the added benefits to the concrete. The most important benefit of adding flyash to concrete is it reduced permeability to water ingress and aggressive chemicals. Properly cured concrete made with flyash creates a denser product because of the reduction in the pore size of the concrete matrix; this increases compressive and flexural strengths.

Low-carbon flyash or the use of a better air-entraining agent at a higher-than-usual addition rate can control the problem of freeze-thaw durability.

3.2 Waste glass as Pozzolan

Glass is one of the most versatile material used around the globe due to its excellent characteristics such as its optical transparency, its inherent strength, it's extremely low permeability and its highly chemical inertness (Lu, Duan, & Poon, 2017).

In developed countries, the commercial glass products are being continuously produced each year and the waste glass collected as municipal solid waste is partially recycled and reused to produce new glass products, there are limitations for the mixed color waste glass in terms of meeting the quality criteria for glass remanufacturing. Therefore, the nonrecyclable fraction is usually discarded and sent to landfills. Each year about 200 million tons of glass is disposed in landfills around the world and the overall recycling rate is quite low (Mirzahosseini et al., 2015).

It is estimated that USA generated 12.25 Million tons (Mt) of waste glass in 2018, out of which 7.55 Mt were sent to landfills (Xiao et al., 2022). In UK, the glass waste collected in 2018 amounted to 2.49 Mt out of which only 1.704 Mt was recycled (UK Statistics on Waste, 2021). In France about 2.30 Mt of household & industrial waste glass and 0.235 Mt of waste flat glass was collected in 2020 (Federec & Recyclage, 2020). Australia produced 1.16 Mt of waste glass in 2018−19, of which only 59% was successfully recycled and the rest went directly to landfill (Dong et al., 2021). Turkey generates 0.12 Mt of waste glass each year with high percentage coming from container glass most of which is disposed on the landfill sites (Al-Kerttani et al., 2018). Some European countries have relatively higher (above 50%) level of waste glass recycling but due to quality limitations, a small fraction can be remanufactured (Mohajerani et al., 2017). In Malaysia about 0.42 Mt of waste glass was generated in 2018 (Chua et al., 2019). In Hong Kong less than 10% of kerbside bottle and container waste were recycled, while 90% were sent to landfills (Lu, Duan, & Poon, 2017). Since glass is nonbiodegradable material, its landfilling can result in contamination of soil and underground water, therefore it is not an environmentally feasible option. Colored glass may contain some heavy metal ions, used as colorants; lead glass, which is used in the manufacturing of display monitors has a higher content of lead oxide hence landfilling of these types of glass can result in contamination of soil and underground water by leaching of these toxic metals (Chen et al., 2006; Song et al., 2015). The mixed color and contaminated glass can only go to the secondary markets or dumped in the landfills.

The huge stockpile of waste glass around the world warrants exploring alternative venues to safely consume and recycle this waste and also improve the circular economy of various countries. The most outstanding problem is that the broken waste glass cannot be sorted by color and the decoloring process of the contaminated waste glass is extremely hard to achieve (Xiao, Ma, et al., 2020).

The common industrially manufactured glass includes soda-lime glass, borosilicate glass, silicate glass, lead glass and e-glass etc., and soda-lime glass is the most common type accounting to 82% of all the glass products (Jiang et al., 2022; Mohajerani et al., 2017). The majority of urban waste glass collected belongs to soda-lime glass category, which include beverage glass bottles, glass containers, food containers, broken window glass, light bulbs etc. Soda-lime glass has highly amorphous microstructure and comes with a relatively fixed chemical composition (about 70% SiO_2, 10% CaO, 15% Na_2O) (Martin et al., 2006).

Since waste glass microstructurally contains high proportion of amorphous silica, it indicates that it has pozzolanic properties and when ground to fine particle size, it can be used in construction materials either as natural sand replacement (fine aggregate, $<$ 4.75 mm) or as cement substitution material, when pulverized to particle size <100 μm. Waste glass is regarded as pozzolanic or cementitious material when it is finely grounded into powder because of the fact that, it contains large amounts of silica and calcium and have amorphous structure conforming to the criteria of

ASTM (C 618). Hence, powdered waste glass has the potential to be used as partial replacement for cement.

In many ways, use of waste glass in the manufacture of concrete is a favorable practice, (1) glass has the ability to undergo pozzolanic reaction when used as partial replacement for cement in the production of mortar or concrete, (2) there is no special treatment needed (sorting the waste glass by color or type) other than pulverizing it to the desired particle size, (3) the risk of leaching out of heavy and toxic metals is almost none, if the pH of the concrete remains usual, (4) with the annual concrete production to the tune of 4.4 Billion tons around the globe (Hilburg et al., 2019), concrete in its production, can consume substantial portion of waste glass and (5) the recycle and reuse of glass in concrete as partial cement replacement will also considerably reduce CO_2 footprint as well. It has been estimated that 1.0 ton of cement production releases 0.9 tons of CO_2 in the atmosphere and concrete production accounts for 8% global CO_2 emissions (Adesina et al., 2020; Sousa et al., 2021). The recycling of waste glass in concrete can be made in various forms such as partial replacement of coarse and/or fine aggregate in concrete production, as partial replacement of cement for concrete production and as precursor raw material in geopolymer production. The most beneficial contribution of glass in concrete can be realized when it is pulverized to powder size, having particle size smaller than cement and used as partial cement replacement for concrete production (Ali-Boucetta et al., 2021; Dong et al., 2021; Khan et al., 2020). Reduction of glass to powder size increases its surface area, which provides more hydration sites and increases the hydration reaction when substituted as partial cement replacement. Furthermore, reduction to powder size increases the pozzolanic reactivity of powdered glass and mitigates and supresses its ability to cause alkali silica reaction (ASR) (Guo et al., 2020; Nassar et al., 2012).

3.2.1 Role of waste glass

Waste glass can be used as partial cement replacement or as aggregate (coarse or fine) in concrete production, however the performance depends on the waste glass particle size, substitution rate and its chemical composition, which must be considered during the concrete mix design. Fig. 3.6 shows the role of particle size of waste glass in concrete.

3.2.2 Physiochemical properties of waste glass

The physicochemical properties of glass can be regarded to a moderate extent, similar to that of ceramic porcelain, because both materials undergo a similar manufacturing process of vitrification at a very high temperature. Glass is considered as a noncrystalline and amorphous solid containing high amount of silica, together with sodium carbonate, dolomite and limestone heated to a temperature of 1600°C, which makes it nonbiodegradable (Omran et al., 2016). Glass products are generally formed through rapid cooling whereby crystallization is avoided (Jani et al., 2014).

The following section highlights the physical and chemical properties of waste glass. It also discusses the suitability and efficiency of waste glass as supplementary cementitious material and highlights its pozzolanic activity to produce durable concrete.

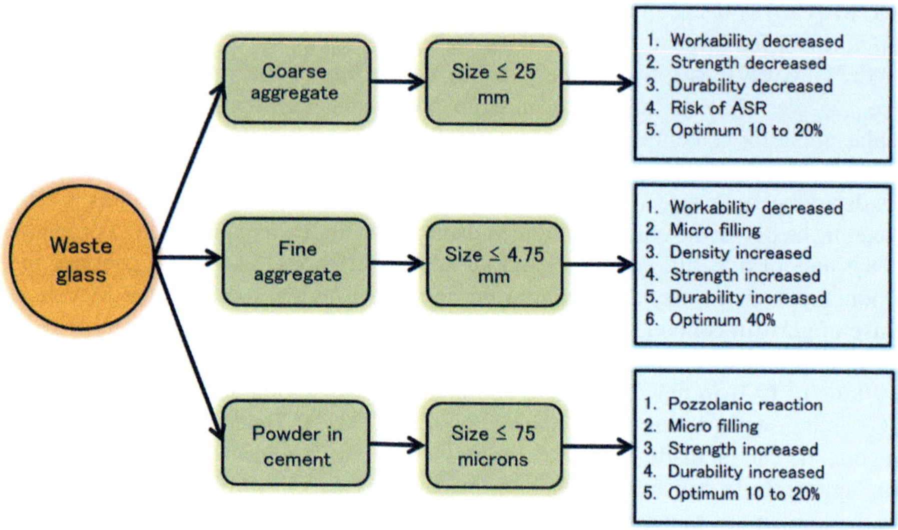

Figure 3.6 Role of particle size of waste glass in concrete (Ahmad & Zhou, 2022).

3.2.2.1 Physical properties

The physical properties of crushed urban waste glass are summarized in Table 3.4. The fineness modulus of powdered glass significantly affects the properties of concrete when used as cement replacement. Studies validate that pulverized/milled glass with particle size similar or lower than cement can act as supplementary cementitious material (Tagnit-Hamou et al., 2018) (Omran et al., 2018). The Los Angeles value of soda-lime glass is approximately 25–28, which suggests that waste glass can be crushed easily in to powder. One study (Xiao Ma et al., 2020) used Los Angeles machine to reduce the particle size of waste glass by putting 40 kg of waste glass in the machine and running it at 33 revolutions for 48 h. The powder obtained had average particle size of 15.4 μm, which has shown very good pozzolanic activity. In addition, waste glass has poor thermal conductivity and high electrical resistivity (Ashby et al., 2012).

3.2.2.2 Chemical properties

The chemical composition of soda-lime glass is relatively stable compared to other glass types. Table 3.5 summarizes the chemical composition based on color and types of glass reported in the literature. As mentioned above, waste glass powder is mainly composed of SiO_2, Na_2O and CaO; however, the quantities of these basic compounds differ according the type of glass being manufactured. Since glass possesses higher amounts of SiO_2 compared to OPC, hence it acts as pozzolan, which reacts with portlandite [$Ca(OH)_2$] released in the cement hydration reaction. Fine glass powder, below the threshold particle size of 75 μm, can fully dissolve into the portlandite alkaline solution and the released silica rich species precipitate to form secondary C–S–H.

Table 3.4 Physical properties of crushed waste glass.

General properties		References
Bulk density	1360 kg/m^3	de Castro and de Brito (2013)
Specific gravity	2.4–2.8	Park and Lee (2004)
Fineness modulus	0.43–3.29	Lam and Poon (2007)
Flakiness index (%)	84.4–94.7	Disfani et al. (2011)
Shape index (%)	30.5	de Castro and de Brito (2013)
Mechanical properties		
California bearing ration (%)	50–75	Ooi et al. (2008)
Los Angelis value (%)	25–28	de Castro and de Brito (2013)
Friction angle	Crit. = 38 (loose state)	Ooi et al. (2008)
Friction angle	Crit. 50–61 (dense state)	Ooi et al. (2008)
Thermal properties		
Thermal conductivity	0.7–1.3 W/m.K	Ashby et al. (2012)
Specific heat capacity	850–950 J/kg.K	Ashby et al. (2012)
Thermal expansion coefficient	9.1–9.5 μstrain/°C	Ashby et al. (2012)
Electrical properties		
Electrical resistivity	7.94×10^{17}–7.94×10^{18} μΩ cm	Ashby et al. (2012)
Dielectric constant	7.0–7.6	Ashby et al. (2012)
Dissipation factor	0.007–0.01	Ashby et al. (2012)

These secondary hydrates have lower Ca/Si ratio = 0.6 to 1.3; (N + K)/S = 0.06 to 0.18, compared to that of primary C–S–H formed from pure Portland cement, which have higher Ca/Si ratio of 1.2 to 2.3. Due to the lower Ca/Si ratio, the secondary C–S–H phases have negative charge promoting the adsorption of alkali cations (Na, K etc.) (Idir et al., 2011; Lu, Duan, & Poon, 2017). Contrarily, the coarse glass powder can only partially dissolve in the alkaline solution and the undissolved particles will be surrounded by the high alkali reaction ring, which may result in the precipitation of reaction (ASR) products. One study (Pelisser et al., 2012) investigated the effect of Ca/Si molar ratio on the micro and nano mechanical properties of synthetic C–S–H measured by nano-indentation, and found that the elastic modulus and hardness increased with decreasing Ca/Si ratio. This shows that the secondary C–S–H phases possessing lower Ca/Si ratios may also have better mechanical properties.

The chemical compositions of commonly used cementitious and supplementary cementitious materials (SCMs) are shown in Fig. 3.7. It can be seen in the figure that soda-lime glass is a silica rich material (SiO_2 ~ 70%) with relatively low calcium and alumina content.

Table 3.5 Chemical composition of soda-lime glass.

Color	Chemical composition (wt.%)											References
	SiO_2	Al_2O_3	Fe_2O_3	S + A + F	Cao	MgO	Na_2O	K_2O	SO_3	TiO_2	Cr_2O_3	
Clear	72.58	1.47	0.85	74.90	10.49	0.61	12.54	0.4	0.2	–	–	Xiao, Ma, et al. (2020)
Clear	68.10	0.9	0.6	69.60	14.50	1.80	12.20	0.8	0.4	–	–	Islam et al. (2017)
Clear	72.42	1.44	0.07	73.93	11.50	0.32	13.64	0.35	0.21	0.035	0.002	Shayan and Xu (2004)
Green	71.12	1.71	0.24	73.07	10.02	3.01	13.17	0.19	0.25	0.07	–	Sobolev et al. (2007)
Green	72.05	2.81	–	74.86	10.26	0.90	14.31	0.52	0.07	0.11	–	Dyer and Dhir (2001)
Brown	71.19	2.38	0.29	73.86	10.38	1.70	13.16	0.70	0.04	0.15	0.041	Sobolev et al. (2007)
Brown	72.214	1.37	0.26	73.84	11.57	0.46	13.75	0.20	0.10	0.026	–	Shayan and Xu (2004)
Amber	2.43	1.24	0.43	44.10	13.43	0.69	10.81	0.45	0.11	–	–	Samarakoon et al. (2020)
Amber	70.01	3.20	–	73.21	10.00	1.46	15.35	0.82	0.06	0.11	–	Dyer and Dhir (2001)
Flint	70.65	1.75	0.45	72.85	10.70	2.45	13.25	0.55	0.45	–	–	Karamberi et al. (2005)
White	70.39	2.41	0.32	73.12	6.43	2.59	16.66	0.23	0.19	0.08	–	Dyer and Dhir (2001)
Mixed	70.30	2.2	0.82	73.32	10.64	0.96	11.21	–	0.33	–	–	Zhang and Poon (2020)
Mixed	69.40	2.0	0.2	71.60	12.5	1.10	13.70	0.6	0.2	–	–	Idir and Cyr (2011)

Type												
Window	71.71	1.26	0.09	73.06	8.44	4.16	13.61	0.4	0.25	0.07	–	Sobolev et al. (2007)
Containers	66–75	0.7–7	–	–	6–12	0.1–5	12–16	–	–	–	–	Mohajerani et al. (2017)
Light bulbs	73.0	1.0	–	–	5.0	4.0	17.0	–	–	–	–	Mohajerani et al. (2017)
Sheet	71–73	0.3	–	–	8–10	1–3.5	12–15	–	–	–	–	Mohajerani et al. (2017)
Float	73–74	–	–	–	8.8	3.7	13–15	0.2	–	–	–	Mohajerani et al. (2017)

$S + A + F = SiO_2 + Al_2O_3 + Fe_2O_3$ (Pozzolanic oxides).

3.2.3 Glass waste as partial cement replacement

Many studies have validated the usefulness of glass as SCM. In most of the studies it was observed that 10%–20% of glass powder can be used as cement replacement (Islam et al., 2017; Omran et al., 2018). Glass powder due to secondary C–S–H formation enhances the pore structure, makes the concrete matrix denser and less permeable due to its denser ITZ, thereby reducing the chloride ion ingress, likewise it also improves the performance of glass concrete in sulfate environment. Chaïd et al. (2015) produced concrete by substituting 30% glass powder (passing 80 μm sieve and approximately 90% of the particles having 25 μm diameter), as cement replacement. The diffusion of chloride ions decreased significantly when glass powder was added, due to its effect on reducing the porosity and microstructure thereby making the concrete matrix denser with low permeability and hence resistant to transport phenomenon. The oxygen permeability of mixes with glass powder was also lower than that of mixes without glass powder.

3.2.4 Glass waste as partial aggregate replacement

The use of waste glass as aggregate in concrete offers a possibility for the elimination of this waste and to reduce the consumption of natural aggregate used in concrete. The glass aggregate also improves the abrasion resistance of concrete together with providing better thermal insulation due to low thermal conductivity of glass compared to limestone aggregate (Polley et al., 1998). However the use of glass waste as aggregate can cause alkali-silica reactions (ASR) that produce expansive products in concrete leading to the deterioration of the mechanical properties of concrete (Lee et al., 2011). Many studies have been carried-out with the objective of using waste glass as fine aggregate and also as coarse aggregate in concrete. Saand et al. (2017) found that compressive strength of concrete increased and gave good performance in mechanical properties, when 12% of waste glass fine aggregate with particle size of 1.19 and 1.71 mm were substituted as replacement of natural fine aggregate. Ibrahim et al. (2017) found that it was possible to use glass waste as partial substitute for sand and the best replacement dose was 15%. The use of glass waste fine aggregate as partial replacement in concrete decreased ASR expansion. Liaqat et al. (2018) observed that with the use of 10% glass waste as coarse aggregate with particle size of 4.75–12.5 mm, was ideal to improve mechanical strength. Comparatively, Nwofor et al. (2017) studied the replacement of both, fine and coarse aggregate by waste glass and found the optimal replacement in both cases was 20%.

The compressive strength of concretes produced with the substitution of waste glass fine and coarse aggregate is related to many parameters, such as replacement percentage, their particle size especially of coarse aggregate.

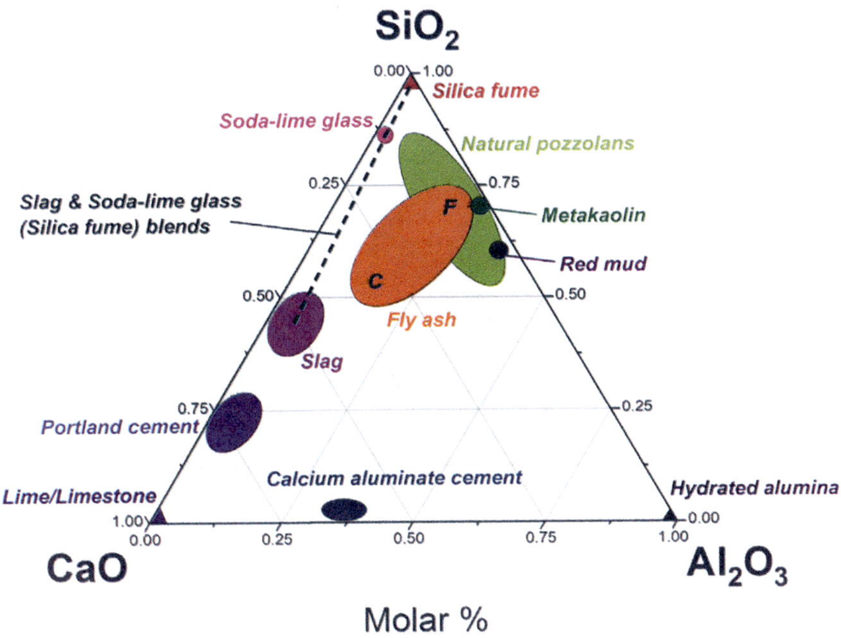

Figure 3.7 Chemical compositions of cementitious and supplementary cementitious materials (SCMs) (Xiao, Jiang, et al., 2020).

3.3 Silica fume

Silica fume is a highly reactive pozzolan material that is used in relatively small amounts to improve or enhance the properties of concrete. It is a by-product of the smelting process in the silicon and ferrosilicon industry. When high-purity quartz is reduced to silicon at temperatures of about 2000°C, silica vapors are produced, which oxidize and condense to tiny particles in the low temperature zone consisting of non-crystalline (amorphous) silica. It is glassy powder of gray color.

The silica fume particles are approximately 100 times smaller than the typical Portland cement particles. The chemical composition and very high fineness of silica particles, make it a very reactive pozzolanic material. It can exhibit both pozzolanic and cementitious properties. Silica fume is also known as micro silica, silica dust and condensed silica fume. Fig. 3.8 shows the schematic diagram of silica fume production process. The silica fume is collected in large bag filters in the baghouse and then stored or transported. It is available in two conditions dry and wet. Dry silica fume can be provided in two forms: (a) as-produced powder and (b) dry, densified powder with or without water reducing admixtures. Slurry of silica fume containing low or high dosages of water reducing admixtures are available.

Silica fume has been recognized as pozzolanic admixture that is effective in enhancing the mechanical properties of concrete substantially. By using silica fume along with superplasticizer in the concrete mix, it is relatively easier to obtain very high compressive strengths of the order of 100 MPa and above in laboratory environment.

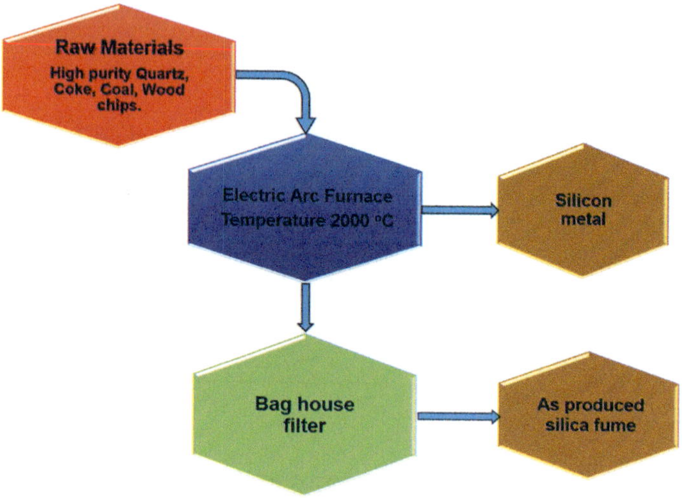

Figure 3.8 Schematic diagram of silica fume production.

3.3.1 Properties of silica fume

3.3.1.1 Physical properties

Silica fume particles are extremely small, the mean primary particle size ranges from 0.1 to 0.2 μm, 95% of the particles are finer than 1 μm. It contains 85%–95% silica (SiO_2). The particles are spherical in shape. SEM micrographs of silica fume particles (a) un-densified (b) densified are presented in Fig. 3.9.

The high surface area of silica fume particles is an important factor affecting the reactivity of the particles and plays very important role in improving the physical and chemical properties of concrete. Silica fume addition benefits concrete in two

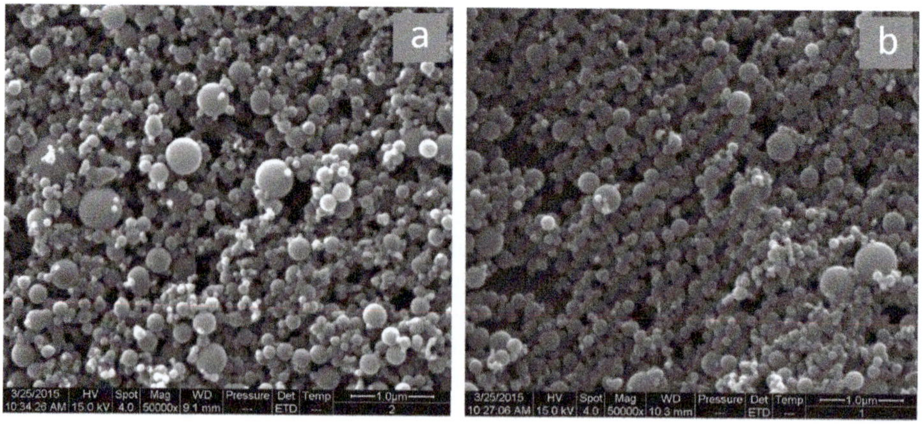

Figure 3.9 SEM micrograph showing silica fume particles (a) un-densified (b) densified (Zhang et al., 2016).

Table 3.6 Typical physical properties of silica fume (Terence, 2005, pp. 1–194).

Property	Value
Particle size (typical)	<1 μm
Surface area (BET)	13,000–30,000 m^2/kg
Specific gravity	2.22
Bulk density	
– As produced	130–430 kg/m^3
– Densified	480–720 kg/m^3
– Slurry	1320–1440 kg/m^3

ways, firstly, the minute particles physically decrease the void space in the cement matrix, and secondly, silica fume works as very effective pozzolan, it chemically reacts to produce calcium silicate hydrate (C–S–H), which results in the increase of compressive strength of concrete. The typical physical properties of silica fume are given in Table 3.6.

Silica fume has specific gravity of about 2.22, which is lower than Portland cement; hence the addition of silica fume to concrete mix will not increase its density.

3.3.1.2 Chemical composition

The chemical composition of silica fume depends on the chemical composition of raw materials used in the furnace. The silica content of silica fume is generally higher than 85%. The other oxides such as alumina, ferric, calcium oxide and alkali content of silica fume is relatively low. The magnesium oxide is also low, while the carbon content varies from 0.5% to 1.5% and usually remains less than 2%. The chemical composition of silica fume samples analyzed by various researchers is given in Table 3.7.

3.3.2 Standards

Various national standards and codes of practice are in use related to silica fume (see Table 3.8). The two globally accepted standards for silica fume are the American ASTM C1240 and the European EN 13263 listed in Table 3.8. Moreover, a number of national and project specific guidelines for the product also exist.

3.3.3 Pozzolanic reaction in cementitious system

Silica fume is very reactive pozzolanic material due to its extreme fineness and very high amorphous silica content. Hydration of Portland cement results in the formation of primary calcium silicate hydrate (C–S–H) and portlandite (alkaline with high pH). The silica fume reacts with portlandite to form secondary (C–S–H), which is similar to primary (C–S–H), this reaction is called pozzolanic reaction (Eqs. 3.1 and 3.2). The combination of both, primary and secondary (C–S–H) refines pore size, reduces

Table 3.7 Chemical composition of silica fume reported by various authors.

References	SiO_2	Al_2O_3	Fe_2O_3	CaO	Na_2O	K_2O	MgO	SO_3	LOI
Banar et al. (2022)	93.54	0.43	0.94	0.49	0.82	1.04	1.0	0.71	0.67
Ashraf et al. (2022)	90.12	1.13	0.31	0.72	2.51	1.81	3.33	—	—
Farrant and Babafemi (2022)	82.00	1.55	5.10	1.62	0.39	2.53	—	—	5.77
Zheng et al. (2022)	94.25	0.43	0.26	0.47	0.32	0.25	0.41	0.53	2.80
Kumar et al. (2022)	92.67	0.88	1.25	0.37	0.43	—	—	0.25	2.10
Singh and Mehta (2022)	96.10	0.75	0.15	0.31	1.25	0.65	0.45	—	0.97
Golewski and Gil (2021)	91.90	0.71	2.54	0.31	—	1.53	1.14	0.45	—
Liu et al. (2021)	94.6	0.13	0.07	0.33	0.29	0.54	0.71	0.07	4.45
Choudhary et al. (2021)	95.58	0.71	0.81	0.90	0.17	0.32	0.60	—	—
Jalal et al. (2015)	93.6	1.30	0.90	0.50	0.45	1.52	1.0	0.4	3.10
Mardani-Aghabaglou et al. (2014)	87.29	0.47	0.63	0.81	1.25	1.28	4.47	0.22	2.70
Güneyisi et al. (2012)	90.36	0.71	1.31	0.45	0.45	1.52	—	0.41	3.11

Table 3.8 Standards of silica fume reported by various countries.

Country	Standards
USA	ASTM C1240 'Silica Fume for use in concrete'
Europe (CEN)	EN 13263—'Silica Fume for Concrete'
Canada	CSA A3001-13 'Cementitious materials for use in concrete'
Japan	JIS A 6207 2000
Australia	AS 3582. Supplementary cementing materials (1992)
China	GB/T 18736
India	IS 15388

permeability, makes the concrete matrix denser, which results in the highly improved physical and durability characteristics of concrete.

$$OPC + H_2O \rightarrow C{-}S{-}H + Ca(OH)_2 \tag{3.1}$$

$$SiO_2 + Ca(OH)_2 + H_2O \rightarrow C{-}S{-}H \tag{3.2}$$

3.3.4 Applications of silica fume in concrete

3.3.4.1 Reduction in heat of hydration

A lower maximum temperature rise and temperature differential takes place in concrete when silica fume is substituted as cement replacement, compared to the concrete of the same strength containing only cement. It has been validated that substitution of silica fume is the most effective way of achieving low heat concrete without compromising the strength at an early age. Its performance is better in comparison to slag and fly ash blends in concrete.

3.3.4.2 Corrosion resistance

The reduction in the permeability of concrete due to silica fume substitution, provides protection against the ingress of chloride ions, thereby increasing the time taken for the chloride ions to reach the steel reinforcement bars and initiate corrosion. In addition, concrete containing silica fume acquires much higher electrical resistivity compared to OPC concrete, thus slowing down the corrosion rate. This combined effect increases the life of a structure by 5–10 times (Sharma et al., 2014).

3.3.4.3 Sulfate resistance

Because of the refinement of pores, silica fume concrete will have higher degree of protection against penetration of sulfate ions, and a high chemical resistance, which provides protection against the sulfate attack, compared to sulfate resisting cements or substitution of flyash or the use of cements containing low Tricalcium aluminate (C_3A) or other cementitious binder systems.

3.3.4.4 Chemical resistance

Silica fume concrete is widely used in the industrial structures that are exposed to various aggressive chemicals. In chemical industry, the exposure of concrete is to mineral acids, phosphates, nitrates, petrochemicals, etc., whereas in the alimentary industry, there is an over exposure to fatty acids, other acids and detergents, etc. Therefore, silica fume concrete is invaluable in both industrial and agricultural sectors.

3.3.4.5 Abrasion resistance

Abrasion resistance of silica fume concrete is very high. The use of silica fume concrete in floor and pavement construction saves time and money. It also improves the operational efficiency for the operator of such facility. The hydraulic abrasion-erosion resistance of silica fume concrete is enhanced, which makes it perfectly suitable for the construction of dam spillways.

3.3.4.6 Waterproofing

Since silica fume concrete has very low permeability, therefore, it can be used for underground structures where some dampness is expected, such as car parks, basements etc.

3.3.4.7 High strength concrete

Condensed silica fume in conjunction with super plasticizers can be used to produce very high strength concrete of about 70–120 MPa. High strength concrete imparts huge economic benefits to the developers with reduced column and wall thickness in tall buildings and improvement of construction schedule. Alkyl micro silica can be used to achieve high strength and high-performance concrete for the construction of tallest buildings such as the Burj Khalifa in Dubai (Bulwark Team, 2021).

3.3.4.8 Bridge construction

In the construction of several of the longest sea bridges in the world, silica fume was used, since it reduces permeability of concrete as well as increases the resistance to sulfates, chloride ingress and reinforcement corrosion. It increases the application thickness per mass and improves the resistance to washouts in marine construction or in wet areas.

3.3.4.9 Shotcrete

Shotcrete is a process of spraying of slurried concrete or mortar on the element or surface with high velocity by pneumatic pressure to give a very dense durable concrete or mortar layer. Silica fume concrete can be used in shotcrete, using either wet or dry process.

3.3.4.10 Tunnel and mining operations

In tunnel and mining operations, application of silica fume is popular in a variety of ways. In sprayed concrete mixtures, particles of micro silica make concrete mixtures more cohesive by reducing rebound and wastage.

3.3.4.11 Disadvantages of micro silica

Its price is comparatively higher compared to cement and fly ash. It increases the shrinkage of the cement paste, which increases the risk of microcracking. Since the workability of silica fume concrete is lower, its application becomes difficult, therefore, the uniformity of surface is compromised.

3.4 Rice husk ash

Rice husk are the hard-natural protective coverings of rice grains, which are separated from the grains in the milling process. Rice husk is an agro-waste abundantly available in all the rice producing countries. The major rice growing countries in Asia are China, Indonesia, Vietnam, Philippines, Malaysia, India and Pakistan. Current rice production in the world in 2021 was 508.84 Million tons (Mt) and in 2022 rice production is expected to increase to 513.02 Mt representing an increase of 4.18 Mt compared to previous year (Habeeb et al., 2010).

Rice husk constitutes about 20%–28% by the weight of rice and is composed of about 50% cellulose, 25%–30% lignin, 15%–20% silica and 10%–15% moisture. Every 100 kg of husk burnt in a boiler will yield about 25 kg of RHA (Singh et al., 2018).

On burning, the volatile organic matter in rice husk i.e., cellulose and lignin evaporate in to the atmosphere, leaving behind essentially silica (SiO_2) and some amount of carbon.

The characteristics of ash are dependent on the (a) composition of rice husk (b) burning temperature (c) burning time (duration) (d) burning in controlled environment. Better quality of rice husk ash (RHA) containing highly reactive amorphous silica can be obtained when the rice husk is burned in controlled environment in reduced environment. From various studies (Fapohunda et al., 2017; Habeeb et al., 2009; Shatat et al., 2016), it was observed that the highest content of amorphous silica could be obtained by burning rice husk (RH) in reduced atmosphere in the temperature range of 500–700°C and then ground to increase its surface area up to 150 m^2/g. Fig. 3.10 depicts the SEM micrographs of rice husk showing changes in the microstructure when burned at 600, 700 and 800°C.

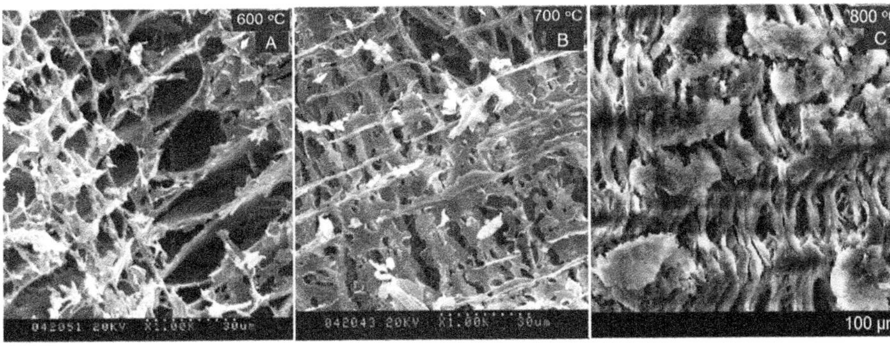

Figure 3.10 Changes in microstructure of RHA burned for 1 h at 600°C, 700 and 800°C. Micrographs (a and b) (Bie et al., 2015), and (c) (Liu et al., 2014).

The RHA having high silica content has been commercially produced in a number of countries including Columbia, Thailand and India (Madandoust et al., 2013), and successfully used as pozzolanic material for partial replacement of cement in concrete (Bie et al., 2015) (Prasittisopin et al., 2012), (Antiohos et al., 2014) (Zain et al., 2011).

3.4.1 Processing of RHA

The physical and chemical characteristics of RHA depends on its processing parameters based on burning methods (a) open field burning (b) controlled burning, as well as (c) separation processes and (d) its pulverization and fineness. The amount of highly needed amorphous silica and carbon content of RHA are dependent on (i) burning temperature and (ii) burning time. Properties of RHA obtained from different burning conditions are shown in Table 3.9.

The pozzolanic activity of RHA is dependent on its amorphous silica content, its surface area and its fineness (Fapohunda et al., 2017).

Generally, RHA containing 85%–95% silica has shown pozzolanic behavior compatible to other SCMs such as flyash, slag, bagasse ash, POFA etc. used to produce blended cements. In certain regions of Asia and South America, rice husk is used as a fuel for parboiling paddy in rice mills, whereas in some places it is used to provide heat to convert soft clay bricks to hard burnt bricks. The calorific value of rice husks is about 50% of that of coal. Assuming that RH contain 8%–10% moisture and no

Table 3.9 Properties of RHA obtained from different burning conditions.

Type of burning	Temperature	Burning duration	Color	LOI	References
Controlled burning	700°C	–	Gray	4.05	Bahri et al. (2018)
Open burning	–	60 min	Gray	3.83	Mayooran et al. (2017)
Muffle furnace	700°C	120 min	–	1.48	Bie et al. (2015)
Pyrolysis	800°C	40 min	Off-white	0.54	Venkatanarayanan and Rengaraju (2015)
Controlled burning in gas	800°C	–	White	–	Jaya et al. (2013)
Controlled burning	500–700°C	30 min	White	3.21	Zain et al. (2011)
Open burning	422°C	72 h	–	–	Rashid et al. (2010)
Incineration	500°C	120 min	Off-white	2.27	Muthadhi et al. (2010)
Controlled burning	650°C	60 min	Whitish	5.91	Ramezanianpour et al. (2009)
Controlled burning	600°C	180 min	White	–	Salas et al. (2009)

bran, the calorific value is estimated to be 15 MJ/kg. The combustion of rice husk in the above cases is not complete and also contributes to air pollution. The uncontrolled open-field burning results in the formation of crystalline silica and high carbon content in RHA (Mboya et al., 2017; Siddika et al., 2018), which can have very adverse impact on the properties of concrete. Conversely, burning of RH under controlled burning at temperatures of 500–700°C has been recommended to be optimum to produce highest content of amorphous silica having a maximum specific surface area of up to 150 m^2/g of RHA particles (Alaneme et al., 2018; Jittin et al., 2020; Siddika et al., 2021). Additionally, after incineration the separation of RHA is required to ensure the elimination of any other impurities present, which could reduce the pozzolanicity of RHA.

The metallic ions present in RHA can be separated using pretreatment such as acid leaching (Adesina et al., 2019; Kwan et al., 2020). To improve the pozzolanic activity, mechanical grinding is important not only to achieve high surface area and fineness of RHA particles, but also to reduce the loss on ignition (LOI) of up to 12% in the grinding of RHA (Van et al., 2013). The recommended minimum grinding time of RHA is 10–30 minutes by mechanical grinder (NBR-12653, 1992; Xu et al., 2015). Moreover, Cordeiro et al. (2009) has reported that grinding time of 120 minutes was sufficient to produce ultra-fine ash having 6.8 μm as average particle size and 33670 m^2/kg BET surface area, which showed Strength Activity Index (SAI) of 109%. Due to its superior pozzolanic activity index, RHA is categorized as pozzolan of Class N as per ASTM C618 (Said et al., 2017). As a result, RHA can be used as a sustainable raw material and SCM in civil engineering construction.

3.4.2 Properties of RHA

3.4.2.1 Physical properties

Rice husk ash is a very fine material having average particle size of 3–10 μm. Completely burnt RHA is gray to white in color and partially burnt RHA is blackish. Due to its highly microporous structure, the specific surface area of RHA determined by Brunauer-Emmett-Teller (BET) using nitrogen adsorption method can range from 20 m^2/g to as high as 270 m^2/g, comparatively, silica fume has specific surface area of 18–23 m^2/g (Singh et al., 2018). Level of carbon detected in RHA before thermal treatment averaged 18.60% and after thermal treatment the level of carbon content decreased considerably to 0.14% when thermally treated at 700°C for 6 h. The physical properties of RHA as reported by some authors are given in Table 3.10.

3.4.2.2 Chemical composition

Rice husk ash is very rich in silica content and contains more than 80% amorphous silica as shown in Table 3.10. Most of the existing research has revealed that the amorphous nature of RHA is very suitable to develop high strength in concrete (Djamaluddin et al., 2018) (Antiohos et al., 2014). The very high silica (amorphous) content and fine particles are very effective in showing pozzolanicity and binding action. Moreover, the other chemical contents of RHA have manifested a contributory effect to

Table 3.10 shows the physical properties of RHA.

Mean particle size (μm)	Specific gravity (g/cm^3)	Fineness passing 45 μm (%)	Specific surface area (m^2/g)	References
6.27	2.08	91	36.47	Kannan et al. (2016)
5.7–15.6	–	–	22.36–25.21	Le et al. (2015)
39.34	2.24	–	0.37	Sua-Iam et al. (2013)
6	2.1	–	2.33	Safiuddin et al. (2012)
11.5–63.8	–	–	25.3–30.4	Abood Habeeb et al. (2010)
3.80	2.06	99	36.47	Ganesan et al. (2008)

its binding property. The oxides of Al_2O_3, Fe_2O_3 have little effect on hydration; however, Al_2O_3 improves the formation of calcium aluminate hydrate (C–A–H) in certain cases, which also helps in enhancing the strength of concrete. By contrast, the alkali content in RHA have shown a negative effect on the setting time (reduces setting time) and the hydration rate (Kashani et al., 2019). The chemical composition of RHA reported in various studies is tabulated in Table 3.11.

Table 3.11 Chemical composition of RHA reported in different studies.

References	Percentage composition (%)				
	SiO_2	S + A + F	CaO	Total alkali	LOI
Kang et al. (2019)	92.00	92.69	0.97	2.75	0.76
Umasabor et al. (2018)	87.22	89.60	2.12	–	1.06
Zabihi et al. (2018)	91.05	91.77	0.41	4.16	0.45
Safari et al. (2018)	91.94	92.48	1.05	1.22	2.65
Molaei Raisi et al. (2018)	88.07	89.64	1.04	2.48	–
Joshaghani et al. (2018)	87.10	87.51	1.96	1.26	–
Jung et al. (2018)	91.79	94.71	1.27	0.64	1.56
Mehta et al. (2018)	89.47	90.83	0.68	0.33	7.61
Huang et al. (2017)	91.56	91.92	1.07	2.63	–
Hossain et al. (2017)	87.12	87.78	1.12	1.83	3.01
Geraldo et al. (2017)	93.30	93.45	0.60	0.69	–
Villaquirán-aicedo et al. (2017)	92.33	92.68	0.63	0.17	2.57
Wei et al. (2016)	94.10	94.20	0.49	1.23	–
Mejía et al. (2016)	90.93	91.23	0.36	1.32	5.10
Le et al. (2016)	87.00	88.20	1.20	2.13	3.70
Tchakouté et al. (2016)	83.05	85.45	0.69	3.85	–

3.4.2.3 Pozzolanic activity

As per ASTM C 618, the combined proportion of SiO_2, Al_2O_3 and Fe_2O_3 in the ash should not be less than 70% and LOI should not exceed 12% as specified in the standard. The rapid method to evaluate the pozzolanic activity of RHA is the measurement of electrical conductivity in a saturated solution of $Ca(OH)_2$ when RHA is added. Initially, the conductivity of saturated solution of $Ca(OH)_2$ (200 ml maintained at 40°C) is measured, then 5 g of RHA sample is added to the solution. After 2 min of continuous stirring, the electrical conductivity is measured. The difference between initial and final conductivities shows the measure of pozzolanicity (Nair et al., 2006). As the pozzolanic reactivity of RHA is much higher than the specified value, after grinding for 8 min, it has the highest potential to be used in concrete.

3.4.3 Utilization of RHA

The RHA utilization as fine aggregate has not only environmental benefits but also will bring revenue to the rice mill owners and rural farmers. Furthermore, incorporation of RHA as SCM will reduce the cost of construction products and also help in reducing the CO_2 emissions. RHA can be utilized as SCM in different types of concrete such as conventional concrete, ultra-high-performance concrete, self-compacting concrete and fiber reinforced concrete.

3.5 Sugarcane bagasse ash (SCBA)

Sugarcane bagasse ash is a by-product waste material of sugar and alcohol industries. After the extraction of sugar juice from sugar cane, the pulp left is recycled and reused as fuel for heat generation in boilers leaving behind 8%–10% ash as waste. The process diagram of sugar cane processing and bagasse ash production is shown in Fig. 3.11. The ash commonly known as sugarcane bagasse ash (SCBA) is generally black in color due to unburnt carbon left in the ash, together with some undesirable components, since bagasse fibers and particles contain some amount of left-over juice and fiber (Savastano et al., 2016). Sugarcane is an important crop, it is lucrative and financially more viable hence, also known as cash crop, grown in more than 100 countries around the world. One ton of sugarcane crushing may produce 0.3 ton of bagasse, which contain good amount of moisture content (50%), press mud (0.03 tons) and molasses (0.04 tons) (Tripathy et al., 2022).

3.5.1 Characterization of bagasse ash (SCBA)

The physical properties and chemical composition of SCBA vary from each other due to many factors such as, sugarcane varieties, its growth, combustion temperature and duration in the boiler, collection location of SCBA in the boiler, type of boiler equipment and method of collection of SCBA (Katare et al., 2017). SBCA collected from bottom of the boiler may be coarser containing irregular particles, while SCBA collected through filtration system contains less carbon (Frías et al., 2011)

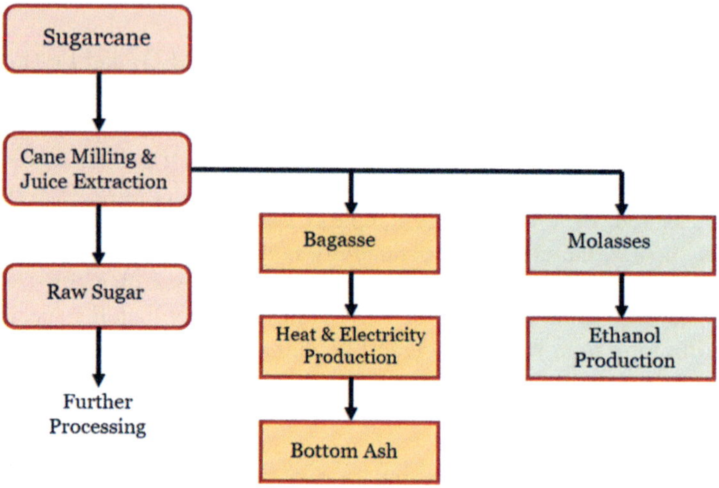

Figure 3.11 Diagram of sugarcane processing and bagasse ash production.

The color of SCBA generated from power plant in sugar factory is dark black to light black. Dark black indicates a higher carbon content due to incomplete combustion (Katare et al., 2017), at high temperature greater than 800°C, SCBA appears gray due to crystallization and decomposition of components contained in the bagasse, but it appears white above 900°C recalcination temperatures (Bahurudeen et al., 2015). In his experimental assessment (Embong et al., 2016) calcined SCBA at different temperatures from 600 to 800°C and for different durations 1 h to 3 h. She found that as calcination temperature and the duration increased, the ash gradually changed its color from char black to white ash. Fig. 3.12 shows the comparative change in color with the increase in temperature and duration of calcination.

Since, SCBA in the power plants of sugar factories is produced in uncontrolled calcination temperatures and durations, therefore SCBA from sugar factories is a mix of different color and morphology.

3.5.1.1 Physical properties

The properties reported by various researchers differ from each other, because the SCBA samples collected by the researchers were collected from different sources and locations, e.g., sugar industries, alcohol industries, from different locations in the boilers etc. These properties are tabulated in Table 3.12.

The specific gravity of SCBA has direct impact on the water requirements for the workability of mortars or concrete mixture proportioning. The specific gravity of SCBA reported by the researchers has a minimum value of 1.13 (Singh et al., 2016) to a maximum value of 2.69 (Moretti et al., 2018), however, these values show that the specific gravity (sp.gr.) of SCBA is lower than that of Portland cement, which have sp.gr = 3.15 and blended cements (e.g., Portland-blast-furnace-slag and Portland-pozzolan cement) may have sp.gr = 2.90 (Civiltoday, 2022).

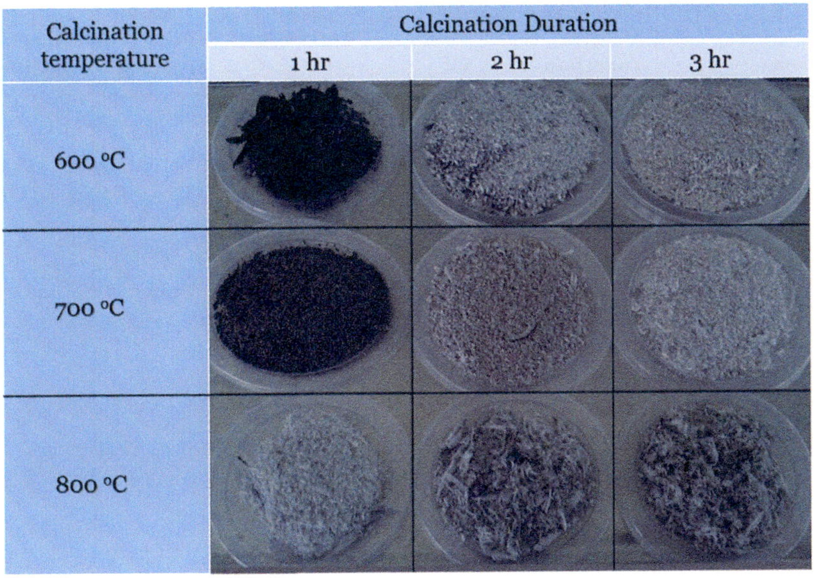

Figure 3.12 Color variation of SCBA with different calcination temperatures and durations (Embong et al., 2016).

Table 3.12 Physical properties of SCBA (collected data).

Authors	Source of sample collection	Mill temp. (°C)	Specific gravity (SG)	Specific surface area (m^2/kg)	Fineness modulus
Cordeiro and Barroso (2018)	Sugar factory (collected during the cleaning of boilers)	900	–	254	–
Deepika et al. (2017)	Sugar mill disposal area	–	2.12	300	–
Joshaghani and Moeini (2017)	Bagasse collected in raw form	–	2.2	–	–
Kazmi et al. (2017)	Sugar factory	–	2.12	–	–
Bahurudeen et al. (2016)	Sugar mill disposal area	–	1.91	169	–
Moretti and Sales et al. (2016)	Sugar factory (collected as bottom ash)	–	2.6	–	1.04

Continued

Table 3.12 Continued

Authors	Source of sample collection	Mill temp. (°C)	Specific gravity (SG)	Specific surface area (m^2/kg)	Fineness modulus
Bahurudeen and Santhanam (2015)	Disposal area	–	2.1	300	–
Almeida et al. (2015)	Sugar & ethanol plants	–	2.57	–	1.15
Bahurudeen and Santhanam (2015)	Disposal area	–	1.91	145	–
Rattanashotinunt et al. (2013)	Sugar factory	600–800	2.27	–	–
Montakarntiwong et al. (2013)	Sugar factory (collected from wet scrubber system with high LOI)	800–1000	2.09 & 2.22	–	–
Cordeiro et al. (2012)	Sugar and alcohol factory (collected during boiler cleaning)	700–900	2.53	893	–
Rukzon and Chindaprasirt (2012)	Sugar factory	–	2.24	1250	–
Amin (2011)	Sugar factory (Mill fired)	–	1.18	900	–
Sales and Lima (2010), Cordeiro et al. (2008)	Sugar factory (collected from 4 different mills during cleaning operation)	–	2.23–2.65	–	0.52–1.46
	Sugar and alcohol factory (collected during cleaning of boilers)	700–900	–	893	–

The specific surface area (SSA) of SCBA usually measured by BET apparatus plays a major role in improving the pozzolanic activity by exposing more surface area of particles for pozzolanic reaction. The pozzolanic reactivity is directly proportional to an increase in SSA depending on temperature threshold (Bahurudeen et al., 2016). There is a great difference in the SSA based on the grinding techniques Cordiero et al. (2017, 2019) have attempted to grind SCBA to ultra-fine particle size. Usually the idea of grinding SCBA is to achieve similar particle size as that of cement. SSA plays an important role on impacting various properties, such as water demand, workability, strength development and durability when added to produce mortars/concretes.

The density of SCBA in the dry form varies from 1.27 g/cm^3 (Anupam et al., 2013) to 2.69 (Moretti et al., 2018). In another investigation Cordeiro et al. (2018) observed increase in density of about 10% from 2.11 to 2.335 g/cm^3 for the same sample due to a simple calcination process, similarly Rodriguez-Diaz et al. (2015), noted that for the same untreated sample of SCBA, the density varied from 1.98 − 2.15 g/cm^3.

Most of the researchers have reported the fineness modulus of SCBA to be greater than 1.0, however, some authors have reported SCBA fineness modulus ranging from 0.52 to 1.46.

3.5.1.2 *Chemical composition*

The chemical composition of SCBA is highly dependent on the source and the quality of SCBA. The major elemental oxides are silica (53%−91%) and alumina (12%−31%). Silica content of SCBA is 3 times higher than Portland cement and plays important role in improving the strength and durability properties of concrete. The other oxides such as ferric oxide and calcium oxide are relatively low. To be classified as natural pozzolan, ASTM C 618 specifies the cumulative silica + alumina + ferric oxide to be greater than 70% (Class F and N), while it should be greater than 50% for (Class C). These three oxides contribute to the pozzolanic properties of SCBA.

The values of total equivalent alkali (Na_2O + 0.658 K_2O) content in SCBA ranges between (0.5%−10%) and its SO_3 content ranges between (0.1%−3.52%). These values conform to the requirements of ASTM C 618 (2014). The loss on ignition (LOI) values depend on the calcination temperature, its duration and the process of calcination (Chusilp et al., 2009a). The LOI is directly proportional to the amount of unburnt particles (carbon content) in the ash. The LOI varies even for the ash collected from the same source (Ríos-Parada et al., 2017). Silica content decreases, whereas the calcium oxide content increases with the rise in LOI (Chusilp et al., 2009a).

Usually, X-ray Fluorescence (XRF) analysis is carried out to determine various oxides available in SCBA. Substantial numbers of studies have reported that almost all the samples of ash meet the specified requirements of ASTM C618.

The chemical and mineralogical composition of SCBA makes it a potential supplementary material in Portland cement blends and also in geopolymeric binders.

The chemical composition data of SCBA from different countries have been summarized in Table 3.13.

Table 3.13 Chemical composition of SCBA from different countries.

Authors	SCBA source	Oxide composition (%)												
		SiO_2	Al_2O_3	Fe_2O_3	CaO	Na_2O	K_2O	SO_3	MgO	TiO_2	P_2O_5	MnO	LOI	S + A + F
Clark et al. (2017)	Australia	70.32	13.01	3.16	1.07	1.13	2.28	–	1.07	0.97	0.30	0.12	6.51	86.54
Agredo and Mejía	Columbia	72.80	6.40	5.50	3.80	–	2.70	–	2.30	–	–	–	3.70	84.7
(2014)	Ethiopia	65.58	5.87	4.32	1.78	1.02	6.41	0.18	1.23	0.25	1.35	0.05	10.48	75.77
Tekle (2011)	Indonesia	63.44	9.44	4.99	4.03	0.15	2.80	–	2.78	0.15	3.52	–	–	80.01
Samadhi et al. (2019)	Iran	64.33	9.08	5.47	8.17	0.87	1.32	–	–	–	–	–	–	78.78
Zareei and Ameri	Mexico	66.12	15.00	7.16	2.57	0.54	3.52	0.26	1.19	1.13	1.14	0.22	9.00	88.28
(2018)	Nigeria	57.95	8.23	3.96	4.52	–	2.41	–	4.47	–	–	–	5.00	70.14
Ríos-Parada, Jiménez-	Pakistan	78.05	3.81	2.01	3.98	0.21	0.42	–	2.24	–	–	–	–	83.87
Quero, et al. (2017)	Brazil (Rio de Janeiro)	54.90	8.90	9.75	11.19	–	10.57	1.93	–	1.06	1.15	0.22	0.5	73.55
Sadeeq and Ochepo (2015)														
Noor-Ul-Amin (2016)	Brazil (Sao Paulo)	36.20	12.30	8.76	7.106.	–	12.8	4.38	4.75	1.98	5.42	0.30	5.37	57.26
Faria and Gurgel (2010)	India (South)	75.90	1.55	2.32	25	0.12	8.4	–	1.77	–	–	–	4.00	79.77
Rodier et al. (2019), Murugesan et al. 2020, Jagadesh et al. (2018)	India (Tamil Nadu)	55.05	5.10	2.62	5.09	0.94	4.00	–	4.82	–	–	–	–	62.77
Hussein and Shafiq	Sudan	77.25	6.37	4.64	4.05	1.36	2.34	0.11	2.61	0.17	0.59	0.27	–	88.08
(2015)	Taiwan	54.4	9.10	5.50	12.4	0.90	1.30	4.10	2.90	–	–	–	9.40	69.00
Cheng et al. (2014)	Thailand	54.10	5.69	3.54	15.37	–	–	0.03	1.41	–	–	–	19.36	63.33
Chusilp (2009b)	Uganda	62.1	5.54	5.42	1.0	–	2.22	–	1.12	–	–	–	–	73.06
Basika and Kigozi (2015)	United States	22.95	2.65	1.65	–	0.30	1.50	0.0	0.45	–	0.30	–	12.28	27.20
Subedi et al. (2022)														

S + A + F= $SiO_2 - Al_2O_3 + Fe_2O_3$ (Pozzolanic oxides), *LOI*, Loss on ignition.

3.5.2 Pozzolanic activity of SCBA

When sugarcane bagasse is burnt as a fuel in the boiler of thermal power plant, reactive amorphous silica is produced and remains in the ash (Chusilp et al., 2009b), the ash however, may contain small amounts of crystalline phases (quartz and cristobalite), which generally come from two sources (i) the sand from soil, which remains stuck with the sugarcane stalks when harvested, chopped and squeezed in the mill and then incinerated (ii) the uncontrolled incineration process.

When the crystalline phases are finely pulverized, the inert particles can serve as fillers in cement mortar/concrete (Sales and Lima, 2010). Conclusively, bagasse ash due to very high content of amorphous silica can be used as pozzolanic material together with amorphous alumina and ferrite (pozzolanic oxides) (De Soares et al., 2016). The main pozzolanic oxides ($SiO_2 + Al_2O_3 + Fe_2O_3$) chemically react with portlandite [$Ca(OH)_2$] released during hydration of cement and this secondary reaction creates additional calcium silicate hydrate (C—S—H) and calcium aluminate hydrate (C—A—H) (Keramatikerman et al., 2019, Arenas-Piedrahita et al., 2016).

Pozzolanic activity can be evaluated by the experimental procedures such as (i) reactivity with slaked lime (ii) electrical conductivity (iii) strength activity index (SAI). SAI is also known as pozzolanic activity index and has been widely used to evaluate the pozzolanic activity of pozzolanic materials, specified in the ASTM C311, (2013) in which relative compressive strength of cement mortars containing 20% bagasse ash, cured for 28 days is compared to that of control mortar. The minimum requirement of SAI is 75%, therefore the pozzolanic activity can be translated to mechanical performance. Fig. 3.13 shows the dependence of SAI and Electrical Conductivity on the amorphous content of SCBA.

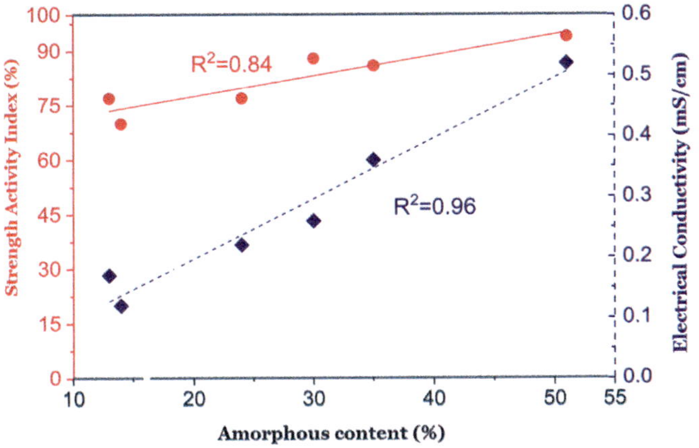

Figure 3.13 Dependence of SAI and electrical conductivity on amorphous content of SCBA (Kolawole et al., 2021).

3.6 Palm oil fuel ash (POFA)

Palm oil is one of the world's most produced and consumed oils. This cheap and highly stable oil is used in a wide variety of food, cosmetic and hygiene products, and can be used as source for bio-fuel or biodiesel (Reeves et al., 1979). Most palm oil is produced in Asia, Africa and South America because the trees require warm temperatures, sunshine and plenty of rain in order to maximize production. Oil palm tree, a native of Africa were planted in countries like Malaysia, Indonesia etc. as an ornamental plant, which later emerged as one of the world's most profitable agricultural commodities with an average life span of 25 years (Thomas et al., 2017).

Palm oil is extracted from reddish pulp (mesocarp) of the fruit of the oil palms. The extraction process and production of POFA are shown in the process flow diagram in Fig. 3.14. The biggest producers of palm oil are Indonesia, Malaysia, Thailand, Columbia and Nigeria. The world production of palm oil in 2021 was 77.047 Million Tons (Mt) (U.S Dept. of Agriculture) and is projected to reach 240 Mt by 2050 (Tullis et al., 2019).

Indonesia the world's largest producer of palm oil produced over 45.5 Mt in 2021 and expects to double its production by the end of 2030 (Gilbert et al., 2012). Malaysia is the world's second largest producer producing (25.8% of the world production), which is about 19.0 Mt. Thailand is the world's third largest producer, producing

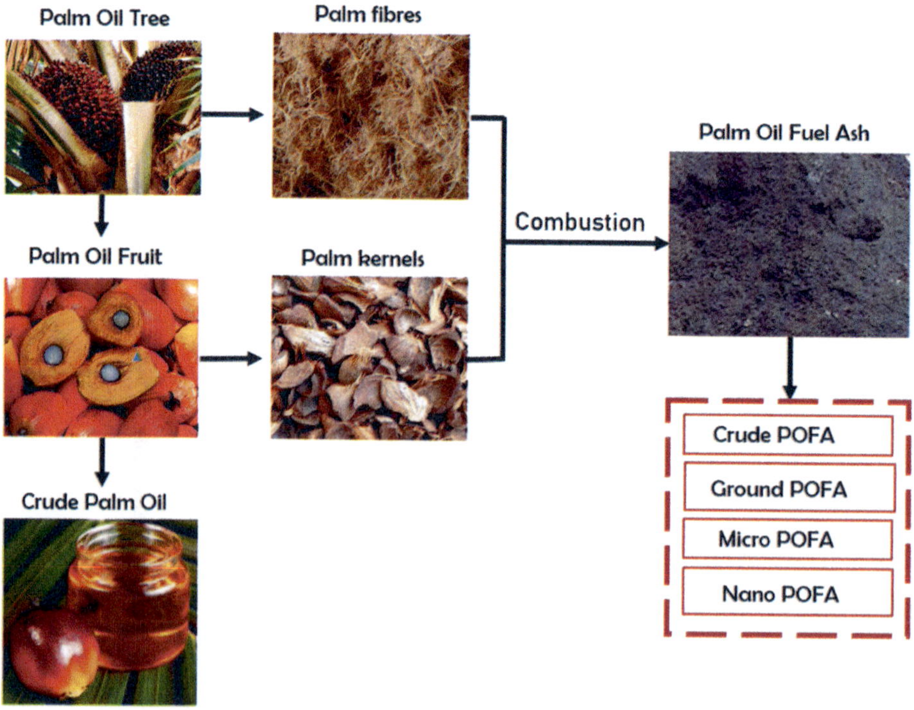

Figure 3.14 Process diagram of palm oil extraction and POFA production.

Table 3.14 Palm oil production in various countries worldwide (U.S Dept. of Agriculture).

Country	Production (million tons)	World production (%)
1. Indonesia	45.5	59.05
2. Malaysia	19.0	24.7
3. Thailand	3.10	4.02
4. Columbia	1.747	2.27
5. Nigeria	1.40	1.82
6. Guatemala	0.88	1.14
7. Honduras	0.60	0.78
8. Papua New Guinea	0.565	0.73
9. Ecuador	0.564	0.73
10. Brazil	0.55	0.71
11. Cote d'Ivoire	0.515	0.67
12. Cameroon	0.465	0.60
13. Ghana	0.365	0.47
14. Congo	0.30	0.39
15. India	0.291	0.38
16. Costa Rica	0.27	0.35
17. Mexico	0.225	0.29
18. Peru	0.221	0.287
19. Philippines	0.09	0.117
20. Sierra Leone	0.075	0.097

approximately 3.1 Mt. The palm oil production around the globe is tabulated in Table 3.14 (U S Dept. Agriculture).

It has been estimated that 1.0 ton of palm oil generates about 4.0 tons of dry biomass waste produced by the processing mills, consisting of kernel shell, empty fruit bunches and fibers, collectively called palm oil waste (POW) (Iskandar et al., 2018; Sadhukhan et al., 2018), which is burned at 800–1000°C in the boilers of thermal power plants for steam and electricity generation. Almost 5% by weight of the biomass incinerates to ash called POFA. About 12 Mt of POFA are generated in the world per annum (Thomas et al., 2017). Disposal of this huge amount of POFA is an issue of paramount concern for the leading palm oil producers (Yusof et al., 2018). It cannot be used as fertilizer because it does not contain nutrients for soil, hence mostly it is disposed-off in the open fields and landfills causing severe harm to human health and environment (Saleh et al., 2014).

3.6.1 Characteristics and properties of POFA

The physical properties, chemical composition of POFA are detailed below.

3.6.1.1 Physical properties

The physical properties of POFA depend upon many factors such as incineration temperature in the boiler and its duration, its treatment process and the part and ratio of different palm oil waste materials being burnt collectively (Huseien et al., 2016; Mohd Salleh & Mustazar, 2018).

3.6.1.1.1 Color

The color of un-ground POFA, in normal state is light gray, but when it is pulverized, color changes to dark gray. Moreover, dark blackish color in the normal state indicates high proportion of un-burned carbon content, which can be reduced substantially after burning further at high temperature, the color then changes back to light gray (Awal et al., 2015; Zeyad et al., 2016).

3.6.1.1.2 Shape

The shape of un-ground POFA particles is moderately spherical and porous, whereas the particle shape of pulverized POFA is irregular and angular (Chindaprasirt et al., 2007). The size of pulverized POFA particles is smaller than cement particles and ranges between 7.2 and 10.1 μm as compared to 10–20 μm of cement particles (Sata et al., 2007).

3.6.1.1.3 Fineness

Fineness is one of the important physical properties of POFA. Due to high surface area, it improves the pozzolanic activity of POFA particles. Usually, POFA is pulverized in ball mills or tube mills, consequently, the porosity of POFA particles is reduced as well. The specific gravity of POFA increases when it is pulverized and the particle size is reduced. Tangchirapat and Jaturapitakkul (2009) pulverized raw POFA in the mill and found that it reduced the particle size from 65.6 to 10.1 μm and increased the specific gravity from 1.97 to 2.33.

The physical properties of POFA containing specific gravity, surface area and Blaine fineness are shown in Table 3.15.

Table 3.15 Physical properties of POFA as reported in literature.

References	**Specific gravity**	**BET fineness (m^2/kg)**	**Blaine fineness (m^2/kg)**	**Retained on sieve no. 325 (%)**	**Median particle size d50 (μm)**
Hamada et al. (2020)	2.52	–	1962	–	0.982
Nagaratnam et al. (2019)	2.2	3833	–	0.3	–
Rajak and Majid (2019)	1.78	1663	–	4	–
Salam and Safiuddin (2018)	2.48	4103	775	5	–

Table 3.15 Continued

References	Specific gravity	BET fineness (m^2/kg)	Blaine fineness (m^2/kg)	Retained on sieve no. 325 (%)	Median particle size d50 (μm)
Alsubari et al. (2018)	–	4000	–	–	–
Mohdhosseini et al. (2017)	2.42	–	493	33	–
Zeyad et al. (2016)	2.59	7670	1136	–	2.06
Ranjbar et al. (2016)	1.81	–	–	96	10
Safiuddin et al. (2016)	2.48	4103	775	5	–
Alsubari and Shafigh (2016)	2.2	7400	–	0.0	–
Lim et al. (2015)	2.42	3046	493	10.5	–
Awal and Shehu (2015)	2.42	–	493	33	–
Hassan et al. (2014)	2.42	2375	–	1.6	–
Alsubari et al. (2014)	2.44	–	–	4	–
Safiuddin and Salam (2014)	2.5	4105	780	5	–

3.6.1.2 Chemical composition

POFA is produced by incinerating different waste parts of palm oil, therefore different samples may vary in their chemical composition. Since the mineral constituents of POFA determine their reactivity, it then becomes imperative to classify them based on their chemical composition. The oxide analysis of these components is preferably done using X-ray Fluorescence (XRF). According to the requirements specified in ASTM C618, the summation of pozzolanic oxides ($SiO_2 + Al_2O_3 + Fe_2O_3$) must be greater than 50% conforming to Class C type pozzolan, while SO_3 should be less than 4.0%. To assess the chemical composition of POFA numerous researchers have done oxide analysis of POFA samples. The notable variations in the chemical composition may be due to different conditions such as burning temperature, its duration, and incineration of different parts of palm oil tree, to produce POFA in different factories. It has been explained by Zeyad et al. (2016) and Johari et al. (2012) that higher silica content is obtained by incinerating the kernel of palm tree, while the other parts of the palm tree produce more carbon in the burning process. Also, with the increase in the fineness of POFA, the amount of silica increases. The strength and durability of POFA concrete is directly proportional to the amount of reactive silica present in the POFA. It can be noted that the K_2O content of POFA is higher due to the absorption of K_2O by palm oil tree from the soil during its cultivation period (Thomas and Kumar, 2017). From the chemical analysis of POFA by various studies, it can be noted that the summation of pozzolanic oxides ($SiO_2 + Al_2O_3 + Fe_2O_3$) is greater than 50%, which according to ASTM C618, conforms to Class C type pozzolan while in some exceptional cases the summation of pozzolanic oxides is greater than 70 conforming to Class F pozzolan. The LOI of POFA is higher than the specified content of 10% as per ASTM C618.

Using waste materials as cement replacement has seen a massive advancement due to their pozzolanic properties being suitable to produce high quality concrete meeting the stringent environmental legislation (Zeyad et al., 2016).

The chemical composition of POFA analyzed and reported by various researchers is shown in Table 3.16.

3.6.2 Pozzolanic activity of POFA

Presently, many types of pozzolan are used worldwide as partial replacement of Portland cement in concrete. It is well known that pozzolan when reacts with cement hydrated portlandite [$Ca(OH)_2$], produce additional calcium silicate hydrate (C—S—H) and calcium aluminate hydrate (C—A—H), which fill the pore spaces in the mortar/concrete matrix, making it denser, thereby, contributing to the increase in mechanical and durability properties of mortar/concrete. The pozzolanic reaction takes place generally in the form shown below:

A_xS_y	+	mCH	+	nH	→	C-S-H	+	C-A-H
Pozzolan		$Ca(OH)_2$		water		Calcium silicate hydrate		Calcium aluminate hydrate

As mentioned in the topic of sugarcane bagasse ash, that there are three methods by which pozzolanic activity index is measured, (i) by reactivity with lime (2) electrical conductivity and (iii) strength activity index (SAI). According to ASTM C311, (2013), the minimum requirement for SAI is 75% for use of pozzolan in concrete. Several investigators have examined the reactivity index of POFA and found that raw POFA showed a reduced SAI, lower than 75%. To attain higher pozzolanicity of POFA, two treatment methods can be used, (a) reduction of particle size by pulverization (b) thermal treatment.

3.6.2.1 Effect of grinding on pozzolanic activity

Grinding has a significant impact on the change in morphology of materials. When ground POFA was observed by scanning electron microscopy (SEM), the researchers (Zeyad et al., 2016; Ranjbar et al., 2016) found broken spherical particles of angular

Table 3.16 Chemical composition of POFA as reported in various studies.

References	SiO_2	Al_2O_3	Fe_2O_3	S + A + F	CaO	MgO	Na_2O	K_2O	SO_3	LOI
Chin et al. (2022)	62.48	18.72	6.54	87.74	4.83	2.65	–	3.18	–	0.48
Hamada et al. (2020)	67.5	4.2	8.12	79.82	3.97	2.72	0.115	8.45	0.535	1.48
Yusof et al. (2020)	76.36	5.89	2.41	84.66	4.43	3.67	–	2.99	2.06	–
Tai et al. (2019)	59.8	4.2	3.1	67.10	3.6	2.1	–	6.2	1.1	17.3
Rajak and Majid (2019)	54.8	7.24	4.47	66.51	14.0	4.14	–	–	0.71	8.5
Adnan et al. (2019)	55.2	4.48	5.44	65.12	4.12	2.25	–	2.28	2.25	–
Yusof et al. (2018)	39.1	1.85	4.50	45.45	3.79	1.73	–	7.58	1.26	–
Tambichik et al. (2018)	62.6	4.65	8.12	75.37	5.7	3.52	–	9.05	1.16	–
Wi, Lee, and Ismail et al. (2018)	70.9	5.63	3.51	80.04	3.78	3.61	0.39	5.66	–	10.1
Huseien et al. (2016)	64.2	4.25	3.13	71.58	10.2	5.90	0.10	8.64	0.09	1.73
Muthusamy and Zamri (2016)	51.55	4.64	8.64	64.83	5.91	2.44	0.07	5.50	0.61	5.0
Mohamad et al. (2016)	43.6	11.4	4.7	59.7	8.4	4.8	0.39	3.5	2.8	18.0
Ul Islam et al. (2016)	71.67	0.94	2.77	75.38	5.61	4.91	–	7.89	1.05	–
Liu et al. (2016)	63.4	5.5	4.2	73.1	4.3	3.7	–	6.3	0.9	6.0
Farzadnia et al. (2015)	47.37	3.53	6.19	57.09	11.83	4.19	–	–	1.22	1.84
Awal and Shehu (2015)	62.6	4.65	8.12	75.37	5.70	3.52	–	9.05	1.16	6.25

LOI is the weight loss of a sample after thermal treatment (burning/furnace) at high temperature, normally 1000°C, for 1 h duration.

$S + A + F = SiO_2 + Al_2O_3 + Fe_2O_3$ (Pozzolanic oxides)

and irregular shape. It was observed that the inclusion of ground POFA in concrete led to considerable improvement in the SAI. The SAI increased above minimum requirement level of 75%. When ultrafine POFA ($d_{50} = 1.1$ μm) was included in concrete, a rise of 32% in SAI was observed compared to ground POFA with particle size ($d_{50} = 14.58$ μm) (Maryam Md Sukaimi et al., 2018) (Karim et al., 2016a). Due to small diameter and high specific surface area, nano-POFA (20–90 nm) improve the silica dissolution which results in the speedy pozzolanic reaction (Tang et al., 2019).

3.6.2.2 Effect of thermal treatment

Thermal treatment has a substantial influence on the pozzolanic reactivity, reduction in unburnt carbon content and change in the morphology of POFA (Alsubari et al., 2016). Reduction in the porosity of particles on the thermal treatment of POFA at 580°C for 3 h, significantly increased the compressive strength of mortar (Karim et al., 2016b). Furthermore, the thermal treatment is accountable for reducing the agglomeration of POFA particles. Substantial reduction in the quantity of unburned carbon was observed when POFA was subjected to calcination at 650°C reducing substantially the LOI as well. The higher amount of unburnt carbon absorbs water through capillary action, reducing the water to cement (W/C) ratio, resulting in the decreased workability of the blended concrete mix. A significant enhancement of 19% in SAI compared to reference sample was observed when thermally treated POFA (500°C for 90 min) was included in the mortar (Sumesh et al., 2019).

3.7 Life cycle assessment (LCA) of recycled concrete containing by-products and plant materials

The LCA of recycled concrete needs to be explicitly discussed by professionals and researchers involved in cement and concrete production. The massive volume of concrete and cement produced worldwide annually transforms this sector into a significant generator of environmental impacts. Currently, it is considered that for each ton of cement produced, almost one ton of dioxide is generated, which directly contributes to the greenhouse effect (Berenguer, Lima, et al., 2020). Besides, other diverse environmental impacts are generated during the whole life cycle of concrete and cement.

In this context, the importance of encouraging the application of life cycle assessment (LCA) in the concrete industry arises. Since 1990, the life cycle assessment method has been used in the construction industry, and this concept is increasingly suitable for analyzing the concrete and cement manufacturing process. The LCA application allows operations to be analyzed respecting their different interrelationships in order to understand the production system systemically and identify critical points in their processes. This is a tool developed to quantitatively assess the environmental impacts of a product, from the extraction of raw materials to the moment the product is discarded. Flows are accounted for and then consolidated into ecological indicators to be presented to interested parties and assist in decision-making.

Although different boundaries can be adopted in an LCA study, a cradle to grave approach is the most suitable for the concrete industry, as the complete cycle includes a diverse range of cement and concrete applications, such as floors, slabs, columns, beams, and bridges. Typically, the LCA approach to evaluating concrete and cement production is the process life cycle analysis (PLCA). It refers to a bottom-up model that describes each relevant process throughout the life cycle of cement in terms of resource inputs and environmental outputs (Ding et al., 2019).

Portland cement production is associated with diverse environmental impacts and greenhouse gas generation. Therefore, incorporating by-products, such as fly ash, slag, silica-fume, rice-husk ash, sugar cane bagasse ash, and Palm-oil fuel ash can mitigate the environmental impacts of recycled aggregate concrete. Therefore, several papers presented in the literature so far (Gursel et al., 2014, 2019; Wang et al., 2017) have conducted LCA studies to evaluate the effect of supplementary cementitious materials (SCM).

Nevertheless, many aspects still need to be better understood in the context of incorporating by-products and SCM. For example, Seto et al. (2017) mentioned the lack of consensus on the upstream production impacts when fly ash replaces Portland cement in concrete. The authors question, among other things, how to attribute the environmental aspects between the production of electricity and the production of fly ash. Besides, it is observed in the literature that most studies focus on replacing natural aggregates with recycled aggregates or Portland cement with SCM, but this replacement is analyzed separately.

Another critical point to note is that the accuracy of LCA results is affected by the selection of functional units. In applying LCA for RCA, the most common functional unit is the volume of concrete, combining volume and strength, or combining volume, strength and durability. Table 3.17 presents commonly used functional unit and system boundaries.

In order to minimize the environmental impacts of the concrete and cement industry, many studies present the application of LCA to compare different concrete mixtures. This is because the selection of materials to be used in the production of concrete is one great possible solution to a sustainable sector; it can represent the minimization of waste and nonrenewable resources consumption. In this context, various plant materials and by-products can be used as recycled concrete aggregates, and each material has its advantages and disadvantages. However, despite the positive aspects on the environment, these materials are still preferred by only a small portion of professionals and consumers connected to the construction industry. It is clear that this sector lacks more public incentives and policies that encourage the use of these materials.

It is also crucial that LCA studies compare concrete mixes with similar mechanical properties, showing that it is possible to obtain optimal results using plant aggregates or by-products in concrete. For example, Bajpai et al. (2020) performed a cradle-to-grave LCA to compare different concrete mixes against the conventional one. In this study, the authors analyzed the use of fly-ash and silica fume. They showed that comparable compressive strength was obtained using silica fume, despite the absence of alkali silicate. In turn, Sáez-Pérez et al. (2020) claim that the mechanical properties of hemp fiber composites are superior to many other materials, with higher values of traction and flexural strength, which can withstand high levels of mechanical strength.

Table 3.17 Examples of the functional unit and system boundaries − RAC + SCM.

Functional unit	System boundary	Country	Reference
1 m^3 concrete Per ton of cement will consume 1.3 t of limestone, 0.3 t of clay and 50 kg of plaster. Recycled aggregate will be produced on-site with a jaw crusher of PE-150×250, the speed of crushing is 0·6∼3 m^3/h, the power of crushing is 0.5 kw	This research object is recycled aggregate concrete containing or not containing fly ash.	China	Ping and Yidong (2011)
1m^3 of each concrete mix for structural concrete	Cradle to gate - Raw material extraction and processing, processing of secondary material input, transport to the manufacturing facility and its manufacture.	Portugal	Kurad et al. (2017)
Scenario 1 with a functional unit of 1 m^3 was assumed for all analyzed concrete mixes (FU based only on the strength requirements) and Scenario 2 with a functional unit of 1.1 m^3 was assumed for RAC, NAC_FA and RAC_FA, and 1.0 m^3 for the other concrete mixes (FU includes strength, serviceability, and durability requirements).	Extraction, transport, electricity production, secondary process. End-of-life was assumed to be comparable (part of the waste was disposed of in landfills and part was recycled), so it was omitted as well.	Serbia	Marinković et al. (2017)

The use of materials such as fly ash, slag, silica-fume, rice-husk ash, sugar cane bagasse ash, and Palm-oil fuel ash can bring benefits not only from an environmental perspective but also influence social and economic aspects associated with the concrete sector. By utilizing locally available plant materials, the skills of the local population can be enhanced, which in turn will prevent them from migrating elsewhere, thus improving the social reality of that place (Shen et al., 2019). Besides, these materials have a reduced associated cost and can improve income generation, influencing economic aspects of the construction sector. Therefore, the use of these materials is closely linked to the goals of sustainable development.

The results found through a comprehensible literature review on materials that can minimize the environmental impacts of the concrete industry are presented below. It is clear that the number of publications on LCA application to evaluate by-products and plant materials is still tiny if we consider all the sustainable potential related to this use in concrete production. However, the concrete industry can already have interesting results to pursue minimizing its impacts, and it should be widely discussed.

- **Fly Ash**

Fly ash is a fine powder from pulverized coal combustion, produced on a large scale. This by-product is used as a partial replacement for Portland cement in concrete, improving its performance and making more economical concrete (Kang et al., 2020).

Zhang et al. (2018) analyzed the compressive strength of ordinary concrete and fly ash concrete with different proportions of water-to-binder and different levels of entrained air. The results obtained in the analysis show that the compressive strength of ordinary concrete without fly ash increased slowly after 28 days. For concrete with fly ash, the compressive strength significantly increased in later stages. Besides, it was proven that fly ash has a strong effect on air entrainment due to its high adsorption and, therefore, this needs to be considered to avoid a high decrease of the concrete mechanical properties.

Kurad et al. (2017) conducted an LCA to analyze the use of supplementary cementing materials in concrete, focusing on using high volumes of fly ash to minimize the environmental impacts. The authors found that the global warming potential (GWP) depends more on the type of materials used than on the strength of concrete mixes. Since the GWP linearly changes with the incorporation ratios of fly ash, it can be concluded that the average GWP for all concrete mixes decreased about 0.91% per unit per cent of fly ash and increased by about 0.20% with the use of 1% superplasticizer.

Another study by Marinković et al. (2017) presented an LCA application of different concrete mixtures for structural use, including a high-volume fly ash concrete with natural and recycled aggregates. The authors found that the concrete mix with fly ash became the best option. Still, it has a limited application due to its characteristic of securing at a temperature of 80°C , which is almost impossible in-situ at a large scale.

- **Silica fume**

Silica fume, which can also be called silica dust, micro silica, condensed silica fume, or volatilized silica, is a by-product of the smelting process in the silicon industry with an average particle size of less than 1 μm (Behnood et al., 2018). Several advantages are discussed in the literature regarding the use of silica-fume in concrete mixtures, such as, a high early compressive strength, a high flexural strength, very low permeability to water intrusion, and enhanced durability and toughness (Siddique et al., 2011). The literature explains these advantages by the spherical shape, high surface area, and superior pozzolanic reactivity of silica fume (Arel et al., 2018).

Khodabakhshian et al. (2018) analyzed the mechanical, economic, and environmental performances of 16 concrete mixes containing silica fume and marble industry waste powder. In this study, the LCA method was applied to compare the different mixtures from an environmental point of view. The authors pointed out that the importance of this study is related to the lack of detailed investigation to determine the optimal combined content of concrete mixes using silica fume as a partial replacement of ordinary Portland cement. Some studies analyze the mechanical issues associated with this replacement but do not consider the environmental aspects. The paper concluded that the mixes containing 10% SF and 5% – 20% waste marble powder showed 25% higher compressive strength compared to control sample at all ages. It also showed best resistance to sulfuric acid upto 1.6 times more as compared to the control sample. The mechanical properties of concrete are improved, greenhouse gas emissions are significantly reduced, and additionally, natural resources are saved.

Dabbaghi et al. (2021) adopted the LCA method to analyze the environmental impacts of producing 30 different lightweight concrete mixtures using silica fume. This paper used a cradle-to-gate approach, considering the following stages of the concrete life cycle: extraction of raw materials, production of ingredients and its transport to the production site, and the concrete manufacturing stage. Besides, the authors applied a genetic algorithm approach in order to minimize the economic cost and environmental life-cycle impacts while maximizing the compressive and tensile strengths. The great advantage of this work is that it uses a mathematical method to help select the best concrete mix considering all these aspects together. In contrast, many studies consider either environmental factors or mechanical characteristics.

- **Rice husk ash**

Rice husk ash is a by-product obtained by burning rice husk. It is already widely used in the concrete industry, minimizing various environmental impacts in this sector and solving pollution problems generated by the disposal of these ashes (Nuaklong et al., 2020). The disposal of this material is an environmental concern due to its high volume, reaching 156 million metric tons of rice husk globally (Gursel et al., 2016). Furthermore, rice husk decomposition is very slow, generating a significant amount of solid waste.

Rice husk ash can increase the compressive strength of concrete when used as a partial cement replacement, and the ideal ratio of cement replacement by this by-product must be between 10% and 30% (Lo et al., 2021). In addition, rice husk ash can increase corrosion resistance and resistance to chloride-ion penetration (Aprianti et al., 2017). Thus, rice husk ash deserves attention among researchers and professionals associated with the construction sector and concrete production.

Gursel et al. (2016) applied LCA to analyze the environmental properties of rice husk ash, also considering its mechanical and durability properties. In this study, interesting results were obtained and discussed. First, it became clear that with decreasing amount of Portland cement and increasing rice husk ash usage per unit volume of concrete mixes, the global warming potential of the concrete mixture decreases. On the other hand, rice husk ash can be associated with a greater amount of energy generation when compared to different types of ash due to the high use of energy for its grinding.

On the other hand, the results obtained from Fernando et al. (2021) showed that the environmental impacts generated by a concrete mix that uses fly ash and alkali-activated binders combined with rice husk ash could be more significant and more cost-effective than Portland cement concrete. The explanation presented by the authors to discuss the results found in their LCA study also refers to the high energy consumption required during the dissolution, processing and filtering processes.

- **Sugarcane bagasse ash**

Sugarcane bagasse is used as biomass to generate energy in sugarcane mills, and this process produces a residue that usually contains unburned bagasse and some sand. Hence, sieving processes are applied to transform this residue into ash for concrete production (Larissa et al., 2020). Sugarcane production is quite representative in Brazil, accounting for more than 760 million tons, while world production corresponds to around 1500 million tons (Sinoh et al., 2020).

A direct advantage of adding sugarcane bagasse ash in the cement matrix is that this by-product currently ends up in landfills; therefore, this is an opportunity for the concrete industry to help minimize waste generation (Berenguer, Lima, et al., 2020). However, publications are still analyzing the other economic, mechanical, and environmental advantages of this use. For the environmental analysis specifically, a few studies have already discussed the application of LCA to compare concrete mixtures using sugarcane bagasse ash. But it is undeniable that the number of publications on this topic are still negligible, especially compared to the number of publications that apply LCA to test other supplementary cementitious materials.

For example, several studies analyze the use of sugarcane bagasse ash from a mechanical perspective without considering the environmental aspects associated with its use. Same is the case of the paper written by Ríos-Parada et al. (2017). They analyzed the results of adding a Mexican sugarcane bagasse ash to binary concrete prepared with blended Portland cement and fly ash. Besides, Moretti, Sales, et al. (2016) analyzed the properties of concrete produced with construction waste and sugarcane bagasse ash sand, determining the compressive strength, modulus of elasticity, void ratio, dry bulk density, water absorption and the carbonation depth of concrete produced with different contents of these materials. Again, the environmental aspects were not considered in this study.

Regarding the environmental aspects of the use of sugarcane bagasse ash, the literature still lacks a deeper discussion. For example, based on a life cycle assessment (LCA) and a life cycle cost analysis (LCCA) application, Khan et al. (2020) showed that the sugarcane bagasse ash is a good substitute if NOx and SO_2 are adequately controlled. However, this study does not discuss comprehensive results about this specific by-product. On the other hand, Silva et al. (2014) evaluated the significant potential environmental impacts of the sugarcane bagasse use in Brazil through an LCA application. Still, this study considered the use of this material for electricity generation and not for concrete production. Few results could be considered by professionals and researchers associated with the concrete and cement industry, as the processes analyzed are very different.

It is clear that research still needs to go a long way in this direction, especially in countries that produce sugarcane on a large scale, such as Brazil, China and India.

- **Palm oil fuel ash**

The palm oil industry produces a significant number of solid wastes, and can occur in the forms of fibers, shells, and empty fruit bunches. These solid waste materials can be used for burning to produce electricity generation, for example. However, another solid waste is generated after combustion called palm oil fuel ash (Sata et al., 2004). The interest in using palm oil fuel ash as part of the concrete mix arose from the perception that the main chemical composition of this by-product is silica, an essential ingredient of pozzolan. Therefore, palm oil fuel ash could be utilized as a pozzolan in standard and high strength concrete (Chindaprasirt et al., 2008).

Thomas et al. (2017) presented a literature review of palm oil fuel ash utilization as supplementary cementitious material and the properties of such concrete at fresh and hardened stages. The authors mentioned that the main problem associated with oil palm cultivation is the massive biomass waste, which must be considered during concrete production. Regarding the research development, the authors mention that the use of this by-product needs to be studied in-depth, as few articles were found on the subject. Furthermore, it is essential to point out that most of the information synthesized by the authors referred to the mechanical aspects of concrete, not observing the potential environmental impacts generated by concrete production using palm oil fuel ash.

- **Waste glass powder**

Glass is a nonbiodegradable material mainly composed of silica (Deschamps et al., 2018). It does not decompose in the environment, causing various environmental impacts. Therefore, waste glass powder started to be used as a partial replacement for cement as a possible solution to glass landfilling. The micron particle size glass milling is an excellent solution for this by-product, as it strengthens the reactions between glass and cement hydrates (Islam et al., 2017). This brings several environmental and economic advantages to produce concrete worldwide. Besides, it has been observed in the literature that the use of glass powder in concrete can improve strength and durability parameters due to the increase in pozzolanic reactivity, and it can also reduce the quantity of cement used (Aygun et al., 2021).

Deschamps et al. (2018) presented an LCA study to compare the business-as-usual concrete production model and a glass powder concrete production process model. This paper uses the postconsumed mixed glass in an open-loop circular approach as supplementary cementitious material. The LCA results showed that the use of this by-product could minimize all environmental indicators, including midpoint and endpoint impact categories.

In turn, Jiang et al. (2014) conducted an LCA analysis in order to compare the environmental impacts of three different concrete mixtures: conventional concrete, glass powder concrete, and alkali-activated slag concrete. The authors adopted a cradle-to-gate approach and chose the following impact categories: greenhouse gas emissions (GHGs), energy use, water use, and potential environmental toxicity. The LCA results proved that it is possible to reduce the environmental impacts of concrete production when using glass powder as a cement replacement in concrete with lower strengths or in mortars with different strengths.

3.8 Summary

Flyash improves the workability, pumpability, cohesiveness, finish, ultimate strength, and durability of concrete as well as solves many problems experienced with concrete today—and all of these benefits at cheaper cost. Flyash, also known as pulverized fuel ash (PFA), has been a popular supplementary cementitious material (SCM) since the mid-1900s. It has been validated that substitution of silica fume is the most effective way of achieving low heat concrete without compromising the strength at an early age. The utilization of Rice Husk Ash (RHA) in construction sector can effectively reduce not only the disposal of RHA but can also reduce the subsequent land pollution. The use of RHA as cement substitution material or as partial replacement of fine aggregate can save natural resource like river sand and can prevent the possible erosion of river bank due to excess sand mining. Regarding the environmental aspects of the use of sugarcane bagasse ash, the literature still lacks a deeper discussion. Although studies on the use of palm oil fuel ash have grown in recent years, it is clear that this by-product has not yet become an object of study for LCA practitioners regarding concrete production. There are some LCA applications in this area, but it is clear that the research is still in its initial phases. Worldwide, efforts have been made to develop different types of glass-based products such as glass fiber, foam glass, sand blasting media; however, these products have a very limited demand and low market value and cannot absorb the very large quantity of waste glass produced worldwide. Hence, it is imperative to develop value added products to fully consume the accumulated waste glass.

References

Abood Habeeb, G., & Bin Mahmud, H. (2010). Study on properties of rice husk ash and its use as cement replacement material. *Materials Research, 13*(2), 185–190.

Adesina, A. (2020). Recent advances in the concrete industry to reduce its carbon dioxide emissions. *Environmental Challenges, 1*(November), 100004. https://doi.org/10.1016/j.envc.2020.100004. Elsevier B.V.

Adesina, P. A., & Olutoge, F. A. (2019). Structural properties of sustainable concrete developed using rice husk ash and hydrated lime. *Journal of Building Engineering, 25*(February), 100804. https://doi.org/10.1016/j.jobe.2019.100804. Elsevier Ltd.

Adjei, S., & Elkatatny, S. (2020). A highlight on the application of industrial and agro wastes in cement-based materials. *Journal of Petroleum Science and Engineering, 195*(June), 107911. https://doi.org/10.1016/j.petrol.2020.107911. Elsevier B.V.

Adnan, S. H., Azemi, N. F. N. M., Osman, M. H., Jeni, M. L. A., Ern, P. A. S., Yassin, N. I. M., … Akasyah, W. M. N. (2019). Influence of Palm oil fuel ash (POFA) towards fire resistance performance of brick. *Journal of Advanced Research in Fluid Mechanics and Thermal Sciences, 64*(2), 316–325. https://www.akademiabaru.com/submit/index.php/arfmts/article/view/2769.

Agredo, J. T., Mejía de Gutiérrez, R., Escandón Giraldo, C. E., & González Salcedo, L. O. (2014). Characterization of sugar cane bagasse ash as supplementary material for Portland cement. *INGENIERÍA E INVESTIGACIÓN, 34*(1), 5–10.

Ahmad, J., Zhou, Z., Usanova, K. I., Vatin, N. I., & El-Shorbagy, M. A. (2022). A step towards concrete with partial substitution of waste glass (WG) in concrete: A review. *Materials, 15*(7), 2525. https://doi.org/10.3390/ma15072525

Al-Kerttani, O. M. G. (2018). Fresh and shrinkage properties of self-compacted concrete when using recycled glass as aggregate. *Structural Concrete, 19*(4), 1245–1254. https://doi.org/10.1002/suco.201700069

Alaneme, K. K., Ekperusi, J. O., & Oke, S. R. (2018). Corrosion behaviour of thermal cycled aluminium hybrid composites reinforced with rice husk ash and silicon carbide. *Journal of King Saud University - Engineering Sciences, 30*(4), 391–397. https://doi.org/10.1016/j.jksues.2016.08.001. King Saud University.

Ali-Boucetta, T., Behim, M., Cassagnabere, F., Mouret, M., Ayat, A., & Laifa, W. (2021). Durability of self-compacting concrete containing waste bottle glass and granulated slag. *Construction and Building Materials, 270*, 121133. https://doi.org/10.1016/j.conbuildmat.2020.121133

Almeida, F. C., Sales, A., Moretti, J. P., & Mendes, P. C. (2015). Sugarcane bagasse ash sand (SBAS): Brazilian agro-industrial by-product for use in mortar. *Construction and Building Materials, 82*, 31–38. https://doi.org/10.1016/j.conbuildmat.2015.02.039

Alsubari, B., Shafigh, P., Jumaat, M. Z., & Alengaram, U. J. (2014). Palm oil fuel ash as a partial cement replacement for producing durable self-consolidating high-strength concrete. *Arabian Journal for Science and Engineering, 39*(12), 8507–8516. https://doi.org/10.1007/s13369-014-1381-3

Alsubari, B., et al. (2018). Properties of eco-friendly self-compacting concrete containing modified treated palm oil fuel ash. *Construction and Building Materials, 158*, 742–754. https://doi.org/10.1016/j.conbuildmat.2017.09.174. Elsevier Ltd.

Alsubari, B., Shafigh, P., Ibrahim, Z., Alnahhal, M. F., & Jumaat, M. Z. (2018). Properties of eco-friendly self-compacting concrete containing modified treated palm oil fuel ash. *Construction and Building Materials, 158*, 742–754. https://doi.org/10.1016/j.conbuildmat.2017.09.174

Amin, N.-U. (2011). Use of bagasse ash in concrete and its impact on the strength and chloride resistivity. *Journal of Materials in Civil Engineering, 23*(5), 717–720. https://doi.org/10.1061/(ASCE)MT.1943-5533.0000227. Dept. of Chemistry, Abdul Wali khan University Mardan, Khyber Pakhtoon Khwa, Pakistan.

AN/NZS 3582.1. (2016). *Australian/New Zealand Standard AS/NZS 3582.1 ' Supplementary cementitious materials.*

Antiohos, S. K., Papadakis, V. G., & Tsimas, S. (2014). Rice husk ash (RHA) effectiveness in cement and concrete as a function of reactive silica and fineness. *Cement and Concrete Research, 61–62*, 20–27. https://doi.org/10.1016/j.cemconres.2014.04.001. Elsevier Ltd.

Anupam, A. K., Kumar, P., & G.D.Ransinchung, R. N. (2013). Use of various agricultural and industrial waste materials in road construction. *Procedia - Social and Behavioral Sciences, 104*, 264–273. https://doi.org/10.1016/j.sbspro.2013.11.119. Elsevier B.V.

Aprianti, S. E. (2017). A huge number of artificial waste material can be supplementary cementitious material (SCM) for concrete production – a review part II. *Journal of Cleaner Production, 142*, 4178–4194. https://doi.org/10.1016/J.JCLEPRO.2015.12.115

Arel, H.Ş., & Shaikh, F. U. A. (2018). Effects of silica fume fineness on mechanical properties of steel fiber reinforced lightweight concretes subjected to ambient and elevated temperatures exposure. *Structural Concrete, 19*, 1829–1837. https://doi.org/10.1002/SUCO.201700281

Arenas-Piedrahita, J. C., Montes-García, P., Mendoza-Rangel, J. M., Calvo, H. L., Valdez-Tamez, P. L., & Martínez-Reyes, J. (2016). Mechanical and durability properties of mortars prepared with untreated sugarcane bagasse ash and untreated fly ash. *Construction and Building Materials, 105*, 69–81. https://doi.org/10.1016/j.conbuildmat.2015.12.047

Ashby, M. F. (2012). *Materials and the environment: Eco-informed material choice*. Elsevier.

Ashraf, M., Iqbal, M. F., Rauf, M., Ashraf, M. U., Ulhaq, A., Muhammad, H., & Liu, Q. F. (2022). Developing a sustainable concrete incorporating bentonite clay and silica fume: Mechanical and durability performance. *Journal of Cleaner Production, 337*, 130315. https://doi.org/10.1016/j.jclepro.2021.130315

ASTM. (2005). *Committee C-9 on Concrete and Concrete Aggregates. Standard specification for coal fly ash and raw or calcined natural pozzolan for use in concrete.* ASTM Int.

ASTM C 618. (2014). *Standard specification for coal fly ash and raw or calcined natural pozzolan for use in concrete, ASTM international, West Conshohocken, PA, 2012.* ASTM International. https://doi.org/10.1520/C0618. www.astm.org

ASTM C311. (2013). *Standard test methods for sampling and testing fly ash or natural pozzolans for use in Portland cement concrete,ASTM-C311/C311M-13.* West Conshohocken, USA: ASTM International.

Aygun, B. F. (2021). An overview of the impact of using glass powder on mechanical, durability properties in self-compacting concrete. *Journal of Sustainable Construction Materials and Technologies, 6*, 116–123. https://doi.org/10.29187/JSCMT.2021.67

Awal, A. S. M. A., & Shehu, I. A. (2015). Performance evaluation of concrete containing high volume palm oil fuel ash exposed to elevated temperature. *Construction and Building Materials, 76*, 214–220. https://doi.org/10.1016/j.conbuildmat.2014.12.001. Elsevier Ltd.

Bajpai, R., Choudhary, K., Srivastava, A., Sangwan, K. S., & Singh, M. (2020). Environmental impact assessment of fly ash and silica fume based geopolymer concrete. *Journal of Cleaner Production, 254*, 120147. https://doi.org/10.1016/J.JCLEPRO.2020.120147

Bahri, S., Mahmud, H. B., & Shafigh, P. (2018). Effect of utilizing unground and ground normal and black rice husk ash on the mechanical and durability properties of high-strength concrete. *Sadhana - Academy Proceedings in Engineering Sciences, 43*(2), 1–12. https://doi.org/10.1007/s12046-018-0795-0. Springer India.

Bahurudeen, A., Wani, K., Basit, M. A., & Santhanam, M. (2016). Assesment of pozzolanic performance of sugarcane bagasse ash. *Journal of Materials in Civil Engineering, 28*(2), 04015095. https://doi.org/10.1061/(ASCE)MT.1943-5533.0001361

Bahurudeen, A., & Santhanam, M. (2015). Influence of different processing methods on the pozzolanic performance of sugarcane bagasse ash. *Cement and Concrete Composites, 56*, 32–45. https://doi.org/10.1016/j.cemconcomp.2014.11.002. Elsevier Ltd.

Banar, R., Dashti, P., Zolfagharnasab, A., Ramezanianpour, A. M., & Ramezanianpour, A. A. (2022). A comprehensive comparison between using silica fume in the forms of water slurry or blended cement in mortar/concrete. *Journal of Building Engineering, 46*, 103802. https://doi.org/10.1016/j.jobe.2021.103802

Basika, E., Kigozi, J., & Kiggundu, N. (2015). Investigation of sugar cane bagasse ash as a binding material for the construction industry. *Journal of Global Ecology and Environment, 2*(4), 205–208. https://doi.org/10.9734/bpi/magees/v3/2101e

Behnood, A., & Golafshani, E. M. (2018). Predicting the compressive strength of silica fume concrete using hybrid artificial neural network with multi-objective grey wolves. *Journal of Cleaner Production, 202*, 54–64. https://doi.org/10.1016/J.JCLEPRO.2018.08.065

Belviso, C., Cavalcante, F., & Fiore, S. (2010). Synthesis of zeolite from Italian coal fly ash: Differences in crystallization temperature using seawater instead of distilled water. *Waste Management, 30*(5), 839–847. https://doi.org/10.1016/j.wasman.2009.11.015. Elsevier Ltd.

Berenguer, R. A., Capraro, A. P. B., Farias de Medeiros, M. H., Carneiro, A. M. P., & Oliveira, R. A. de (2020). Sugar cane bagasse ash as a partial substitute of Portland cement: Effect on mechanical properties and emission of carbon dioxide. *Journal of Environmental Chemical Engineering, 8*, 103655. https://doi.org/10.1016/J.JECE.2020.103655

Berenguer, R., Lima, N., Valdés, A. C., Medeiros, M. H. F., Lima, N. B. D., Delgado, J. M. P. Q., Silva, F. A. N., Azevedo, A. C., Guimarães, A. S., & Rangel, B. (2020). *Durability of concrete structures with sugar cane bagasse ash, advances in materials science and engineering.* https://doi.org/10.1155/2020/6907834, 2020.

Bie, R. S., Song, X. F., Liu, Q. Q., Ji, X. Y., & Chen, P. (2015). Studies on effects of burning conditions and rice husk ash (RHA) blending amount on the mechanical behavior of cement. *Cement and Concrete Composites, 55*, 162–168. https://doi.org/10.1016/j.cemconcomp.2014.09.008

Binal, A., Bas, B., & Karamut, O. R. (2016). Improvement of the strength of Ankara clay with self-cementing high alkaline fly ash. *Procedia Engineering, 161*, 374–379. https://doi.org/10.1016/j.proeng.2016.08.577. The Author(s).

Blissett, R. S., & Rowson, N. A. (2012). A review of the multi-component utilisation of coal fly ash. *Fuel, 97*, 1–23. https://doi.org/10.1016/j.fuel.2012.03.024. Elsevier Ltd.

BP Energy. (2018). *BP energy Outlook*. https://doi.org/10.1088/1757-899X/342/1/012091, 2018.

BS EN 14227-4. (2013). *Hydraulically bound Specifications. Fly ash for hydraulically bound mixtures*. https://doi.org/10.3403/30240493

Bulwark Team. (2021). *Micro silica (silica fume): Production & applications*. Available at: https://www.bulwarkconchem.com/micro-silica-silica-fume-production-applications/.

de Castro, S., & de Brito, J. (2013). Evaluation of the durability of concrete made with crushed glass aggregates. *Journal of Cleaner Production, 41*, 7–14. https://doi.org/10.1016/j.jclepro.2012.09.021. Elsevier Ltd.

Chaïd, R., Kenaï, S., Zeroub, H., & Jauberthie, R. (2015). Microstructure and permeability of concrete with glass powder addition conserved in the sulphatic environment. *European Journal of Environmental and Civil Engineering, 19*(2), 219–237. https://doi.org/10.1080/19648189.2014.939310

Chen, D., Masui, H., Miyoshi, H., Akai, T., & Yazawa, T. (2006). Extraction of heavy metal ions from waste colored glass through phase separation. *Waste Management, 26*(9), 1017–1023. https://doi.org/10.1016/j.wasman.2005.09.019

Cheng, A., Lin, W. T., Chao, S. J., & Hsu, H. M. (2014). Variation in fineness of cement-based composites containing sugarcane bagasse ashes. In *Advanced materials research* (Vol. 894, pp. 13–17). Trans Tech Publications, Ltd. https://doi.org/10.4028/www.scientific.net/amr.894.13

Chindaprasirt, P., Rukzon, S., & Sirivivatnanon, V. (2008). Resistance to chloride penetration of blended Portland cement mortar containing palm oil fuel ash, rice husk ash and fly ash. *Construction and Building Materials, 22*, 932–938. https://doi.org/10.1016/J.CONBUILDMAT.2006.12.001

Chindaprasirt, P., De Silva, P., Sagoe-Crentsil, K., & Hanjitsuwan, S. (2012). Effect of SiO_2 and Al_2O_3 on the setting and hardening of high calcium fly ash-based geopolymer systems. *Journal of Materials Science, 47*(12), 4876–4883. https://doi.org/10.1007/s10853-012-6353-y

Chindaprasirt, P., Homwuttiwong, S., & Jaturapitakkul, C. (2007). Strength and water permeability of concrete containing palm oil fuel ash and rice husk-bark ash. *Construction and Building Materials, 21*(7), 1492–1499. https://doi.org/10.1016/j.conbuildmat.2006.06.015

Chin, W. Q., Lee, Y. H., Amran, M., Fediuk, R., Vatin, N., Kueh, A. B. H., & Lee, Y. Y. (2022). A sustainable reuse of agro-industrial wastes into green cement bricks. *Materials, 15*(5), 1713. https://doi.org/10.3390/ma15051713

Choudhary, R., Gupta, R., Alomayri, T., Jain, A., & Nagar, R. (October 2021). Permeation, corrosion, and drying shrinkage assessment of self-compacting high strength concrete comprising waste marble slurry and fly ash, with silica fume. In *Structures* (Vol. 33, pp. 971–985). Elsevier. https://doi.org/10.1016/j.istruc.2021.05.008

Chua, H. S., Bashir, M. J. K., Tan, K. T., & Chua, H. S. (July 2019). A sustainable pyrolysis technology for the treatment of municipal solid waste in Malaysia. In *AIP conference proceedings* (Vol. 2124, p. 020016). AIP Publishing LLC. https://doi.org/10.1063/1.5117076 (1).

Chusilp, N., Jaturapitakkul, C., & Kiattikomol, K. (2009a). Effects of LOI of ground bagasse ash on the compressive strength and sulfate resistance of mortars. *Construction and Building Materials, 23*(12), 3523–3531. https://doi.org/10.1016/j.conbuildmat.2009.06.046. Elsevier Ltd.

Chusilp, N., Jaturapitakkul, C., & Kiattikomol, K. (2009b). Utilization of bagasse ash as a pozzolanic material in concrete. *Construction and Building Materials, 23*(11), 3352–3358. https://doi.org/10.1016/j.conbuildmat.2009.06.030. Elsevier Ltd.

Civiltoday. (2022). *'Properties of Cement'. Civil Engineering*. Available at: https://civiltoday.com/civil-engineering-materials/cement/111-properties-of-cement-physical-chemical-properties?highlight=WyJzcGVjaWZpYyIsImdyYXZpdHkiLCJvZiIsImNlbWVud CIsImNlbWVudCdzIiwiY2VtZW50JyIsInNwZWNpZmljIGdyYXZpdHkiLCJzcGVja WZpYyBncmF2aXR5IG9mIiwiZ.

Clark, M. W., Despland, L. M., Lake, N. J., Yee, L. H., Anstoetz, M., Arif, E., Parr, J. F., & Doumit, P. (2017). High-efficiency cogeneration boiler bagasse-ash geochemistry and mineralogical change effects on the potential reuse in synthetic zeolites, geopolymers, cements, mortars, and concretes. *Heliyon, 3*(4), e00294. https://doi.org/10.1016/j.heliyon.2017.e00294

Cokca, E., & Yilmaz, Z. (2004). Use of rubber and bentonite added fly ash as a liner material. *Waste Management, 24*(2), 153–164. https://doi.org/10.1016/j.wasman.2003.10.004

Cook, D. J., & Cao, T. (1992). Fly ash, slag and silica fume. In *Australian concrete technology* (pp. 67–98). Melbourne, VIC: Addison-Wesley Longman.

Cordeiro, G. C., Andreão, P. V., & Tavares, L. M. (2019). Pozzolanic properties of ultrafine sugar cane bagasse ash produced by controlled burning. *Heliyon, 5*(10), e02566. https://doi.org/10.1016/j.heliyon.2019.e02566

Cordeiro, G. C., Barroso, T. R., & Toledo Filho, R. D. (2018). Enhancement the properties of sugar cane bagasse ash with high carbon content by a controlled re-calcination process. *KSCE Journal of Civil Engineering, 22*(4), 1250–1257. https://doi.org/10.1007/s12205-017-0881-6

Cordeiro, G. C., Toledo Filho, R. D., Tavares, L. M., & Fairbairn, E. M. R. (2008). Pozzolanic activity and filler effect of sugar cane bagasse ash in Portland cement and lime mortars. *Cement and Concrete Composites, 30*(5), 410–418. https://doi.org/10.1016/j.cemconcomp.2008.01.001

Cordeiro, G. C., Toledo Filho, R. D., Tavares, L. M., & Fairbairn, E. M. R. (2012). Experimental characterization of binary and ternary blended-cement concretes containing ultrafine residual rice husk and sugar cane bagasse ashes. *Construction and Building Materials, 29*, 641–646. https://doi.org/10.1016/j.conbuildmat.2011.08.095

Cordeiro, G. C., & Kurtis, K. E. (2017). Effect of mechanical processing on sugar cane bagasse ash pozzolanicity. *Cement and Concrete Research, 97*, 41–49. https://doi.org/10.1016/j.cemconres.2017.03.008. Elsevier Ltd.

Cordeiro, G. C., Toledo Filho, R. D., & De Moraes Rego Fairbairn, E. (2009). Use of ultrafine rice husk ash with high-carbon content as pozzolan in high performance concrete. *Materials and Structures/Materiaux et Constructions, 42*(7), 983–992. https://doi.org/10.1617/s11527-008-9437-z

Dabbaghi, F., Tanhadoust, A., Nehdi, M. L., Nasrollahpour, S., Dehestani, M., & Yousefpour, H. (2021). Life cycle assessment multi-objective optimization and deep belief network model for sustainable lightweight aggregate concrete. *Journal of Cleaner Production, 318*, 128554. https://doi.org/10.1016/J.JCLEPRO.2021.128554

Dash, S. K., Kar, B. B., Mukherjee, P. S., & Mustakim, S. M. (2016). A comparison among the physico-chemical-mechanical of three potential aggregates fabricated from fly ash. *Journal of Civil and Environmental Engineering, 6*, 243. https://doi.org/10.4172/2165-784x.1000243

Das, P., & Kumaran, S. (2015). Monitoring and control of an in-house steam power plant: a support of the idea of distributed power generation. *International Journal of Advancements in Technology, 06*(02), 6–8. https://doi.org/10.4172/0976-4860.1000149

de Soares, M. M., Garcia, D. C., Figueiredo, R. B., Aguilar, M. T. P., & Cetlin, P. R. (2016). Comparing the pozzolanic behavior of sugar cane bagasse ash to amorphous and crystalline SiO_2. *Cement and Concrete Composites, 71*, 20–25. https://doi.org/10.1016/j.cemconcomp.2016.04.005

Deepika, S., Anand, G., Bahurudeen, A., & Santhanam, M. (2017). Construction products with sugarcane bagasse ash binder. *Journal of Materials in Civil Engineering, 29*(10), 04017189. https://doi.org/10.1061/(ASCE)MT.1943-5533.0001999

Deschamps, J., Simon, B., Tagnit-Hamou, A., & Amor, B. (2018). Is open-loop recycling the lowest preference in a circular economy? Answering through LCA of glass powder in concrete. *Journal of Cleaner Production, 185*, 14–22. https://doi.org/10.1016/J.JCLEPRO.2018.03.021

Dhir, R. K. (1986). Pulverized fuel ash. *Cement Replacement Materials, 3*, 197–255. Surrey University Press Guildford, Surrey, UK.

Ding, N., Liu, J., Kong, Z., Yan, L., & Yang, J. X. (2019). Life cycle greenhouse gas emissions of Chinese urban household consumption based on process life cycle assessment: Exploring the critical influencing factors. *Journal of Cleaner Production, 210*, 898–906. https://doi.org/10.1016/J.JCLEPRO.2018.10.242

Disfani, M. M., Arulrajah, A., Bo, M. W., & Hankour, R. J. W. M. (2011). Recycled crushed glass in road work applications. *Waste Management, 31*(11), 2341–2351. https://doi.org/10.1016/j.wasman.2011.07.003

Djamaluddin, A. R., Caronge, M. A., Tjaronge, M. W., Rahim, I. R., & Noor, N. M. (2018). Abrasion resistance and compressive strength of unprocessed rice husk ash concrete. *Asian Journal of Civil Engineering, 19*(7), 867–876. https://doi.org/10.1007/s42107-018-0069-5

Dodson, V. H. (2013). *Concrete Admixtures. Springer US (VNR structural engineering series)*. Available at: https://books.google.com.au/books?id=9NEFCAAAQBAJ.

Dong, W., Li, W., & Tao, Z. (2021). A comprehensive review on performance of cementitious and geopolymeric concretes with recycled waste glass as powder, sand or cullet. *Resources, Conservation and Recycling, 172*(May), 105664. https://doi.org/10.1016/j.resconrec.2021.105664. Elsevier B.V.

Dyer, T. D., & Dhir, R. K. (2001). Chemical reactions of glass cullet used as cement component. *Journal of Materials in Civil Engineering, ASCE, 13*(6), 412–417.

ECOBA. (2016). *Utilisation of flyash in construction industry & underground mining in Europe*. Available at: https://www.ecoba.com/ecobaccputil.html.

Embong, R., Shafiq, N., Kusbiantoro, A., & Nuruddin, M. F. (2016). Effectiveness of low-concentration acid and solar drying as pre-treatment features for producing pozzolanic sugarcane bagasse ash. *Journal of Cleaner Production, 112*, 953–962. https://doi.org/10.1016/j.jclepro.2015.09.066

EN 450. (2012). *EN 450-1: 2012, fly ash for concrete - Part 1: Definition, specifications and conformity criteria*. Comite Europeen de Normalisation.

Fapohunda, C., Akinbile, B., & Shittu, A. (2017). Structure and properties of mortar and concrete with rice husk ash as partial replacement of ordinary Portland cement – a review.

International Journal of Sustainable Built Environment, 6(2), 675–692. https://doi.org/10.1016/j.ijsbe.2017.07.004. The Gulf Organisation for Research and Development.

Faria, K. C., Gurgel, R. F., & Holanda, J. N. F. (2010). Characterization of sugarcane bagasse ash for use in ceramic bodies. In *Materials science forum* (Vol. 660, pp. 1049–1052). Trans Tech Publications Ltd. https://doi.org/10.4028/www.scientific.net/MSF.660-661.1049

Farrant, W. E., Babafemi, A. J., Kolawole, J. T., & Panda, B. (2022). Influence of sugarcane bagasse ash and silica fume on the mechanical and durability properties of concrete. *Materials, 15*(9), 3018. https://doi.org/10.3390/ma15093018

Farzadnia, N., Noorvand, H., Yasin, A. M., & Aziz, F. N. A. (2015). The effect of nano silica on short term drying shrinkage of POFA cement mortars. *Construction and Building Materials, 95*, 636–646. https://doi.org/10.1016/j.conbuildmat.2015.07.132

Federec, & Recyclage. (2020). *Le marche Du recyclage 2020.*

Fernando, S., Gunasekara, C., Law, D. W., Nasvi, M. C. M., Setunge, S., & Dissanayake, R. (2021). Life cycle assessment and cost analysis of fly ash–rice husk ash blended alkali-activated concrete. *Journal of Environmental Management, 295*, 113140. https://doi.org/10.1016/J.JENVMAN.2021.113140

Feuerborn, H. J. (June 2009). Calcareous ash in Europe-a reflection on technical and legal issues. In *Proceedings of the 2nd hellenic conference on utilisation of industrial by-products in construction, Aiani Kozani, Greece. June 1–3, 2009* (Vol. 1).

Franus, W., Wiatros-Motyka, M. M., & Wdowin, M. (2015). Coal fly ash as a resource for rare earth elements. *Environmental Science and Pollution Research, 22*(12), 9464–9474. https://doi.org/10.1007/s11356-015-4111-9

Frías, M., Villar, E., & Savastano, H. (2011). Brazilian sugar cane bagasse ashes from the cogeneration industry as active pozzolans for cement manufacture. *Cement and Concrete Composites, 33*(4), 490–496. https://doi.org/10.1016/j.cemconcomp.2011.02.003

Ganesan, K., Rajagopal, K., & Thangavel, K. (2008). Rice husk ash blended cement: assessment of optimal level of replacement for strength and permeability properties of concrete. *Construction and Building Materials, 22*(8), 1675–1683. https://doi.org/10.1016/j.conbuildmat.2007.06.011

GB/T 1596. (2017). *Fly ash used for cement and concrete., National Standards of China*, 10.1016/j.gecco.2019.e00539%0A10.1016/j.foreco.2018.06.029%0Ahttp://www.cpsg.org/sites/cbsg.org/files/documents/Sunda Pangolin National Conservation Strategy and Action Plan %28LoRes%29.pdf%0A10.1016/j.forec. Available at:.

Geraldo, R. H., Ouellet-Plamondon, C. M., Muianga, E. A., & Camarini, G. (2017). Alkali-activated binder containing wastes: A study with rice husk ash and red ceramic. *Cerâmica, 63*, 44–51. https://doi.org/10.1590/0366-69132017633652057

Gilbert, N. (2012). Palm-oil boom raises conservation concerns. *Nature, 487*(7405), 14–15. https://doi.org/10.1038/487014a

Golewski, G. L., & Gil, D. M. (2021). Studies of fracture toughness in concretes containing fly ash and silica fume in the first 28 Days of curing. *Materials, 14*(2), 319. https://doi.org/10.3390/ma14020319

Gollakota, A. R. K., Volli, V., & Shu, C. M. (2019). Progressive utilisation prospects of coal fly ash: a review. *Science of the Total Environment, 672*, 951–989. https://doi.org/10.1016/j.scitotenv.2019.03.337. Elsevier B.V.

Gursel, A. P., Maryman, H., & Ostertag, C. (2016). A life-cycle approach to environmental, mechanical, and durability properties of "green" concrete mixes with rice husk ash. *Journal of Cleaner Production, 112*, 823–836. https://doi.org/10.1016/j.jclepro.2015.06.029

Gursel, A. P., & Ostertag, C. (2019). Life-cycle assessment of high-strength concrete mixtures with copper slag as sand replacement. *Advances in Civil Engineering*. https://doi.org/10.1155/2019/6815348, 2019.

Güneyisi, E., Gesoğlu, M., Karaoğlu, S., & Mermerdaş, K. (2012). Strength, permeability and shrinkage cracking of silica fume and metakaolin concretes. *Construction and Building Materials, 34*, 120–130. https://doi.org/10.1016/j.conbuildmat.2012.02.017

Guo, P., Meng, W., Nassif, H., Gou, H., & Bao, Y. (2020). New perspectives on recycling waste glass in manufacturing concrete for sustainable civil infrastructure. *Construction and Building Materials, 257*, 119579. https://doi.org/10.1016/j.conbuildmat.2020.119579

Habeeb, G. A., & Fayyadh, M. M. (2009). Rice Husk Ash concrete: the effect of RHA average particle size on mechanical properties and drying shrinkage. *Australian Journal of Basic and Applied Sciences, 3*(3), 1616–1622.

Hamada, H. M., Alya'a, A., Yahaya, F. M., Muthusamy, K., Tayeh, B. A., & Humada, A. M. (2020). Effect of high-volume ultrafine palm oil fuel ash on the engineering and transport properties of concrete. *Case Studies in Construction Materials, 12*, e00318. https://doi.org/10.1016/j.cscm.2019.e00318

Harris, D., Heidrich, C., & Feuerborn, J. (2019). Global aspects on coal combustion products. *VGB Power Technology*, 25–33. Available at: https://www.vgb.org/vgbmultimedia/PT202010HARRIS-p-16422.pdf.

Hasan, M., Khan, M. A., & Alam, J. (2014). Permeability characteristic of fly ash treated with bentonite. In *Proceedings of Indian geotechnical conference IGC, December* (pp. 18–22).

Hassan, I. O., et al. (2014). Flow characteristics of ternary blended self-consolidating cement mortars incorporating palm oil fuel ash and pulverised burnt clay. *Construction and Building Materials, 64*, 253–260. https://doi.org/10.1016/j.conbuildmat.2014.04.057. Elsevier Ltd.

Hassett, D. J., & Eylands, K. E. (1999). Mercury capture on coal combustion fly ash. *Fuel, 78*(2), 243–248. https://doi.org/10.1016/S0016-2361(98)00150-1

Hilburg, J. (2019). *Concrete production produces eight percent of the world's carbon dioxide emissions*. Architecture, International, News, Sustainability. Available online: https://www.archpaper.com/2019/01/concrete-production-eight-percent-co2-emissions/. (Accessed 20 August 2020).

Hossain, M. M., Karim, M. R., & Zain, M. F. M. (2017). Acid and sulfate resistance of alkali-activated ternary blended composite binder. *Journal of Materials in Civil Engineering, 30*(2), 04017276. https://doi.org/10.1061/(asce)mt.1943-5533.0002131

Huang, H., Gao, X., Wang, H., & Ye, H. (2017). Influence of rice husk ash on strength and permeability of ultra-high performance concrete. *Construction and Building Materials, 149*, 621–628. https://doi.org/10.1016/j.conbuildmat.2017.05.155

Huseien, G. F., Mirza, J., Ismail, M., & Hussin, M. W. (2016). Influence of different curing temperatures and alkali activators on properties of GBFS geopolymer mortars containing fly ash and palm-oil fuel ash. *Construction and Building Materials, 125*, 1229–1240. https://doi.org/10.1016/j.conbuildmat.2016.08.153

Hussein, A. A. E., Shafiq, N., & Nuruddin, M. F. (May 2015). Compressive strength and interfacial transition zone of sugar cane bagasse ash concrete: A comparison to the established pozzolans. In *AIP conference proceedings* (Vol. 1660, p. 070002). AIP Publishing LLC. https://doi.org/10.1063/1.4915720 (1).

Ibrahim, K. I. M. (2017). The effect of using waste glass [WG] as partial replacement of sand on concrete. *IOSR Journal of Mechanical and Civil Engineering, 14*(2), 41–45. https://doi.org/10.9790/1684-1402024145

Idir, R., Cyr, M., & Tagnit-hamou, A. (2011). Cement & Concrete Composites Pozzolanic properties of fine and coarse color-mixed glass cullet. *Cement and Concrete Composites, 33*(1), 19–29. https://doi.org/10.1016/j.cemconcomp.2010.09.013. Elsevier Ltd.

Iskandar, M. J., Baharum, A., Anuar, F. H., & Othaman, R. (2018). Palm oil industry in South East Asia and the effluent treatment technology—a review. *Environmental Technology & Innovation, 9*, 169–185. https://doi.org/10.1016/j.eti.2017.11.003

Islam, G. M. S., Rahman, M. H., & Kazi, N. (2017). Waste glass powder as partial replacement of cement for sustainable concrete practice. *International Journal of Sustainable Built Environment, 6*(1), 37–44. https://doi.org/10.1016/j.ijsbe.2016.10.005. The Gulf Organisation for Research and Development.

Jagadesh, P., Ramachandramurthy, A., & Murugesan, R. (2018). Evaluation of mechanical properties of sugar cane bagasse ash concrete. *Construction and Building Materials, 176*, 608–617. https://doi.org/10.1016/j.conbuildmat.2018.05.037. Elsevier Ltd.

Jalal, M., Pouladkhan, A., Harandi, O. F., & Jafari, D. (2015). Comparative study on effects of Class F fly ash, nano silica and silica fume on properties of high performance self-compacting concrete. *Construction and Building Materials*, (94), 90–104. https://doi.org/10.1016/j.conbuildmat.2015.07.001

Jani, Y., & Hogland, W. (2014). Waste glass in the production of cement and concrete - a review. *Journal of Environmental Chemical Engineering, 2*(3), 1767–1775. https://doi.org/10.1016/j.jece.2014.03.016. Elsevier B.V.

Jaya, R. P., Nor, M. A. A. M., Ahmad, Z. A., & Amin, Z. M. (2013). Properties of mortar containing rice husk ash at different temperature and exposed to aggressive environment. In *Advanced materials research* (Vol. 620, pp. 87–93). Trans Tech Publications Ltd. https://doi.org/10.4028/www.scientific.net/AMR.620.87

Jiang, M., Chen, X., Rajabipour, F., & Hendrickson, C. T. (2014). Comparative life cycle assessment of conventional, glass powder, and alkali-activated slag concrete and mortar. *Journal of Infrastructure Systems, 20*, 04014020. https://doi.org/10.1061/(ASCE)IS.1943-555X.0000211

Jiang, X., Xiao, R., Bai, Y., Huang, B., & Ma, Y. (2022). Influence of waste glass powder as a supplementary cementitious material (SCM) on physical and mechanical properties of cement paste under high temperatures. *Journal of Cleaner Production, 340*, 130778. https://doi.org/10.1016/j.jclepro.2022.130778

JIS A 6201. (2019). *Flyash for use in concrete* (15th ed.).

Jittin, V., Bahurudeen, A., & Ajinkya, S. D. (2020). Utilisation of rice husk ash for cleaner production of different construction products. *Journal of Cleaner Production, 263*, 121578. https://doi.org/10.1016/j.jclepro.2020.121578. Elsevier Ltd.

Joshaghani, A., & Moeini, M. A. (2017). Evaluating the effects of sugar cane bagasse ash (SCBA) and nanosilica on the mechanical and durability properties of mortar. *Construction and Building Materials, 152*, 818–831. https://doi.org/10.1016/j.conbuildmat.2017.07.041. Elsevier Ltd.

Joshaghani, A., & Moeini, M. A. (2018). Evaluating the effects of sugarcane-bagasse ash and rice-husk ash on the mechanical and durability properties of mortar. *Journal of Materials in Civil Engineering, 30*(7), 04018144. https://doi.org/10.1061/(ASCE)MT.1943-5533.0002317

Jung, S. H., Saraswathy, V., Karthick, S., Kathirvel, P., & Kwon, S. J. (2018). Microstructure characteristics of fly ash concrete with rice husk ash and lime stone powder. *International Journal of Concrete Structures and Materials, 12*(1), 1–9. https://doi.org/10.1186/s40069-018-0257-4

Kang, S. H., Hong, S. G., & Moon, J. (2019). The use of rice husk ash as reactive filler in ultra-high performance concrete. *Cement and Concrete Research, 115*(March 2018), 389–400. https://doi.org/10.1016/j.cemconres.2018.09.004. Elsevier.

Kang, S., Lloyd, Z., Kim, T., & Ley, M. T. (2020). Predicting the compressive strength of fly ash concrete with the Particle Model. *Cement and Concrete Research, 137*, 106218. https://doi.org/10.1016/J.CEMCONRES.2020.106218

Kannan, V., & Ganesan, K. (2016). Effect of tricalcium aluminate on durability properties of self-compacting concrete incorporating rice husk ash and metakaolin. *Journal of Materials in Civil Engineering, 28*(1), 04015063. https://doi.org/10.1061/(ASCE)MT.1943-5533.0001330

Karamberi, A., & Moutsatsou, A. (2005). Participation of coloured glass cullet in cementitious materials. *Cement and Concrete Composites, 27*(2), 319–327. https://doi.org/10.1016/j.cemconcomp.2004.02.021

Karim, M. R., Hashim, H., & Abdul Razak, H. (2016a). Assessment of pozzolanic activity of palm oil clinker powder. *Construction and Building Materials, 127*, 335–343. https://doi.org/10.1016/j.conbuildmat.2016.10.002. Elsevier Ltd.

Karim, M. R., Hashim, H., & Abdul Razak, H. (2016b). Thermal activation effect on palm oil clinker properties and their influence on strength development in cement mortar. *Construction and Building Materials, 125*, 670–678. https://doi.org/10.1016/j.conbuildmat.2016.08.092. Elsevier Ltd.

Kashani, A., Ngo, T. D., & Hajimohammadi, A. (2019). Effect of recycled glass fines on mechanical and durability properties of concrete foam in comparison with traditional cementitious fines. *Cement and Concrete Composites, 99*(January), 120–129. https://doi.org/10.1016/j.cemconcomp.2019.03.004

Katare, V. D., & Madurwar, M. V. (2017). Experimental characterization of sugarcane biomass ash – A review. *Construction and Building Materials, 152*, 1–15. https://doi.org/10.1016/j.conbuildmat.2017.06.142. Elsevier Ltd.

Kazmi, S. M. S., Munir, M. J., Patnaikuni, I., & Wu, Y. F. (2017). Pozzolanic reaction of sugarcane bagasse ash and its role in controlling alkali silica reaction. *Construction and Building Materials, 148*, 231–240. https://doi.org/10.1016/j.conbuildmat.2017.05.025

Keramatikerman, M., Chegenizadeh, A., & Terzaghi, S. (2019). Review on effect of sugarcane bagasse ash as an additive in construction industry. *EJGE, 24*, 453–470. Available at: https://www.scopus.com/inward/record.uri?eid=2-s2.0-85091010255&partnerID=40&md5=d47257b5ace0c5869d1d30c8903fa23f.

Khan, M. N. N., Saha, A. K., & Sarker, P. K. (2020). Reuse of waste glass as a supplementary binder and aggregate for sustainable cement-based construction materials: A review. *Journal of Building Engineering, 28*(June 2019), 101052. https://doi.org/10.1016/j.jobe.2019.101052. Elsevier Ltd.

Khodabakhshian, A., de Brito, J., Ghalehnovi, & Asadi Shamsabadi, M. E. (2018). Mechanical, environmental and economic performance of structural concrete containing silica fume and marble industry waste powder. *Construction and Building Materials, 169*, 237–251. https://doi.org/10.1016/J.CONBUILDMAT.2018.02.192

Kim, B., Prezzi, M., & Salgado, R. (2005). Geotechnical properties of fly and bottom ash mixtures for use in Highway Embankments. *Journal of Geotechnical and Geoenvironmental Engineering, 131*(7), 914–924. https://doi.org/10.1061/(asce)1090-0241(2005)131:7(914)

Kizhakkumodom Venkatanarayanan, H., & Rangaraju, P. R. (2015). Effect of grinding of low-carbon rice husk ash on the microstructure and performance properties of blended cement concrete. *Cement and Concrete Composites, 55*, 348–363. https://doi.org/10.1016/j.cemconcomp.2014.09.021. Elsevier Ltd.

Kolawole, J. T. A., Babafemi, A. J., Ebenezer Fanijo, E., Paul, S. C., & Riaan Combrinck, R. (2021). State-of-the-art review on the use of sugarcane bagasse ash in cementitious materials. *Cement and Concrete Composites, 118*, 103975. https://doi.org/10.1016/j.cemconcomp.2021.103975. Elsevier Ltd.

Kolbe, J. L., Lee, L. S., Jafvert, C. T., & Murarka, I. P. (2011). Use of alkaline coal ash for reclamation of a former strip mine. In *World of coal ash (WOCA) conference, USA* (pp. 1–15).

Kolias, S., Kasselouri-Rigopoulou, V., & Karahalios, A. (2005). Stabilisation of clayey soils with high calcium fly ash and cement. *Cement and Concrete Composites, 27*(2), 301–313. https://doi.org/10.1016/j.cemconcomp.2004.02.019

Koukouzas, N. K., Zeng, R., Perdikatsis, V., Xu, W., & Kakaras, E. K. (2006). Mineralogy and geochemistry of Greek and Chinese coal fly ash. *Fuel, 85*(16), 2301–2309. https://doi.org/10.1016/j.fuel.2006.02.019

Külaots, I., Hurt, R. H., & Suuberg, E. M. (2004). Size distribution of unburned carbon in coal fly ash and its implications. *Fuel, 83*(2), 223–230. https://doi.org/10.1016/S0016-2361(03)00255-2

Kumar, A., Bheel, N., Ahmed, I., Rizvi, S. H., Kumar, R., & Jhatial, A. A. (2022). Effect of silica fume and fly ash as cementitious material on hardened properties and embodied carbon of roller compacted concrete. *Environmental Science and Pollution Research, 29*(1), 1210–1222. https://doi.org/10.1007/s11356-021-15734-0

Kurad, R., Silvestre, J. D., de Brito, J., & Ahmed, H. (2017). Effect of incorporation of high volume of recycled concrete aggregates and fly ash on the strength and global warming potential of concrete. *Journal of Cleaner Production, 166*, 485–502. https://doi.org/10.1016/J.JCLEPRO.2017.07.236

Kwan, W. H., & Wong, Y. S. (2020). Acid leached rice husk ash (ARHA) in concrete: A review. *Materials Science for Energy Technologies, 3*(KeAi Communications Co., Ltd), 501–507. https://doi.org/10.1016/j.mset.2020.05.001

Lam, C. S., Poon, C. S., & Chan, D. (2007). Enhancing the performance of pre-cast concrete blocks by incorporating waste glass - ASR consideration. *Cement and Concrete Composites, 29*(8), 616–625. https://doi.org/10.1016/j.cemconcomp.2007.03.008

Larissa, L. C., Marcos, M. A., Maria, M., de Souza, N. S. L., & de Farias, E. C. (2020). Effect of high temperatures on self-compacting concrete with high levels of sugarcane bagasse ash and metakaolin. *Construction and Building Materials, 248*, 118715. https://doi.org/10.1016/J.CONBUILDMAT.2020.118715

Lee, S. H., et al. (2003). Effect of particle size distribution of fly ash-cement system on the fluidity of cement pastes. *Cement and Concrete Research, 33*(5), 763–768. https://doi.org/10.1016/S0008-8846(02)01054-2

Lee, G., et al. (2011). Effects of crushed glass cullet sizes, casting methods and pozzolanic materials on ASR of concrete blocks. *Construction and Building Materials, 25*(5), 2611–2618. https://doi.org/10.1016/j.conbuildmat.2010.12.008

Le, H. T., & Ludwig, H. M. (2016). Effect of rice husk ash and other mineral admixtures on properties of self-compacting high performance concrete. *Materials and Design, 89*, 156–166. https://doi.org/10.1016/j.matdes.2015.09.120

Le, H. T., Siewert, K., & Ludwig, H. M. (2015). Alkali silica reaction in mortar formulated from self-compacting high performance concrete containing rice husk ash. *Construction and Building Materials, 88*, 10–19. https://doi.org/10.1016/j.conbuildmat.2015.04.005

Liaqat, M., Shah, M. L., & Baig, M. A. (2018). Effect of waste glass as partial replacement for coarse aggregate in concrete. *International Journal of Technical Innovation in Modern Engineering & Science, 4*(6), 2455−2585.

Lim, N. H. A. S., et al. (2015). The effects of high volume nano palm oil fuel ash on microstructure properties and hydration temperature of mortar. *Construction and Building Materials, 93*, 29−34. https://doi.org/10.1016/j.conbuildmat.2015.05.107

Liu, S.-T., et al. (2014). Electromagnetic performance of rice husk ash. *BioResources, 9*(2), 2328−2340. https://doi.org/10.15376/biores.9.2.2328-2340

Liu, M. Y. J., et al. (2016). Microstructural investigations of palm oil fuel ash and fly ash based binders in lightweight aggregate foamed geopolymer concrete. *Construction and Building Materials, 120*, 112−122. https://doi.org/10.1016/j.conbuildmat.2016.05.076

Liu, H., et al. (2021). High strength flowable lightweight concrete incorporating low C3A cement, silica fume, stalite and macro-polyfelin polymer fibres. *Construction and Building Materials, 281*, 122410. https://doi.org/10.1016/j.conbuildmat.2021.122410

Lo, F. C., Lee, M. G., & Lo, S. L. (2021). Effect of coal ash and rice husk ash partial replacement in ordinary Portland cement on pervious concrete. *Construction and Building Materials, 286*, 122947. https://doi.org/10.1016/J.CONBUILDMAT.2021.122947

Lopes Silva, D. A., Delai, I., Delgado Montes, M. L., & Roberto Ometto, A. (2014). Life cycle assessment of the sugarcane bagasse electricity generation in Brazil. *Renewable and Sustainable Energy Reviews, 32*, 532−547. https://doi.org/10.1016/J.RSER.2013.12.056

Lu, G. Q., & Do, D. D. (1991). Adsorption properties of fly ash particles for NOx removal from flue gases. *Fuel Processing Technology, 27*(1), 95−107. https://doi.org/10.1016/0378-3820(91)90011-Z

Lu, J., Duan, Z., & Poon, C. S. (2017). Combined use of waste glass powder and cullet in architectural mortar. *Cement and Concrete Composites, 82*, 34−44. https://doi.org/10.1016/j.cemconcomp.2017.05.011

Lu, J., et al. (2017). Using glass powder to improve the durability of architectural mortar prepared with glass aggregates. *Materials & Design, 135*, 102−111. https://doi.org/10.1016/j.matdes.2017.09.016

Madandoust, R., & Ghavidel, R. (2013). Mechanical properties of concrete containing waste glass powder and rice husk ash. *Biosystems Engineering, 116*(2), 113−119. https://doi.org/10.1016/j.biosystemseng.2013.07.006. IAgrE.

Mahvash, S., López-Querol, S., & Bahadori-Jahromi, A. (2017). Effect of class F fly ash on fine sand compaction through soil stabilization. *Heliyon, 3*(3), e00274. https://doi.org/10.1016/j.heliyon.2017.e00274. Elsevier Ltd.

Mardani-Aghabaglou, A., İnan Sezer, G., & Ramyar, K. (2014). Comparison of fly ash, silica fume and metakaolin from mechanical properties and durability performance of mortar mixtures view point. *Construction and Building Materials, 70*, 17−25. https://doi.org/10.1016/j.conbuildmat.2014.07.089

Marinković, S., Dragaš, J., Ignjatović, I., & Tošić, N. (2017). Environmental assessment of green concretes for structural use. *Journal of Cleaner Production, 154*, 633−649. https://doi.org/10.1016/j.jclepro.2017.04.015

Martin, J. (2006). *Materials for engineering*. Woodhead Publishing. Available at: http://160592857366.free.fr/joe/ebooks/Mechanical Engineering Books Collection/MATERIAL SCIENCE/Materials for engineering, 3rd Edition - (Malestrom).pdf.

Maryam Md Sukaimi, A., et al. (2018). Effect of grinding period on physical properties of modified bitumen using palm oil fuel ash (POFA). *Journal of Physics: Conference Series, 1049*(1). https://doi.org/10.1088/1742-6596/1049/1/012004

Mayooran, S., Ragavan, S., & Sathiparan, N. (2017). Comparative study on open air burnt low- and high-carbon rice husk ash as partial cement replacement in cement block production. *Journal of Building Engineering, 13*(July), 137–145. https://doi.org/10.1016/j.jobe.2017.07.011. Elsevier Ltd.

Mboya, H. A., et al. (2017). Measurement of pozzolanic activity index of Scoria, pumice, and rice husk ash as potential supplementary cementitious materials for Portland cement. *Advances in Civil Engineering*. https://doi.org/10.1155/2017/6952645

Megat Johari, M. A., et al. (2012). Engineering and transport properties of high-strength green concrete containing high volume of ultrafine palm oil fuel ash. *Construction and Building Materials, 30*(Elsevier Ltd), 281–288. https://doi.org/10.1016/j.conbuildmat.2011.12.007

Mehta, A., & Siddique, R. (2018). Sustainable geopolymer concrete using ground granulated blast furnace slag and rice husk ash: Strength and permeability properties. *Journal of Cleaner Production, 205*, 49–57. https://doi.org/10.1016/j.jclepro.2018.08.313

Mejía, J. M., Mejía De Gutiérrez, R., & Montes, C. (2016). Rice husk ash and spent diatomaceous earth as a source of silica to fabricate a geopolymeric binary binder. *Journal of Cleaner Production, 118*, 133–139. https://doi.org/10.1016/j.jclepro.2016.01.057

Memphisedu (no date). Fly ash, Slag, silica fume, and natural Pozzolans. In: *Design and control of concrete mixtures*, Chapter 3, pp. 57–72.

Mirzahosseini, M., & Riding, K. A. (2015). Influence of different particle sizes on reactivity of finely ground glass as supplementary cementitious material (SCM). *Cement and Concrete Composites, 56*, 95–105. https://doi.org/10.1016/j.cemconcomp.2014.10.004. Elsevier Ltd.

Mohajerani, A., et al. (2017). Practical recycling applications of crushed waste glass in construction materials: A review. *Construction and Building Materials, 156*, 443–467. https://doi.org/10.1016/j.conbuildmat.2017.09.005. Elsevier Ltd.

Mohamad, M. E., et al. (2016). A review of the mechanical properties of concrete containing Biofillers. *IOP Conference Series: Materials Science and Engineering, 160*(1). https://doi.org/10.1088/1757-899X/160/1/012064

Mohd Salleh, I. S., Mustazar, N. A., & Yussof, H. W. (2018). Mercury removal from wastewater using palm oil fuel ash. *MATEC Web of Conferences, 150*, 10–14. https://doi.org/10.1051/matecconf/201815002007

Molaei Raisi, E., Vaseghi Amiri, J., & Davoodi, M. R. (2018). Mechanical performance of self-compacting concrete incorporating rice husk ash. *Construction and Building Materials, 177*, 148–157. https://doi.org/10.1016/j.conbuildmat.2018.05.053. Elsevier Ltd.

Mollamahmutoğlu, M., & Yilmaz, Y. (2001). Potential use of fly ash and bentonite mixture as liner or cover at waste disposal areas. *Environmental Geology, 40*(11–12), 1316–1324. https://doi.org/10.1007/s002540100355. Gazi University, Fac. of Engineering and Architecture, Civil Engineering Department, 06570 Ankara, Turkey.

Montakarntiwong, K., et al. (2013). Strength and heat evolution of concretes containing bagasse ash from thermal power plants in sugar industry. *Materials and Design, 49*, 414–420. https://doi.org/10.1016/j.matdes.2013.01.031. Elsevier Ltd.

Moreno, N., et al. (2005). Physico-chemical characteristics of European pulverized coal combustion fly ashes. *Fuel, 84*(11), 1351–1363. https://doi.org/10.1016/j.fuel.2004.06.038

Moretti, J. P., et al. (2016). Joint use of construction waste (CW) and sugarcane bagasse ash sand (SBAS) in concrete. *Construction and Building Materials, 113*, 317–323. https://doi.org/10.1016/j.conbuildmat.2016.03.062. Elsevier Ltd.

Moretti, J. P., Nunes, S., & Sales, A. (2018). Self-compacting concrete incorporating sugarcane bagasse ash. *Construction and Building Materials, 172*, 635–649. https://doi.org/10.1016/j.conbuildmat.2018.03.277. Elsevier Ltd.

Moretti, J. P., Sales, A., Almeida, F. C. R., Rezende, M. A. M., & Gromboni, P. P. (2016). Joint use of construction waste (CW) and sugarcane bagasse ash sand (SBAS) in concrete. *Construction and Building Materials, 113*, 317–323. https://doi.org/10.1016/J.CONBUILDMAT.2016.03.062

Mukherjee, A. B., et al. (2008). Mercury flow via coal and coal utilization by-products: A global perspective. *Resources, Conservation and Recycling, 52*(4), 571–591. https://doi.org/10.1016/j.resconrec.2007.09.002

Murugesan, T., Vidjeapriya, R., & Bahurudeen, A. (2020). Development of sustainable alkali activated binder for construction using sugarcane bagasse ash and marble waste. *Sugar Tech, 22*(5), 885–895. https://doi.org/10.1007/s12355-020-00825-y. Springer India.

Muthadhi, A., & Kothandaraman, S. (2010). Optimum production conditions for reactive rice husk ash. *Materials and Structures/Materiaux et Constructions, 43*(9), 1303–1315. https://doi.org/10.1617/s11527-010-9581-0

Muthusamy, K., & Zamri, N. A. (2016). Mechanical properties of oil palm shell lightweight aggregate concrete containing palm oil fuel ash as partial cement replacement. *KSCE Journal of Civil Engineering, 20*(4), 1473–1481. https://doi.org/10.1007/s12205-015-1104-7

Nagaratnam, B. H., et al. (2019). Strength and microstructural characteristics of palm oil fuel ash and fly ash as binary and ternary blends in Self-Compacting concrete. *Construction and Building Materials, 202*, 103–120. https://doi.org/10.1016/j.conbuildmat.2018.12.139. Elsevier Ltd.

Nair, D. G., Jagadish, K. S., & Fraaij, A. (2006). Reactive pozzolanas from rice husk ash: An alternative to cement for rural housing. *Cement and Concrete Research, 36*(6), 1062–1071. https://doi.org/10.1016/j.cemconres.2006.03.012

Nairn, J., Blackburn, D., & Wilson, M. (2001). Research and development for fly ash: Opportunity or alchemy. In *18th annual International Pittsburg coal Conference, Newcastle, 2001*.

Nassar, R., & Soroushian, P. (2012). Strength and durability of recycled aggregate concrete containing milled glass as partial replacement for cement. *Construction and Building Materials, 29*, 368–377. https://doi.org/10.1016/j.conbuildmat.2011.10.061. Elsevier Ltd.

NBR-12653, ABNT (1992) NBR-12653. (1992). *Pozzolanic materials: (Brazilian standard). Rio de Janeiro*. Available at: https://qdoc.tips/nbr-12653-materiais-pozolanicos-pdf-free.html.

Nhan, C. T., Graydon, J. W., & Kirk, D. W. (1996). Utilizing coal fly ash as a landfill barrier material. *Waste Management, 16*(7), 587–595. https://doi.org/10.1016/S0956-053X(96)00108-0

Noor-Ul-Amin, et al. (2016). Synthesis and characterization of geopolymer from bagasse bottom ash, waste of sugar industries and naturally available China clay. *Journal of Cleaner Production, 129*, 491–495. https://doi.org/10.1016/j.jclepro.2016.04.024. Elsevier Ltd.

Nuaklong, P., Jongvivatsakul, P., Pothisiri, T., Sata, V., & Chindaprasirt, P. (2020). Influence of rice husk ash on mechanical properties and fire resistance of recycled aggregate high-calcium fly ash geopolymer concrete. *Journal of Cleaner Production, 252*, 119797. https://doi.org/10.1016/J.JCLEPRO.2019.119797

Nwofor, T. C., & Ukpaka, C. (2017). Predictive models for compressive strength of concrete produced with waste glass as replacement for aggregates. *International Journal of Civil Engineering, 4*(1), 31–34. https://doi.org/10.14445/23488352/ijce-v4i1p104

Omran, A., et al. (2018). Performance of ground-glass pozzolan as a cementitious material-a review. *Advances in Civil Engineering Materials, 7*(1), 237–270. https://doi.org/10.1520/ACEM20170125

Omran, A., & Tagnit-Hamou, A. (2016). Performance of glass-powder concrete in field applications. *Construction and Building Materials, 109*, 84–95. https://doi.org/10.1016/j.conbuildmat.2016.02.006. Elsevier Ltd.

Ooi, P. S. K., et al. (2008). Shear strength characteristics of recycled glass. *Transportation Research Record, 2059*, 52–62. https://doi.org/10.3141/2059-06

Pandian, N. S. (2004). Fly ash characterization with reference to geotechnical applications. *Journal of the Indian Institute of Science, 84*(6), 189–216. Department of Civil Engineering, Indian Institute of Science, Bangalore 560 012, India https://www.scopus.com/inward/record.uri?eid=2-s2.0-20044373111&partnerID=40&md5=05aaf1af48c1321f4d02e0670d35ca9f. Available at:.

Park, S., & Lee, Bong-chun (2004). Studies on expansion properties in mortar containing waste glass and fibers. *Cement and Concrete Research, 34*(34), 1145–1152. https://doi.org/10.1016/j.cemconres.2003.12.005

Pelisser, F., Gleize, P. J. P., & Mikowski, A. (2012). Effect of the Ca/Si molar ratio on the micro/nanomechanical properties of synthetic C-S-H measured by nanoindentation. *Journal of Physical Chemistry C, 116*(32), 17219–17227. https://doi.org/10.1021/jp302240c

Petek Gursel, A., Masanet, E., Horvath, A., & Stadel, A. (2014). Life-cycle inventory analysis of concrete production: A critical review. *Cement and Concrete Composites, 51*, 38–48. https://doi.org/10.1016/j.cemconcomp.2014.03.005

Ping, W., & Yidong, X. (2011). Life cycle assessment of recycled aggregate concrete containing fly ash. In *Second International Conference on mechanic Automation and control engineering* (pp. 2287–2290). https://doi.org/10.1109/MACE.2011.5987436

Polley, C., Cramer, S. M., & Cruz, R. V. de la (1998). Potential for using waste glass in Portland cement concrete. *Journal of Materials in Civil Engineering, 10*(4), 210–219. https://doi.org/10.1061/(asce)0899-1561(1998)10:4(210)

Pragyaditya, D., Kumaran, S., & Kanna, B. C. (2015). Monitoring and control of an in-house steam power plant: a support of the idea of distributed power generation. *International Journal of Advanced Manufacturing Technology, 6*(149), 2. https://doi.org/10.4172/0976-4860.1000149

Prashanth, J. P., Sivapullaiah, P. V., & Sridharan, A. (2001). Pozzolanic fly ash as a hydraulic barrier in land fills. *Engineering Geology, 60*(1–4), 245–252. https://doi.org/10.1016/S0013-7952(00)00105-8

Prasittisopin, L., & Trejo, D. (2012). *Characterization of chemical treatment method for rice husk ash cementing materials*. American Concrete Institute, ACI Special Publication (294 SP).

Rajak, M. A. A., Majid, Z. A., & Ismail, M. (2019). Pozzolanic activity of nanosized palm oil fuel ash: A comparative assessment with various fineness of palm oil fuel ash. *IOP Conference Series: Earth and Environmental Science, 220*(1), 012061. https://doi.org/10.1088/1755-1315/220/1/012061

Ramachandran, V. S. (1996). *Concrete admixtures Handbook, 2nd ed.: Properties, science and technology*. Elsevier Science (Building materials science series). Available at: https://books.google.com.au/books?id=aSjXbx9H9WAC.

Ramezanianpour, A. A. (2014). Cement replacement materials. *Springer Geochemistry/Mineralogy, 10*, 973–978. Springer.

Ramezanianpour, A. A., Mahdi khani, M., & Ahmadibeni, Gh (2009). The effect of rice husk ash on mechanical properties and durability of sustainable concretes. *International Journal of Civil Engineering, 7*(2), 83–91.

Ram, L. C., & Masto, R. E. (2014). Fly ash for soil amelioration: A review on the influence of ash blending with inorganic and organic amendments. *Earth-Science Reviews, 128*, 52–74. https://doi.org/10.1016/j.earscirev.2013.10.003. Elsevier B.V.

Ranjbar, N., et al. (2016). Durability and mechanical properties of self-compacting concrete incorporating palm oil fuel ash. *Journal of Cleaner Production, 112*, 723–730. https://doi.org/10.1016/j.jclepro.2015.07.033. Elsevier Ltd.

Rashid, M. H., Ali Molla, M. K., & Ahmed, T. U. (2010). Durability of mortar in presence of rice husk ash. *World Academy of Science, Engineering and Technology, 43*, 736–739.

Rattanashotinunt, C., et al. (2013). Use of calcium carbide residue and bagasse ash mixtures as a new cementitious material in concrete. *Materials and Design, 46*, 106–111. https://doi.org/10.1016/j.matdes.2012.10.028. Elsevier Ltd.

Reeves, J. B., & Weihrauch, J. L. (1979). *Composition of foods: Fats and oils: Raw, processed, prepared. Agriculture. US department of agriculture, science and Education Administration.* Available at: https://books.google.com.au/books?hl=en&lr=&id=e6U_jcvb-lQC&oi=fnd&pg=PA1&ots=MbZ6NUnEkZ&sig=VJPycTHFQcPp20xzHtxuks5o_5Y#v=onepage&q&f=false.

Ríos-Parada, V., et al. (2017). Characterization and use of an untreated Mexican sugarcane bagasse ash as supplementary material for the preparation of ternary concretes. *Construction and Building Materials, 157*, 83–95. https://doi.org/10.1016/j.conbuildmat.2017.09.060. Elsevier Ltd.

Rodier, L., et al. (2019). Potential use of sugarcane bagasse and bamboo leaf ashes for elaboration of green cementitious materials. *Journal of Cleaner Production, 231*, 54–63. https://doi.org/10.1016/j.jclepro.2019.05.208

Rodríguez-Díaz, J. M., et al. (2015). Comprehensive characterization of sugarcane bagasse ash for its use as an adsorbent. *Bioenergy Research, 8*(4), 1885–1895. https://doi.org/10.1007/s12155-015-9646-6

Rukzon, S., & Chindaprasirt, P. (2012). Utilization of bagasse ash in high-strength concrete. *Materials and Design, 34*, 45–50. https://doi.org/10.1016/j.matdes.2011.07.045. Elsevier Ltd.

Saand, A., et al. (2017). Utilization of waste glass as partial replacement of fine aggregate in concrete. *Engineering Science and Technology International Research Journal, 1*(1), 28–32.

Sadeeq, J., Ochepo, J., & Salahudeen, A. (2015). Assessment of bagasse ash effect on the California bearing ratio of used oil contaminated lateritic soils. *Nigerian Journal of Technology, 34*(2), 223. https://doi.org/10.4314/njt.v34i2.2

Sadhukhan, J., et al. (2018). Role of bioenergy, biorefinery and bioeconomy in sustainable development: Strategic pathways for Malaysia. *Renewable and Sustainable Energy Reviews, 81*(April 2017), 1966–1987. https://doi.org/10.1016/j.rser.2017.06.007

Sáez-Pérez, M. P., Brümmer, M., & Durán-Suárez, J. A. (2020). A review of the factors affecting the properties and performance of hemp aggregate concretes. *Journal of Building Engineering, 31*, 101323. https://doi.org/10.1016/j.jobe.2020.101323

Safari, J., et al. (2018). Effect of rice husk ash and macro-synthetic fibre on the properties of self-compacting concrete. *Construction and Building Materials, 175*, 371–380. https://doi.org/10.1016/j.conbuildmat.2018.04.207. Elsevier Ltd.

Safiuddin, M., et al. (2016). Modeling of compressive strength for self-consolidating high-strength concrete incorporating palm oil fuel ash. *Materials, 9*(5), 396'.

Safiuddin, M., Salam, M. A., & Jumaat, M. Z. (2014). Key fresh properties of self-consolidating high-strength POFA concrete. *Journal of Materials in Civil Engineering, 26*(1), 134–142. https://doi.org/10.1061/(asce)mt.1943-5533.0000782

Safiuddin, M., West, J. S., & Soudki, K. A. (2012). Properties of freshly mixed self-consolidating concretes incorporating rice husk ash as a supplementary cementing

material. *Construction and Building Materials, 30*, 833–842. https://doi.org/10.1016/j.conbuildmat.2011.12.066. Elsevier Ltd.
Sata, V., Jaturapitakkul, C., & Kiattikomol, K. (2004). Utilization of palm oil fuel ash in high-strength concrete. *Journal of Materials in Civil Engineering, 16*, 623–628. https://doi.org/10.1061/(ASCE)0899-1561(2004)16:6(623)
Said, I., et al. (2017). Utilization of rice husk ash as a pozzolan in self compacting concrete. *International Journal of Civil Engineering and Technology, 8*(1), 98–108. Department of Civil Engineering, Iqra National University, Peshawar, Pakistan: IAEME Publication https://www.scopus.com/inward/record.uri?eid=2-s2.0-85011067818&partnerID=40&md5=74ee9bba663fb128e051bb5ae32511fd. Available at:.
Salam, M. A., Safiuddin, M., & Jumaat, M. Z. (2018). Durability indicators for sustainable self-consolidating high-strength concrete incorporating palm oil fuel ash. *Sustainability (Switzerland), 10*(7). https://doi.org/10.3390/su10072345
Salas, A., et al. (2009). Comparison of two processes for treating rice husk ash for use in high performance concrete. *Cement and Concrete Research, 39*(9), 773–778. https://doi.org/10.1016/j.cemconres.2009.05.006. Elsevier B.V.
Saleh, A. M., et al. (2014). Utilization of palm oil fuel ash and rice husks in unfired bricks for sustainable construction materials development. *MATEC Web of Conferences, 15*, 1–8. https://doi.org/10.1051/matecconf/20141501032
Sales, A., & Lima, S. A. (2010). Use of Brazilian sugarcane bagasse ash in concrete as sand replacement. *Waste Management, 30*(6), 1114–1122. https://doi.org/10.1016/j.wasman.2010.01.026. Elsevier Ltd.
Samadhi, T. W., et al. (2019). Synthesis of geopolymer from blends of tropical biomass ashes. *IOP Conference Series: Materials Science and Engineering, 622*(1). https://doi.org/10.1088/1757-899X/622/1/012034
Samarakoon, M. H., et al. (2020). Properties of one-part fly ash/slag-based binders activated by thermally-treated waste glass/NaOH blends: A comparative study. *Cement and Concrete Composites, 112*(December 2019), 103679. https://doi.org/10.1016/j.cemconcomp.2020.103679. Elsevier Ltd.
Sata, V., Jaturapitakkul, C., & Kiattikomol, K. (2007). Influence of pozzolan from various by-product materials on mechanical properties of high-strength concrete. *Construction and Building Materials, 21*(7), 1589–1598. https://doi.org/10.1016/j.conbuildmat.2005.09.011
Savastano, H., et al. (2016). Chap-19 - sustainable use of vegetable fibres and particles in civil construction. In J. M. B. T.-S. Of C. M. Khatib (Ed.), *Sustainability of construction materials, Woodhead publishing Series in civil and structural engineering* (Second E., pp. 477–520). Woodhead Publishing. https://doi.org/10.1016/B978-0-08-100370-1.00019-6
Seto, K. E., Churchill, C. J., & Panesar, D. K. (2017). Influence of fly ash allocation approaches on the life cycle assessment of cement-based materials. *Journal of Cleaner Production, 157*, 65–75. https://doi.org/10.1016/j.jclepro.2017.04.093
Sharma, U., Khatri, A., & Kanoungo, A. (2014). Use of micro-silica as additive to concrete-state of art. *International Journal of Civil Engineering Research, 5*(1), 9–12. Available at: http://www.ripublication.com/ijcer.htm.
Shen, L., Yang, J., Zhang, R., Shao, C., & Song, X. (2019). The benefits and barriers for promoting bamboo as a green building material in China—an integrative analysis. *Sustainability, 11*, 2493. https://doi.org/10.3390/SU11092493
Shatat, M. R. (2016). Hydration behavior and mechanical properties of blended cement containing various amounts of rice husk ash in presence of metakaolin. *Arabian Journal of Chemistry, 9*, S1869–S1874. https://doi.org/10.1016/j.arabjc.2013.12.006. King Saud University.

Shayan, A., & Xu, A. (2004). Value-added utilisation of waste glass in concrete. *Cement and Concrete Research, 34*(1), 81–89. https://doi.org/10.1016/S0008-8846(03)00251-5

Siddika, A., et al. (2021). State-of-the-art-review on rice husk ash: A supplementary cementitious material in concrete. *Journal of King Saud University - Engineering Sciences, 33*(5), 294–307. https://doi.org/10.1016/j.jksues.2020.10.006. King Saud University.

Siddique, R., & Chahal, N. (2011). Use of silicon and ferrosilicon industry by-products (silica fume) in cement paste and mortar. *Resources, Conservation and Recycling, 55*, 739–744. https://doi.org/10.1016/J.RESCONREC.2011.03.004

Siddika, A., Mamun, M. A. Al, & Ali, M. H. (2018). Study on concrete with rice husk ash. *Innovative Infrastructure Solutions, 3*(1), 1–9. https://doi.org/10.1007/s41062-018-0127-6. Springer International Publishing.

Siddique, R., & Khan, M. I. (2011). Supplementary cementing materials. In M. I. K. Siddique (Ed.), *Rafat. Berlin, Heidelberg: Springer Berlin Heidelberg (Engineering materials)* (1st ed.). https://doi.org/10.1007/978-3-642-17866-5

Singh, B. (2018). 13 - Rice husk ash. In R. Siddique, P. B. T.-W. Cachim, & S. C. M. in C. (Eds.), *Woodhead Publishing Series in civil and structural engineering* (pp. 417–460). Woodhead Publishing. https://doi.org/10.1016/B978-0-08-102156-9.00013-4

Singh, A., Mehta, P. K., & Kumar, R. (2022). Performance of binary admixtures (fly ash and silica fume) on self compacting concrete. *Materials Today: Proceedings, 58*, 970–977. https://doi.org/10.1016/j.matpr.2022.02.489. Elsevier Ltd.

Singh, D., & Singh, J. (2016). Use of Agrowaste in concrete construction. *International Journal of Environment, Ecology, Family and Urban Studie, 6*(1), 119–130.

Sinoh, S. S., Ibrahim, Z., Othman, F., & Kuang, Zaki L. M. A. (2020). Life cycle assessment of sugarcane bagasse ash as partial cement replacement in concrete. In *Proceedings of the 4th International Conference on sustainable Innovation 2020–technology, engineering and agriculture (ICoSITEA 2020)* (pp. 144–150).

Sobolev, K., et al. (2007). Utilization of waste glass in ECO-cement: Strength properties and microstructural observations. *Waste Management, 27*(7), 971–976. https://doi.org/10.1016/j.wasman.2006.07.014

Sokolar, R., & Nguyen, M. (2018). The fly ash of class C for ceramic technology. *IOP Conference Series: Materials Science and Engineering, 385*(1), 012053. https://doi.org/10.1088/1757-899X/385/1/012053

Song, Q., et al. (2015). Environmental risk assessment of CRT and PCB workshops in a mobile e-waste recycling plant. *Environmental Science and Pollution Research, 22*(16), 12366–12373. https://doi.org/10.1007/s11356-015-4350-9

Soomro, M., & Hung, Y. T. (2014). Use of selective recycled materials in civil engineering construction. In *Handbook of environment & waste management* (1st ed., pp. 635–684). Singapore: World Scientific.

Sousa, V., & Bogas, J. A. (2021). Comparison of energy consumption and carbon emissions from clinker and recycled cement production. *Journal of Cleaner Production, 306*. https://doi.org/10.1016/j.jclepro.2021.127277

Sua-Iam, G., & Makul, N. (2013). Utilization of limestone powder to improve the properties of self-compacting concrete incorporating high volumes of untreated rice husk ash as fine aggregate. *Construction and Building Materials, 38*, 455–464. https://doi.org/10.1016/j.conbuildmat.2012.08.016. Elsevier Ltd.

Subedi, S., et al. (2022). Feasibility of ECC with high contents of post-processed bagasse ash as partial cement replacement. *Construction and Building Materials, 319*(May 2021), 126023. https://doi.org/10.1016/j.conbuildmat.2021.126023. Elsevier Ltd.

Sumesh, M., et al. (2019). Microstructural and strength characteristics of high-strength mortar using nontraditional supplementary cementitious materials. *Journal of Materials in Civil Engineering, 31*(4), 04019017. https://doi.org/10.1061/(asce)mt.1943-5533.0002626

Tagnit-Hamou, A., et al. (2018). Ground glass pozzolan in conventional, high, and ultra-high performance concrete. *MATEC Web of Conferences, 149*, 1–7. https://doi.org/10.1051/matecconf/201714901005

Tai, Z. S., et al. (2019). Influence of pre-treatment temperature of palm oil fuel ash on the properties and performance of green ceramic hollow fiber membranes towards oil/water separation application. *Separation and Purification Technology, 222*(December 2018), 264–277. https://doi.org/10.1016/j.seppur.2019.04.046

Takada, T., et al. (1995). Utilization of coal ash from fluidized-bed combustion boilers as road base material. *Resources, Conservation and Recycling, 14*(2), 69–77. https://doi.org/10.1016/S0921-3449(95)80001-8

Tambichik, M. A., et al. (2018). Utilization of construction and agricultural waste in Malaysia for development of green concrete: A review. *IOP Conference Series: Earth and Environmental Science, 140*(1). https://doi.org/10.1088/1755-1315/140/1/012134

Tangchirapat, W., Jaturapitakkul, C., & Chindaprasirt, P. (2009). Use of palm oil fuel ash as a supplementary cementitious material for producing high-strength concrete. *Construction and Building Materials, 23*(7), 2641–2646. https://doi.org/10.1016/j.conbuildmat.2009.01.008. Elsevier Ltd.

Tang, W. Le, et al. (2019). Optimization of micro and nano palm oil fuel ash to determine the carbonation resistance of the concrete in accelerated condition. *Materials, 12*(1). https://doi.org/10.3390/ma12010130

Tchakouté, H. K., et al. (2016). Geopolymer binders from metakaolin using sodium waterglass from waste glass and rice husk ash as alternative activators: A comparative study. *Construction and Building Materials, 114*, 276–289. https://doi.org/10.1016/j.conbuildmat.2016.03.184

Tekle, B. (2011). *Bagasse ash as a cement replacing material*. https://doi.org/10.13140/RG.2.1.2257.8166

Terence, C. H. (2005). *Silica fume User's manual*. Federal Highway Administration.

Thomas, B. S., Kumar, S., & Arel, H. S. (2017). Sustainable concrete containing palm oil fuel ash as a supplementary cementitious material – a review. *Renewable and Sustainable Energy Reviews, 80*(July 2016), 550–561. https://doi.org/10.1016/j.rser.2017.05.128. Elsevier Ltd.

Tripathy, A., & Acharya, P. K. (2022). Characterization of bagasse ash and its sustainable use in concrete as a supplementary binder – a review. *Construction and Building Materials, 322*(July 2021), 126391. https://doi.org/10.1016/j.conbuildmat.2022.126391. Elsevier Ltd.

Tullis, Paul (2019). How the world got hooked on palm oil. *The Guardian*. https://doi.org/10.1007/978-90-481-8661-7_46

UK Statistics on Waste. (2021). Dept.of environment., food and rural Affairs UK.

Ul Islam, M. M., et al. (2016). Durability properties of sustainable concrete containing high volume palm oil waste materials. *Journal of Cleaner Production, 137*, 167–177. https://doi.org/10.1016/j.jclepro.2016.07.061. Elsevier Ltd.

Umasabor, R. I., & Okovido, J. O. (2018). Fire resistance evaluation of rice husk ash concrete. *Heliyon, 4*(12), e01035. https://doi.org/10.1016/j.heliyon.2018.e01035. Elsevier Ltd.

Van, V. T. A., et al. (2013). Mesoporous structure and pozzolanic reactivity of rice husk ash in cementitious system. *Construction and Building Materials, 43*, 208–216. https://doi.org/10.1016/j.conbuildmat.2013.02.004. Elsevier Ltd.

Vichaphund, S., et al. (2014). Characteristic of fly ash derived-zeolite and its catalytic performance for fast pyrolysis of Jatropha waste. *Environmental Technology (United Kingdom), 35*(17), 2254−2261. https://doi.org/10.1080/09593330.2014.900118

Villaquirán-Caicedo, M. A., Mejía De Gutiérrez, R., & Gallego, N. C. (2017). A novel MK-based geopolymer composite activated with rice husk ash and KOH: Performance at high temperature. *Materiales de Construccion, 67*(326), 1−13. https://doi.org/10.3989/mc.2017.02316

Wang, J., Wang, Y., Sun, Y., Tingley, D. D., & Zhang, Y. (2017). Life cycle sustainability assessment of fly ash concrete structures. *Renewable and Sustainable Energy Reviews, 80*, 1162−1174. https://doi.org/10.1016/j.rser.2017.05.232

Wang, S., & Wu, H. (2006). Environmental-benign utilisation of fly ash as low-cost adsorbents. *Journal of Hazardous Materials, 136*(3), 482−501. https://doi.org/10.1016/j.jhazmat.2006.01.067

Wei, J., & Meyer, C. (2016). Utilization of rice husk ash in green natural fiber-reinforced cement composites: Mitigating degradation of sisal fiber. *Cement and Concrete Research, 81*, 94−111. https://doi.org/10.1016/j.cemconres.2015.12.001. Elsevier Ltd.

Wesche, K. (1991). *Fly ash in concrete: Properties and performance*. CRC Press.

Wi, K., et al. (2018). Effect of using Micropalm oil fuel ash as partial replacement of cement on the properties of cement mortar. *Advances in Materials Science and Engineering*, 1−8. https://doi.org/10.1155/2018/5164030, 2018.

Xiao, R., et al. (2022). A state-of-the-art review of crushed urban waste glass used in OPC and AAMs (geopolymer): Progress and challenges. *Cleaner Materials, 4*(February), 100083. https://doi.org/10.1016/j.clema.2022.100083

Xiao, R., Jiang, X., et al. (2020). Analytical investigation of phase assemblages of alkali-activated materials in CaO-SiO_2-Al_2O_3 systems: The management of reaction products and designing of precursors. *Materials and Design, 194*, 108975. https://doi.org/10.1016/j.matdes.2020.108975. The Authors.

Xiao, R., Ma, Y., et al. (2020). Strength, microstructure, efflorescence behavior and environmental impacts of waste glass geopolymers cured at ambient temperature. *Journal of Cleaner Production, 252*, 119610. https://doi.org/10.1016/j.jclepro.2019.119610. Elsevier Ltd.

Xu, W., et al. (2015). Effect of rice husk ash fineness on porosity and hydration reaction of blended cement paste. *Construction and Building Materials, 89*, 90−101. https://doi.org/10.1016/j.conbuildmat.2015.04.030. Elsevier Ltd.

Yao, Z. (2014). Part A: Sources, collection and characterisation of flyash. In *Flyash:Sources, applications and potential environments impacts*. Nova Science Publishers, Inc.

Yao, Z. T., et al. (2015). A comprehensive review on the applications of coal fly ash. *Earth-Science Reviews, 141*, 105−121. https://doi.org/10.1016/j.earscirev.2014.11.016

Yusof, M. S. M., et al. (2018). Feasibility study of cadmium adsorption by palm oil fuel ash (POFA)-based low-cost hollow fibre zeolitic membrane. *Environmental Science and Pollution Research, 25*(22), 21644−21655. https://doi.org/10.1007/s11356-018-2286-6. Environmental Science and Pollution Research.

Yusof, M. S. M., et al. (2020). Arsenic adsorption mechanism on palm oil fuel ash (POFA) powder suspension. *Journal of Hazardous Materials, 383*(September 2019), 121214. https://doi.org/10.1016/j.jhazmat.2019.121214. Elsevier.

Zabihi, S. M., Tavakoli, H., & Mohseni, E. (2018). Engineering and microstructural properties of fiber-reinforced rice husk−ash based geopolymer concrete. *Journal of Materials in Civil Engineering, 30*(8), 04018183. https://doi.org/10.1061/(asce)mt.1943-5533.0002379

Zain, M. F. M., et al. (2011). Production of rice husk ash for use in concrete as a supplementary cementitious material. *Construction and Building Materials, 25*(2), 798–805. https://doi.org/10.1016/j.conbuildmat.2010.07.003. Elsevier Ltd.

Zareei, S. A., Ameri, F., & Bahrami, N. (2018). Microstructure, strength, and durability of eco-friendly concretes containing sugarcane bagasse ash. *Construction and Building Materials, 184*, 258–268. https://doi.org/10.1016/j.conbuildmat.2018.06.153

Zeyad, A. M., et al. (2016). Efficiency of treated and untreated palm oil fuel ash as a supplementary binder on engineering and fluid transport properties of high-strength concrete. *Construction and Building Materials, 125*, 1066–1079. https://doi.org/10.1016/j.conbuildmat.2016.08.065. Elsevier Ltd.

Zhang, B., He, P., & Poon, C. S. (2020). Optimizing the use of recycled glass materials in alkali activated cement (AAC) based mortars. *Journal of Cleaner Production, 255*, 120228. https://doi.org/10.1016/j.jclepro.2020.120228. Elsevier Ltd.

Zhang, Z., Zhang, B., & Yan, P. (2016). Comparative study of effect of raw and densified silica fume in the paste, mortar and concrete. *Construction and Building Materials, 105*, 82–93. https://doi.org/10.1016/j.conbuildmat.2015.12.045

Zhang, P., Li, D., Qiao, Y., Zhang, S., Sun, C., & Zhao, T. (2018). Effect of air entrainment on the mechanical properties, chloride migration, and microstructure of ordinary concrete and fly ash concrete. *Journal of Materials in Civil Engineering, 30*, 04018265. https://doi.org/10.1061/(ASCE)MT.1943-5533.0002456

Zheng, S., et al. (2022). Experimental investigation on the effect of nano silica fume on physical properties and microstructural characteristics of lightweight cement slurry. *Construction and Building Materials, 329*(February), 127172. https://doi.org/10.1016/j.conbuildmat.2022.127172

Recycled aggregates and their properties

4

Fernando L. Gayarre [1], *Carlos López-Colina* [1], *Lara Carral* [2], *Miguel A. Serrano* [1], *Jesús M. Suárez* [1] *and Rebeca Martínez* [3]
[1]University of Oviedo, Gijón/Xixón, Asturias, Spain; [2]Autonomous University of Aguascalientes, Aguascalientes, Ags., Mexico; [3]University of León, León, Spain

4.1 Introduction

Circular economy in construction is vital in light of the huge amount of construction and demolition waste (CDW) generated all over the world. CDW is estimated to represent approximately 30% of urban solid wastes in some countries (Rodríguez et al., 2015). Because of the prevalence of concrete structures over steel construction, the support and enhancement of all the reuse and recycling practices involving this material must be a priority. Moreover, as recycling is a very common practice in the production of structural steel, concrete must become more sustainable in order to compete.

In terms of weight and volume, aggregates, understood as the mainly inert granular materials commonly named as sand (fine aggregate) and gravel (coarse aggregate) are the most important part of traditional concretes. The global production of aggregates for construction, including aggregates for concrete and for other applications, is about 40,000 million tonnes, and manufacturers are planning an increase up to more than 55,000 million tonnes by 2030 (O'Brien, 2016). According to the most recent studies (Grand View Research, 2020; Persistence Market Research, 2015), the global construction aggregates market had a value of 463,300 million US dollars in 2020, reaching a compound annual growth rate of about 6.0% since 2015 in terms of market value and a compound annual growth rate of 4.6% in terms of volume. The last forecast for the period from 2020 to 2027 predicts a compound annual growth rate of 3.3% despite the COVID-19 restrictions in the construction activities at the beginning of this period. All these numbers remark the necessity of a global plan for the use of some recycled materials as aggregates in concrete instead of the natural aggregates specifically extracted from quarries for this purpose.

The origin of the recycled aggregate used in new concrete can be very diverse. However, one of the most extended points of view of recycling is to treat an apparently useless old material turning it into raw material, obtaining then an equivalent to the initial one. Following this viewpoint, the most studied group of recycled aggregates for concrete has been the construction and demolition waste. Among these, the one that has been more deeply studied due to its intrinsic similarities with the original aggregates is the result of crushing old concrete (Chen et al., 2019). This has been probably followed by ceramic demolition waste (Costa et al., 2013), including bricks (Elçi, 2016), tiles (Baradaran-Nasiri & Nematzadeh, 2017), and all the concrete-ceramic mixtures.

Recycled Concrete. https://doi.org/10.1016/B978-0-323-85210-4.00003-5

According to the last data provided by the European Aggregates Association (European Aggregates Association, 2020), the recycled aggregates, understood in a restrictive way exclusively as recycled and reused construction and demolition waste, currently represent 10.5% of the European production of aggregates. However, the differences between countries are very important. They are presented in Fig. 4.1, with values varying from 26% of France to the almost inappreciable values in the many European countries that are not represented in the graph.

The European revised Waste Framework (European Union, 2008) set a very ambitious 70% level of recycling and reuse of nonhazardous construction and demolition waste generated at buildings and civil engineering constructions by 2020. However, Europe did not reach this objective and it is still far from achieving it, leaving clear that its low levels of production and use of recycled aggregates are one of the main causes of the failed goal.

Although CDW is the main source of recycled aggregates, there are many other residues that can be included as mainly inert granular material in the fabrication of concrete. Therefore, the recycled aggregates can be divided into two main groups according to their importance: the ones obtained from construction and demolition waste and the rest.

4.2 Aggregates from CDW

The recycled aggregates from Construction and Demolition Waste (CDW) can be defined as aggregates obtained by processing the inorganic material previously used in construction (CEN, 2013). The activities of demolition, construction, refurbishment,

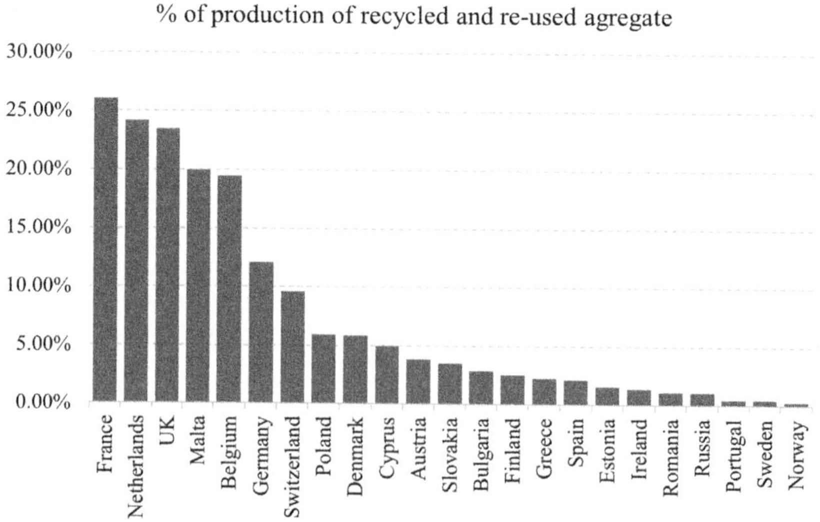

Figure 4.1 Recycled and reused aggregate in Europe Percentage of recycled and reused aggregate produced in European countries.

and maintenance of buildings and infrastructures produce waste that, properly crushed and selected, can be used as fine and/or coarse aggregate in concrete. Of course, the implementation of selective demolition practices is highly necessary to obtain aggregates with adequate properties. In addition, some broken or discarded pieces from the building materials industry, like bricks, blocks, tiles, etc. are usually assimilated as CDW despite their different origin (GEAR Project, 2012).

The very different construction practices all around the world give as a result a huge variety in the composition of CDW. Furthermore, countries apply different definitions of construction and demolition waste, which difficulties comparisons. Even in the same country, the architectural techniques, demolition practices, and raw materials can vary from one region to another (Cabral et al., 2012). However, there are groups of countries or geographical regions that can be grouped by general trends and more or less similar practices (Mediterranean countries, Scandinavian countries, big areas of South America, USA and Canada, etc.). The properties described in this section are based on all the research work and its comprehensive reviews that can be found in the literature (González-Taboada et al., 2016a; Silva et al., 2014). It is essential to note that they have been mainly carried out in the countries that have begun to recycle their CDW or have advanced plans to do it. Some compilations of standardization proposals of requirements for CDW have also been considered (ACHE C 2 GT 2/5, 2006).

By assuming one of the most generalized classifications, in this section we will sort out the CDW aggregates into three categories:

Recycled Masonry Aggregate (RMA) is the one understood coming primarily from masonry rubble (at least 90% by mass). The composition of this group of recycled aggregates can vary hugely depending on the local construction practices. While in some countries RMA is almost a synonym of "ceramic recycled aggregate" coming from ceramic bricks and tiles Fig. 4.2A in others, it can mainly include concrete blocks (lightweight, aerated, etc.) or even sand-lime or blast-furnace slag bricks in some cases. RILEM classifies this as Type I aggregate (RILEM TC 121-DRG, 1994a).

Recycled Concrete Aggregate (RCA) is commonly agreed to have a 90% by mass of recycled concrete-based fragments: cement and natural aggregates Fig. 4.2B. This coincides with the definition of Type II aggregate by RILEM (RILEM TC 121-DRG, 1994a).

(a) (b)

Figure 4.2 RMA and RCA Ceramic RMA (A) and RCA (B).

Mixed Recycled Aggregate (MRA) is a mixture of the previous two categories: concrete aggregate and masonry aggregate. Generally, a recycled aggregate is considered MRA, when it has less than 90% of recycled concrete-based fragments and the other portion, is mainly composed of masonry-based materials (BS-8500, 2006; NBR-15.116, 2005).

RILEM calls Type III aggregate to the mix of at least 80% of natural aggregate and recycled aggregate. But the inclusion into this category also requires the recycled masonry aggregate (RMA or RILEM Type I aggregate) to be less than the 10% in mass.

The size, physical and chemical properties of RMA are very heterogeneous since they depend on the construction materials, demolition techniques, contaminants, etc. In the following sub-sections, the most representative or common values and requirements for the chemical and physical characteristics of RMA are presented.

4.3 Composition of RMA and RCA

Fig. 4.3 presents some indicative values for the usual ranges of chemical composition recycled concrete and mainly ceramic recycled masonry debris (CEDEX, 2014). Of course, the values depend mainly on the composition of the source material. In the case of masonry, they depend on the main composition of the ceramic bricks and tiles, but for other types of masonry (concrete blocks, for example), the results would vary considerably. However, the most studied parameters of the chemical composition of aggregates for new concrete are not the majority components, but the ones that could affect the performance and durability of the new concrete or the rebars. The most studied are the sulfate, chloride, and alkali content.

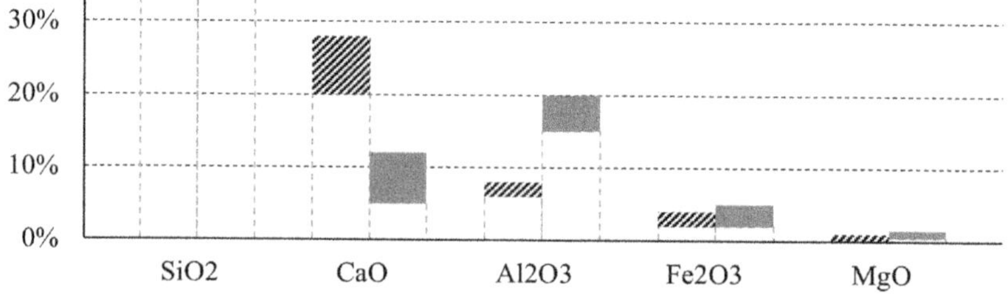

Figure 4.3 Composition of RCA and RMA Typical composition ranges of debris for RCA and RMA.

4.3.1 Sulfate, chloride, and alkali content

The sulfate content in the RMA and the RCA coming from building demolitions strongly depends on the presence of gypsum plaster in the source material. The water-soluble sulfates could cause expansive reactions and it is generally admitted that it should not exceed 1% in mass of the dry aggregate (CEN, 2002). One of the possible measures to reduce the gypsum content is to discard the finer fraction of aggregate, in which the gypsum dust is concentrated.

In addition to the clear addition of sulfates caused by gypsum, some researchers (Barbudo et al., 2012; de Juan & Gutiérrez, 2009) also found a strong influence of the amount of crushed clay brick and the adhered mortar on the presence of these harmful chemical compounds. Therefore, when an RMA is considered, in general, we should expect higher sulfate content than in RCA.

Since chloride content is usually associated with recycled aggregates from structures subjected to marine environments or ice melt salt for roads, this property is usually associated with civil engineering structures and, therefore, mainly with RCA. It is difficult to find specific studies of this chemical parameter for RMA. However, the recommendation to avoid the harmful effect of chlorides in RMA from structures exposed to some saline environment is the same to all recycled aggregate from CDW: to deeply wash them or to immerse them in water for at least 14 days (Debieb et al., 2010). If not, the chloride could cause the corrosion of the reinforcement bars of the new concrete.

In general, the statutory requirements for the chloride content in the recycled aggregates for concrete are the same as the ones stated for natural aggregates. The typical value among the existing standards for the maximum chloride content allowed is 0.05%. However, the DIN 4223 standard (DIN 4226−100:2002−02, 2002) allows 0.15% of chloride content when RMA or MRA are considered. RILEM originally gave some detailed requirements (RILEM, 1992) for the chloride content in RMA that allowed until 0.1% in the finer fraction (0−4 mm) when the aggregate was to be used in reinforced concrete (see Table 4.1).

The expansive reaction between the alkalis from cement and silica in aggregates is usually considered to be more possible when recycled aggregates are used. This is because of the additional presence of alkali-rich adhered mortar in the aggregate, and the less-controlled composition of the aggregates, which could include more

Table 4.1 Maximum chloride content.

	Maximum chloride content (% in mass of dry aggregate)		
	Unreinforced concrete	**Reinforced concrete**	**Prestressed concrete**
0−4 mm fraction	1.00	0.10	0.015
(>4 mm)	1.00	0.05	0.007

reactive silica than the well-known natural aggregates. Therefore, some standards consider any CDW recycled aggregate as potentially reactive regarding this problem. In general, the reactivity of recycled aggregates is expected to be higher than the one of natural aggregates. However, no difference in advance can be made between RMA and RCA regarding the alkali-silica reactivity, since it depends on the control of the crushed waste, the amount of adhered mortar, and grading. According to the RILEM recommendations (RILEM TC 121- DRG, 1994b), for any wet or highly moisturized environment, it is necessary to assess the resistance of the resulting recycled concrete to the alkali-silica reaction through some additional tests.

4.3.2 Attached mortar

One of the main properties that are usually quantified and the one which dramatically determines the properties of RMA, RCA, and MRA is the amount of adhered mortar in the coarse aggregate. It is not only important for the reactivity (sulfates and alkali-silica expansive reactions) but also for other physical properties and qualities: the presence of mortar causes higher absorption, less density, and less freeze-thaw resistance of the aggregate. These properties worsen the stiffness, durability, resistance, creep, and shrinkage of the concrete obtained with the recycled aggregate. In addition, the presence of attached mortar and its inherent higher water absorption affect the workability of the fresh new concrete.

The adhered mortar in recycled masonry (RMA) should be removed as much as possible and the ideal situation arises when the aggregates come from faulty, discarded, or broken pieces from the brick, block, or tile industry (see Fig. 4.4).

In RCA, the amount of adhered mortar should also be reduced to improve the quality of the aggregate. However, the industrialized processes of recycling aggregates

(a) (b)

Figure 4.4 RMA from ceramic brick industry RMA from ceramic brick industry without attached mortar. Far view (A). Close view (B).

trend to remove only the more deteriorated and less adherent mortar after the primary crushing. The removal and discard of all the attached cementitious material to obtain the original coarse natural aggregate would lead to inefficient and low rates of recycling.

The values of adhered mortar found in literature reviews about RCA show high variability, varying from 20% to 60%, being one of its causes the different methods used in the literature to determine them (González-Taboada et al., 2016a). A general observation is that the finer fraction (5–10 mm) presents more adhered mortar than the coarse one (10–20 mm). However, successive crushing processes could reduce the amount of adhered mortar in recycled coarse aggregates by discarding the finest part after each stage. Fig. 4.5 schematically shows the two main examples found in the literature (Nagataki et al., 2004; Yagisita et al., 1993).

The quality of the source concrete does not have much influence on the amount of attached mortar of the recycled aggregate, but slightly higher values are found in those concretes with higher water-cement ratio, which are in general concretes with higher resistance. However, the dependence of other essential properties on the quality of the original concrete is very clear and intuitive.

The presence of attached old mortar in the recycled aggregate implies having two phases and a transition zone only in the aggregate, adding one interfacial transition zone (ITC) to the recycled concrete. Therefore, there will be two ITC's: one older, between the original aggregate and the old mortar and the other one newer, between the attached old mortar and the new mortar matrix (see Fig. 4.6). The New interfacial

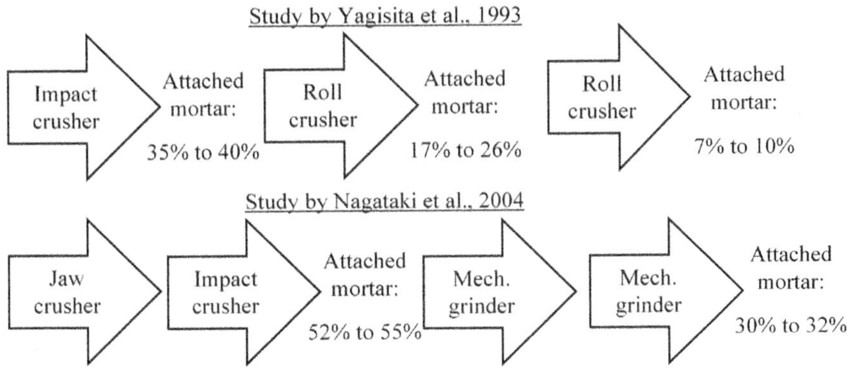

Figure 4.5 Effect of successive crushing Effect of successive crushing on the amount of attached mortar.

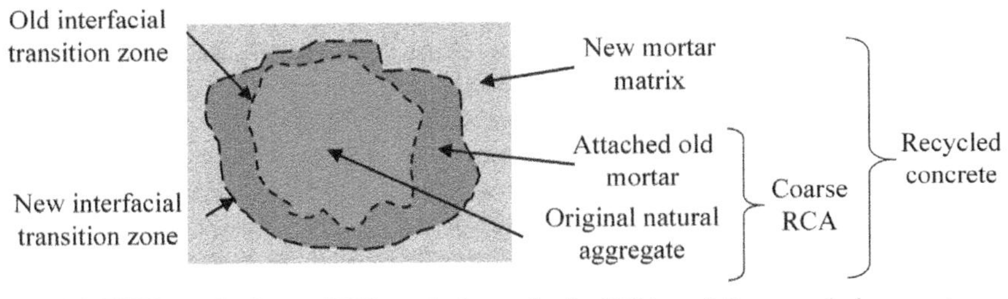

Figure 4.6 ITZ's and phases ITZ's and phases in the RCA and the recycled concrete.

transition zone, between the attached mortar and the new mortar matrix, is weaker than the old one, because of the worse properties of the adhered mortar compared with the ones of the natural aggregate.

4.3.3 Other contaminants

Despite the fact it is detrimental to the fabrication of new concretes, mortar cannot be considered as a contaminant in RCA and RMA since it is an essential part of the source material that is wanted to be recycled. There are other many materials that are actually contaminants, because they come from different parts of the demolishes structures that could not be separated of the concrete or masonry. These impurities are very varied: asphalt, soil, rubber, wood, glass, gypsum, aluminum, steel, organic material, etc. But they all have something in common: they cause a decrease of strength in the new concrete.

The amounts and type of contaminants depend on the recycled aggregate origin and processing. As a rule, the RCA have less contaminants, especially when it comes from civil engineering structures. The aggregates from masonry coming from buildings considerably increase the number of contaminants. Therefore, RMA and MRA are expected to have more contaminating material.

As is the case with the attached mortar, the contaminants are concentrated in the finest fraction. Therefore, a good practice is to discard this for the use as aggregate in new concrete.

Table 4.2 shows the most common contaminants, the maximum values allowed in the most representative recommendations and the main damage mechanism identified that make them harmful for the new concrete. The table also presents for each maximum value one of the references where these limitations can be found, but it is worth noting that most of them are adopted in many other standards, codes and recommendations.

Table 4.2 Common contaminants, associated degradation mechanism and maximum allowed.

Contaminant	Degradation mechanism	Some representative limitations
Gypsum	Sulfate attack	Indirectly: SO_3 limited to 1% [RILEM]
Clay soil	Shrinkage	0.6% [EHE-08][a]
Bricks (in RCA)	Freeze-thaw	10% (in RCA) [RILEM]
Wood, cardboard, and paper	Surface flaking	0.1% [EN 12620]
Asphalt	Low strength	10% [EN 12620]
Glass	Brittleness and alkali	1% [EN 12620]

[a]The limit is valid for concretes with a maximum of 20% of recycled aggregate. If not, the limitation for natural aggregates of 0.25% should be adopted.

Gypsum as source of sulfates has been discussed above. There are not direct limitations for this contaminant, but for the maximum sulfate content. After applying selective demolition, an effective procedure to avoid gypsum is to discard the finest aggregate where the gypsum dust is found.

The presence of clay soil in the aggregates can cause increased shrinkage in concrete. Soil and clay particles can be found stuck to stone, concrete, and brick pieces. Washing the recycled aggregates is the best practice to avoid this undesired material and its effects.

Bricks are a main component of RMA and MRA, but they are undesirable in RCA. Its limits were established when the three main CDW aggregates were defined. Like all other contaminants, the presence of brick, especially clay brick, affects the compressive strength of the recycled concrete made with RCA due to its low strength. It also affects the freeze-thaw resistance of the resulting material. The presorting of CDW is the main way to avoid this mixing of materials when using recycled concrete in the aggregate.

As much paper, cardboard, plastic and wood as possible should be removed from the debris. Then air blowing or water processing can be used to clean the remaining small pieces from the aggregate. Apart from their obvious effect on the resistance of concrete by creating small hollows and discontinuities, these contaminants can cause surface-flaking. The recommendations usually mention floating nonstone materials, which have a very strict limit of 0.1% in mass.

4.3.4 Leaching

When water comes into contact with a solid material like recycled aggregate, some substances are dissolved and removed. This is effect sought when washing out contaminants. However, if the recycled aggregate has not been washed, water can cause leaching of contaminants when the new concrete is mixed or even after it has hardened.

Some of the contaminants mentioned above (metals, wood, gypsum) as well as other organic materials, can cause leaching contamination. This is especially worrying when the recycled aggregate is to be used in road surfaces, but it cannot be ignored when the aggregate is for recycled concrete. Due to the higher proportion of contaminants, leaching is worse in recycled aggregates from buildings than in those from common civil engineering structures. Therefore, it is worse in RMA or MRA than in RCA.

As an example, the European Standard for aggregates EN 12,620 en (CEN, 2002) forbids the leaching of heavy metals, polyaromatic hydrocarbons and other dangerous substances, but does not specify any compulsory leaching test. This may be due to the common idea of aggregates as inert materials. However, this supposed lack of chemical activity or contaminant leaching should be proved through tests such as the "Compliance test for leaching of granular waste materials and sludges", EN 12,457 (CEN, 2003). In practice, the majority of CDW is, in fact, inert. Only in special cases of recycled concrete or masonry from highly contaminated areas where residual contamination in the rubble is expected, do leaching tests seem advisable.

4.3.5 Physical properties of RMA and RCA

As it happens with the chemical properties, the physical properties of the recycled aggregates coming from construction and demolition waste depend mainly on the quality of the original material and the recycling procedure. In this group of properties we can include size, shape, density, water absorption and strength. Of course, most of them depend directly on the composition or on the crushing and sieving process of the CDW. In addition, these properties are clearly interrelated.

4.3.5.1 Size

The product obtained from the process of selection, washing and crushing of concrete or masonry is a mixture of fine (≤4 mm) and coarse (>4 mm) aggregates. The coarse aggregate constitutes between 70% and 90% of the weight of the original concrete in the case of RCA (CEDEX, 2014).

Successive crushing is common in the production of recycled aggregates. The first stage is usually a jaw crushing, followed by some cone or impact crushing process to adjust the grain-size distribution. The size specifications for coarse aggregates for recycled concrete can be obtained by simply adjusting the crusher aperture (Hansen, 1992).

With successive crushing the coarse fraction of the RCA can be adjusted to the adequate grading curve for the concrete production. The main disadvantage is the greater proportion of fine declassified aggregates (ACHE C 2 GT 2/5, 2006). The grading curves are the main tool to determine the size of aggregates. There are various standards and recommendations that give detailed limits for the grading curves of recycled aggregates of different sizes. Fig. 4.7 shows the upper and lower limits proposed for different commercial sizes of aggregates according to the standard (Vincke & Rousseau, 1994). Fig. 4.8 presents the limits according to the Japanese requirements

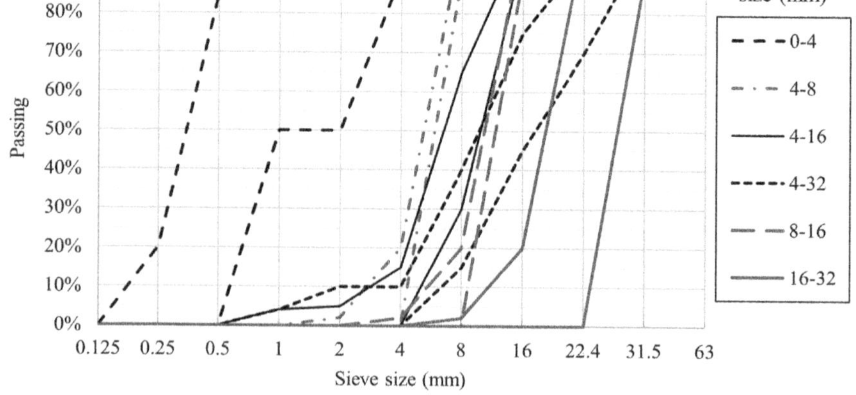

Figure 4.7 Belgian grading limits for RCA Grading limits for the different sizes of RCA according to the Belgian recommendations.

(Kasai, 1994), which classify the aggregates according to their maximum size and differentiate between use for buildings or for civil engineering structures.

The fineness modulus of recycled aggregates from CDW varies depending on the original material and the crushing method. However, a general rule for the fineness modulus of recycled aggregates from CDW is that the recycled aggregates obtained from materials with higher strength have slightly higher fineness moduli than those with a lower strength (Kikuchi et al., 1993).

According to a study of a database of 71 different datasets of RCA [14], an average value of the fineness modulus of 6.6 was found for the coarse recycled aggregate, with a minimum value of 5.7 (corresponding to a maximum grain size of 10 mm) and a maximum of 7.4 (for a maximum grain size of 25 mm). For the fine RCA, an average value of 3.1 was obtained for the fineness modulus, with a maximum of 3.8 (for aggregates up to 6 mm) and a minimum of 2.4 (for aggregates up to 1.25 mm).

Fines are usually defined as particles smaller than 0.063 mm. The fines content of aggregates should be limited due to the bad qualities of this fraction: low density, high absorption and high content of mortar, gypsum, clay soil, sulfate, chloride, and contaminants in general. In addition, fines can adversely affect the bond between the recycled aggregate and the new cement paste and increase the need for water to achieve good workability. It is true that recycled aggregates generate more fines than natural aggregates when they are handled and transported because small particles of mortar become detached. Therefore, strict limitations on fines content could restrict the use of many recycled aggregates. For this reason the restrictions on fines content is lighter for recycled aggregates from CDW than for natural aggregates, allowing for 2%–5% of fines (see examples in Table 4.3).

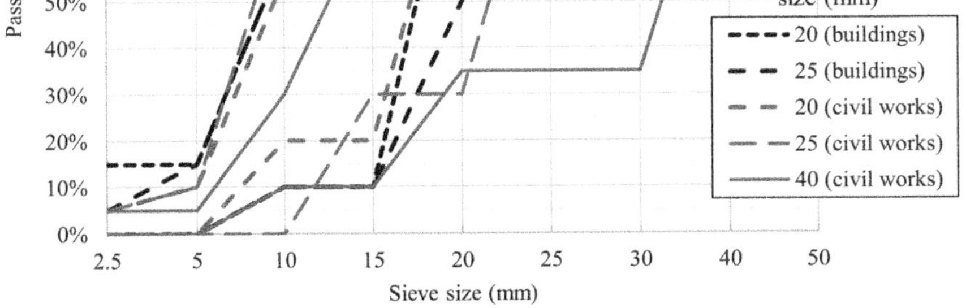

Figure 4.8 Japanese grading limits for RCA Grading limits for the different sizes of RCA according to the Japanese recommendations.

Table 4.3 Fines content. Restriction on fines content according to some recommendations.

	RILEM			Belgium		Germany	United Kingdom	
Fines size	0.063 mm			0.08 mm		0.063 mm	0.063 mm	
Max. fines content	RMA	RCA	MRA	RMA	RCA		RCA	MRA
	3%	2%	2%	5%	3%	4%	5%	3%
Max. sand (<4 mm)	5%			–		–	–	

The same circumstances that cause a high content of particles smaller than 0.063 mm in the coarse recycled aggregates from CDW (the unbounding of adhered mortar during manipulation) also cause an increase on particles smaller than 4 mm. It is estimated that after manipulation of recycled coarse aggregates from CDW, the increase on sand content can reach the 2% of the total weight.

The RILEM recommendations limit the sand content in coarse aggregate from CDW to 5%. This is lower than the typical values for natural aggregates because the bad properties of the small particles of mortar that have been mentioned before. However, this is not a very strict limitation because it does not mean that the coarse aggregate cannot be used with a higher content of small particles. It can be used, but these particles should be considered as sand in the mix design and calculations.

4.3.5.2 Shape

As with the most of properties described above, the shape and texture of recycled aggregates from CDW depend on the proportion of attached mortar. Mortar causes increased roughness and porosity. Of course, shape also depends on the crushing process: rounder particles are produced in the secondary crushing stages, after the primary jaw crushing.

The main difference in shape between natural and recycled aggregates are in the flakiness index. RCAs have a lower flakiness index because of the effect of adhered mortar, decreasing the number of flat particles and causing a positive effect on the shape (ACHE C 2 GT 2/5, 2006). However, no significant differences were found in the shape index when comparing natural and recycled aggregates from CDW.

The best crushers to obtain more spherical shapes and lower flakiness indexes in recycled aggregates from CDW are cone and impact crushers when they are used in secondary stages (Silva et al., 2014).

Fine recycled aggregates obtained by crushing are coarser and more angular than the standard sands for concrete (ACI Committee 555, 2002). This could negatively affect the workability of fresh recycled concrete.

4.3.5.3 Density

Density or specific gravity are the most used properties to classify the recycled aggregates from construction and demolition waste since they are also the key properties to characterize aggregates in general.

RCA are usually classified as "normal weight" aggregates (>2000 kg/m^3) in a three-group classification (heavy, normal, and light-weight aggregates). As guide values, common density of RCA ranges from 2100 to 2400 kg/m^3, or from 2300 to 2500 kg/m^3 if the Saturated Surface-Dried Density (SSDD) is considered. This makes them comparable to the most common natural aggregates such as limestone, sandstone, granite and basalt. However, it is true that the densities of RCAs made with different stone materials as natural aggregates are lower than those of the corresponding natural aggregates (NA). This is mainly due to the presence of adhered mortar, with a density between 1000 and 1600 kg/m^3, and other impurities. This drop in the density of RCA compared with the NA is between 5% and 10%. A review of 292 coarse RCAs (Silva et al., 2014) gave a mean oven-dried density of 2327 kg/m^3, and a mean SSDD of 2442 kg/m^3 with relative standard deviations of 5.0% and 3.4% respectively.

Exceptionally, recycled RMA can be near the lower limit adopted for "normal-weight", but they are usually under it. This is caused by the low density of the masonry materials and its higher porosity (e.g., ceramic bricks and concrete blocks). The most common guide values for the original ceramic bricks range between 1200 and 1800 kg/m^3. The density of RMA made with these bricks has guide values between 1000 and 1500 kg/m^3 because of the contaminants. However, RMA made with dense bricks, low level of contaminants, and a level of crushed concrete near the limit of 10%, can have a density closer to 2000 kg/m^3. A review of 35 RMAs (Silva et al., 2014) gave a mean oven-dried density of 1885 kg/m^3, and a mean SSDD of 2152 kg/m^3 with relative standard deviations of 11.1% and 6.4% respectively.

The importance of a recycling procedure to remove part of the adhered mortar and, of course, the density of the original masonry materials have been proved in the literature (Silva et al., 2014). In turn, density affects the compressive strength of the original bricks since this property increases with the brick density.

The density of MRA depends on the amount of RMA mixed with the RCA. It can be easily estimated by using the RCA and RMA densities and the composition ratios: a higher proportion of recycled masonry in the mix will decrease the resulting density.

The higher proportion of detached mortar in the finer fractions of the recycled aggregates from CDW indicates a clear relation between size and density (Evangelista & de Brito, 2007) that has been proved in the main systematic reviews (González-Taboada et al., 2016b; Silva et al., 2014): the finer the recycled aggregate (RCA, RMA or MRA), the lower the density. An illustrative example can be found in the mentioned literature reviews with average densities 11.3% lower in the fine than in the coarse aggregate when the oven-dried density is considered and 5.8% lower when the SSDD is considered.

The density requirements according to RILEM recommendations are presented in Table 4.4 as a representative example of the existing values in the international framework.

4.3.5.4 Water absorption

Water absorption (WA) of recycled aggregates from CDW is probably the property that most strongly differentiates them from natural aggregates, with notably higher values due to the inherent porosity and WA of:

Adhered mortar in all recycled aggregates from CDW.

Table 4.4 Density requirements.

Recycled aggregate (rilem type)	Dry density (kg/m^3)	Material with SSDD>2200 kg/m^3	Material with SSDD >1800 kg/m^3	Material with SSDD>1800 kg/m^3
RMA (I)	≥1500	–	≤10%	≤1%
RCA (II)	≥2000	≤10%	≤1%	≤0.5%

Masonry bricks and blocks (ceramic, lightweight concrete, etc.) in the case of RMA.

Other porous contaminants in all recycled aggregates from CDW.

The most common natural aggregates have water absorption values between 0.5% and 1.5%. These low values are negligible in the calculations of the water-cement ratio for common concrete. However, the recycled aggregates from CDW have much higher values, which, depending on their quality and size, can reach up to 15%.

The high water-absorption of the aggregates affects the workability of the fresh concrete: the water absorbed by the aggregates is subtracted from the hydration reaction of the new cement. Thus it is necessary to adjust the water-cement ratio of the mix.

RMA aggregates have higher water absorption than RCA due to the high porosity of bricks and blocks. The literature reviews have registered a mean value of 13.4% for coarse RMA while the WA for coarse RCA is 4.9% (Silva et al., 2014). The absorption of mixed aggregates depends on the mix ratio, and can be predicted using a weighted average.

The quality of the original material of RMA has a strong influence on porosity and, therefore, on the WA of the resulting aggregate. The quality of clay bricks depends on the raw clay used and the temperature and duration of the manufacturing process. Less porous bricks are denser and stronger, reducing water absorption.

Particle size also influences water absorption: the finer the aggregate, the higher the absorption. This is because of the high level of mortar and porous contaminants in the fine fractions caused by crushing. As a result, the average WA of coarse RCA (4.9%) is almost half of that of fine RCA (9.5%). The difference is less dramatic in MRA because the masonry increases the absorption of the coarse fraction while the WA of the fine fraction maintains its high value due to the mortar effect. The mean value of WA for coarse MRA is 7.2% and for fine MRA 9.3%.

The limits proposed by RILEM for water absorption allow up to 20% in RMA and up to 10% in RCA (RILEM TC 121-DRG, 1994b). These values are similar in most of recommendations regarding recycled aggregates from CDW.

Density and water absorption are also interrelated: they depend on size because of their relationship with the amount of mortar in the fine fraction. There is also a clear relationship between porosity, density, and water absorption: higher porosity is associated with lower density and higher water absorption.

4.3.5.5 Resistance

The strength of an aggregate is usually indirectly measured as its hardness and abrasion resistance through the Los Angeles coefficient, which represents the weight loss after the standardized Los Angeles (LA) abrasion test.

The recycled aggregates from CDW give worse results than the natural aggregates in the abrasion test, which means higher LA values (weight loss). The main cause of this inferior performance of RCA is the attached mortar, which implies a weak porous layer that is detached from the recycled aggregate in the abrasion test. In the case of RMA, in addition to the attached mortar, the weakness of the masonry causes an even higher LA coefficient.

The LA coefficient of RCA ranges between 25% and 45% (CEDEX, 2014), with an average value of 32.5%. This means that most coarse RCA is within the requirement of less than 40%, which is often considered a limit for the abrasion coefficient in aggregate for concrete (EHE-08, 2010). The LA coefficient of RMAs varies more, ranging from 20% to 55% of weight loss, but with a higher average than the RCAs.

Like all the properties described above, aggregate size also influences abrasion resistance due to the lower quality of the fine fractions caused by the high proportion of detached mortar. However, the harmful effect of the maximum aggregate size on the LA coefficient is not so clear as with other properties (González-Taboada et al., 2016b).

Density and water absorption have opposite relationships with abrasion resistance: while denser aggregates give lower LA coefficients, the water absorption increases when the LA weight loss increases.

4.4 Relationship between physical properties

The clear relationship between the aforementioned properties of aggregates can lead to predictions about their general performance.

For example, the systematic review carried out by Silva et al. (2014) lead to a third-grade polynomial approximation to the relationship between oven-dry density and water absorption of recycled CDW aggregates. Fig. 4.9 shows this model between two other polynomial curves that represent the boundaries of the 95% confidence interval. The authors proposed a classification of recycled aggregates from construction and demolition waste based on three limits for dry density (2400 kg/m^3, 2100 kg/m^3 and 1800 kg/m^3) and three limits for water absorption (3.5%, 8.5% and 15%). These limits make it possible to classify them into four categories from the best to the worst performance: A, B, C and D.

Fig. 4.9 also presents the limits proposed by RILEM for water absorption and dry density of RMA and RCA, showing that most RCAs will be in zones A and B while the RMA could be in zones C and D.

Regarding the relationship between WA, density and abrasion resistance, the literature indicates that water absorption values in RCA under 7% very usually correspond to dry densities over 2000 kg/m^3 and lead to LA coefficients under 40% (ACHE C 2

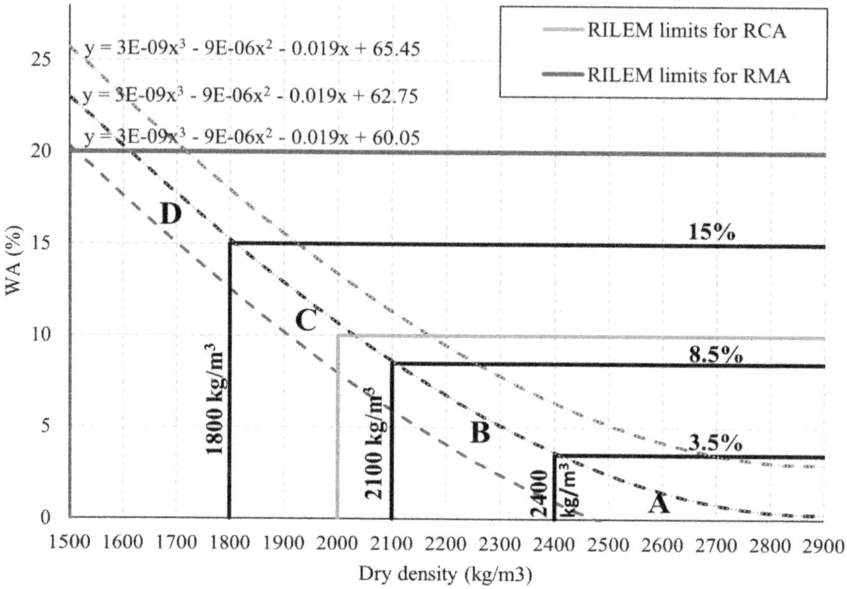

Figure 4.9 Proposed classification Model for the relationship WA-density and proposed classification by Silva et al. (2014).

GT 2/5, 2006; González-Taboada et al., 2016b; Silva et al., 2014). All these values are considered adequate for concrete fabrication.

4.5 Blast furnace slag (BFS)

Blast-furnace slag, BFS, is a by-product of iron when submitted to a blast furnace. It is mainly made up of silicate and aluminosilicate of melted calcium from the original iron ore combined with oxides from the limestone that is periodically removed from the blast furnace.

The cooling process can be either rapid or slow and will determine the resultant particle structure. Crystallized slag is obtained through a slow cooling process and only a small part of it remains in a glassy state. Small particles are then ground to <45 μm with a surface area varying between 400 and 600 m^2/kg.

Crystallized slag is a stony material and similar to igneous rocks. Its texture is rough, predominantly cubic and shows abundant external and internal porosity. Crystallized slag is considered suitable for use as a substitute for conventional aggregates.

4.5.1 Physical and chemical properties

BFS is a coarse material with particles similar in size to sand. This material has cementitious properties and hydraulic behavior in its finer ground form. Its chemical composition depends on the raw materials used in the production of the iron while the physical properties depend on the process used to cool the molten materials.

An average particle size of between one and 1.5 mm can be established, as seen in Fig. 4.10, which shows the gradation curve of a BFS sample (Maghool et al., 2020). Table 4.5 shows the other main physical characteristics of BFS (Özkan et al., 2007).

The fineness of BFS is a very important parameter that is sometimes measured by specific surface area and which controls its reactivity (Pal et al., 2003). This parameter usually ranges from 375 to 450 m^2/kg Blaine fineness.

The chemical makeup of BFS, shown in Table 4.6 (Kosmatka & Wilson, 2016), closely resembles that of Ordinary Portland Cement (OPC). BFS has a mildly alkaline character with a pH in solution in the range of 8–10. The leachate tends to be slightly alkaline and does not present a corrosion risk.

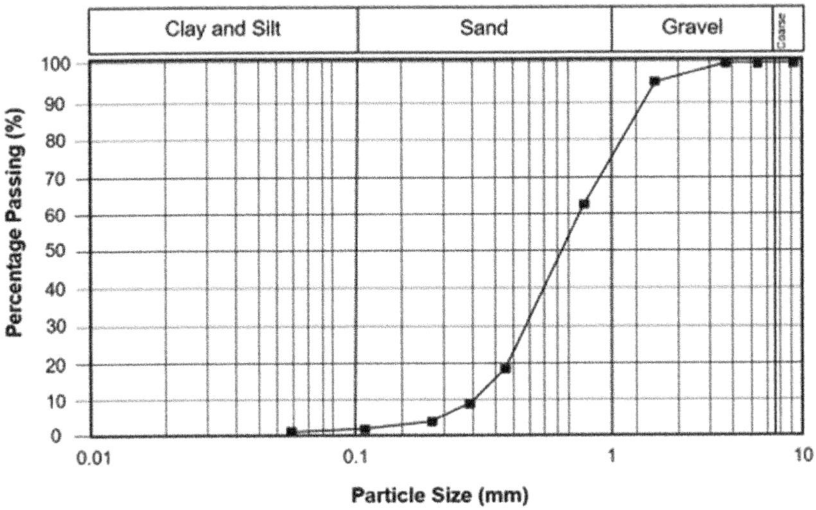

Figure 4.10 Particle size distribution of BFS.

Table 4.5 Physical properties of BFS (Özcan et al., 2007).

Property	Value
Loose bulk density (kg/m^3)	1052
Compact bulk density (kg/m^3)	1236
Specific gravity (g/cm^3)	2.08
Water absorption (%)	8.3
Amount of clay (%)	1
Organic impurities (clay lumps and friable particles) (%)	–
Loss on ignition (%)	1.8
Lightweight particles (proportion) (%)	3

Table 4.6 Chemical composition of BFS.

Oxides (wt.%) in GBFS	CaO	SiO_2	Al_2O_3	Fe_2O_3	SO_3	Na_2O	K_2O	L.O.I.
	40.0	35.0	12.0	1.0	9.0	0.30	0.4	1

Several factors including chemical composition and glass content affect the reactivity index of BFS that contributes to its cementitious performance and ASTM C989 classifies slag by its increasing level of reactivity (ASTM-C989, 2018). Glass content of slag shall be not less than 66% according to EN 15,167-1: 2006. More specifically, the glass content of the slag should be in excess of 90% to show satisfactory properties (Siddique & Kaur, 2012) as cementitious material. However, if the use as aggregate is considered, crystallized slag is a nonplastic material, not susceptible to frost and with good draining behavior, so it can be used as an aggregate for mortar and concrete, after checking that it is free of compounds that affect volumetric stability, such as excess Fe and Mg oxides (BSI, 1983; CEN, 2002).

4.6 Electric arc furnace slag (EAF-slag)

The basic feed materials of an electric arc furnace are steel residue and pig iron. The EAF slag obtained in the process of refining of molten steel is a nonmetallic by-product composed mainly of silicates and oxides.

The main function of these slags is the absorption of deoxidation products and impurities from the molten steel (Shi, 2004), but EAF slags also play other important roles in the molten steel refining process. The first is to protect the electrodes and furnace refractories from thermal radiation by inhibiting oxidation (Luz et al., 2018). EAF slags also perform as heat insulators, avoiding possible heat losses and preventing reoxidation of the molten steel.

Electric arc slag can be added in bituminous mixtures, can be used as aggregates for mortar reaching an optimum compressive strength when a 30%–40% of slag is used, and also as aggregates for concrete improving its density up to 2770 kg/m^3 (Manso & Marquínez, 2006).

4.6.1 Physical and chemical properties

This type of slag usually appears as lumpy masses in greyish or blackish tones and with a rough surface texture. The color is linked to the ferrous oxide content (Pellegrino & Gaddo, 2009). The variability in the chemical composition of the scrap used as raw material as well as the steel grade, the refining process and the furnace type determine the differences between EAF slags produced internationally (Grubeša et al., 2016). However, regardless of these differences in chemical composition, EAF slags exhibit a common range of Mohs hardness value of 6–7.

EAF slag show an average density of 2.8 g/cm^3 in a range of 1.5–3.8 g/cm^3 (see Table 4.7) (Faleschini et al., 2016; Monosi et al., 2016; Sekaran et al., 2015; Skaf et al., 2017) whereas the water absorption level is around 0.95–4, with an average % of 2.6% (Adegoloye et al., 2016).

It is essential to know the water absorption capacity of EAF slags as there are known cases of degradation by penetration of acidic fluids (i.e., CO_2 or H_2CO_3) in slags with a high alkaline composition (i.e., Ca and Mg oxides). As a result of this acid neutralization process, a decomposed carbonate phase is formed (Baciocchi et al., 2010).

Since the EAF steel production process is essentially a scrap recycling process, the chemical composition of these slags will largely depend on the properties and characteristics of the recycled steel. Consequently, the main chemical constituents of the EAF slags can vary widely (see Table 4.8).

The most common mineral phases in these slags are $CaCO_3$, FeO, MgO, Fe_2O_3, $Ca_2Al(AlSiO_7)$, Ca_2SiO_4 (Barra et al., 2001) and since EAF slags have high iron oxide contents, solid solutions of FeO (Wustite) are typically observed as one of the main mineral phases.

4.7 Coal fly ash

The United States Environmental Protection Agency (EPA) defines Coal ash as a coal combustion residue (CCR) produced primarily from the burning of coal in coal-fired thermal power plants. It is removed from the exhaust gases of fossil fuel power plants before they are discharged into atmosphere by dust collection systems as very fine, predominantly spherical glassy particles. As per ASTM C 618 (1993) specification for "Fly ash and raw or calcined natural pozzolan for use as mineral admixture in Portland cement concrete," pozzolans are defined as "silicious and aluminous materials, which in themselves possess little or no cementitious value but will, in finely divided form and in the presence of moisture, chemically react with Ca $(OH)_2$ at normal temperatures to form compounds possessing cementitious properties."

ASTM C 618 categorizes fly ashes into the following two categories based on matrix components (see Table 4.9) (ASTM, 1993; BN-79, 2009, pp. 6722–09).

Class F: Fly ash normally produced from burning anthracite or bituminous coal. This class of fly ash exhibits pozzolanic properties but rarely, if ever, self-hardening properties.

Table 4.7 Physical properties of EAFS.

Particle size range (mm)	**5–40**	
Surface pore diameter (μm)	0.01–10	Monosi et al. (2016)
Water absorption (%)	0.95–<4	Faleschini et al. (2016), Skaf et al. (2017)
Density (g/cm^3)	1.54–3.85	Sekaran et al. (2015), Faleschini et al. (2016)

Table 4.8 Chemical composition of EAFS.

	Oxide composition (wt.%)											
Reference	CaO	SiO_2	Al_2O_3	MgO	FeO	Fe_2O_3	Fe_{total}	SO_3	MnO	TiO_2	P_2O_5	Free CaO
Manso and Marquínez (2006)	23.9	15.3	7.4	5.1	—	—	42.5	0.1	4.5	—	—	0.5
Tsakiridis et al. (2008)	35.7	17.5	6.3	6.5	—	26.4	—	—	2.5	0.8	—	—

Table 4.9 Fly Ashes classification using the standards ASTM C618 (USA) and BN-79/6722-09 (Poland).

Properties	Fly ash class (ASTM C618, 1993)		Fly ash class (BN-79/6722-09)		Content (wt%)			
	Class F	Class C			SiO_2	Al_2O_3	CaO	SO_3
$SiO_2 + Al_2O_3 + Fe_2O_3$ (% min)	70	50	Silicate	k	>40	<30	≤10	<4
SO_3 (% max)	5	5	Aluminum	g	>40	≥30	≤10	<3
Moisture content (% max)	3	3	Calcium	w	>30	<30	>10	≥3
LOI (% max)	6[a]	6						

[a]The use of Class F containing up to 12% LOI may be approved if acceptable performance results are available (Manz, 1999).

Class C: Fly ash normally produced from lignite or sub-bituminous coal. This class of fly ash has both pozzolanic and varying degrees of self-cementitious properties.

Matrix compounds, i.e., aluminosilicates, are considered essential because they determine the properties of an ash and its weak water solubility, as well as the classification according to the standard BN-79/6722−09 (see Table 4.9).

The basic difference between F and C ashes is the CaO content, which determines whether they will be calcium ashes or aluminum ashes. Silico-aluminous ash is used in the manufacture of lime gravel for base courses, reaching strengths of up to 20 MPa. It can also be used as a filler in bituminous mixtures and as a substitute for aggregates in concrete, improving its workability and resistance to sulfate attack. Another recognized and regulated use is the manufacture of lightweight aggregates.

4.7.1 Physical and chemical properties

Normally fly ash particles consist of clear glassy spheres and a spongy aggregate (Mehta, 1988). The exterior surfaces of the solid and hollow spherical particles of low calcium oxide fly ashes are generally smooth and well defined. Typical Physical characteristics of FA are shown in Table 4.10 (EPRI, 1987; Joshi, 1970).

Fly ash generally has a crystalline phase of 15%−45%, accentuated in calcium ashes, Class C, with intervals of 25%−45% of crystalline matter. Although these ashes have a much smaller amorphous phase, it is much more reactive due to the large amount of calcium on the particle surface that made them strongly strained and probably this is why this Class C fly ashes are highly reactive. Table 4.11 presents the chemical analysis of different fly ashes as reported by EPRI (1987).

4.8 Scrap tires

Approximately one billion scrap tires are generated every year, with around four billion currently stockpiled across the world. However, the recycling industry processes only 100 million tires a year (Goldstein Market Intelligence, 2020). The tire has an intricate design with several complex manufacturing processes. It is virtually indestructible by natural processes and difficult to recycle. Thus, their use as aggregates in cement pastes and mortars is a significant step forward in the management

Table 4.10 Physical characteristics of Fly Ashes.

Parameter	Joshi (1970)	EPRI (1987)
Average particle size (μm)	7−12	
Specific gravity	1.3−4.8	2.14−2.69
Moisture content (%)		0.0−0.38
Retained on #325 sieve (%)		3.55−36.90
Blaine fineness (cm_2/g)		1579−5550

Table 4.11 Chemical characterization of Fly Ashes.

Oxide composition (wt%)									
SiO_2	**Al_2O_3**	**Fe_2O_3**	**CaO**	**MgO**	**SO_3**	**Na_2O**	**K_2O**	**Available alkalis (Na_2O equivalent)**	**LOI**
30.92–62.76	12.30–26.95	2.84–24.43	1.10–30.53	0.69–6.69	0.31–3.85	0.20–2.04	0.22–3.03	0.23–1.54	<0.01–16.60

Table 4.12 Scrap tire disposal in Europe and the United States of America.

Market or disposal (thousands of tons)	U.S.A. (2015)	Europe (2018)
Ground rubber	1020.75	1505
Scrap tires used in civil engineering applications	274.92	96.12
Scrap tires generated	4038.8	3573.9
Treated (%)	99.1	91

of this waste. There are now specific regulations, such as the ASTM D 6270-98 (ASTM D 6270-98, 2002), which develops the standard practice for use of scrap tires in civil engineering applications, and the EU Waste Directive 2008/98/EC (EU, 2008), which includes a definition of waste for end-of-life tire-derived products.

Construction and cement are the major end-use industries of the tire recycling downstream products market. New data consolidated by the European Tire and Rubber Manufacturers' Association (ETRMA) on the management of End-of-Life Tires (ELTs) shows that 91% of ELTs were collected and treated for material recycling and energy recovery in 2018 (European Tire & Rubber Manufacturers Association, 2018). Table 4.12 shows comparative data between the United States and Europe on scrap tire disposal. However, these tires must be processed in order to be suitable for use as aggregates.

4.8.1 Composition and physical properties

Scrap tires can be sourced either from cars, weighing approximately 9 kg, or trucks weighing 44–45 kg. Table 4.13 shows the percentage of the total finished element weight of the major materials employed in the manufacture of tires (Rubber Manufacturer's Association, 2000).

Scrap tires are recycled using techniques such as pyrolysis and shredding. Shredding includes mechanical, cryogenics, and other methods to reduce tires to shreds/chips (Raghavan et al., 1998), treads (Albano et al., 2005), ground rubber, crumbs rubber (Khatib & Bayomy, 1999) to be used as a replacement of coarse and fine aggregates. These materials are call "Tyre Derived Aggregate" (TDA). Table 4.14 show some of the main characteristics of TDA that justify its use as aggregates.

Table 4.13 Typical composition of manufactured tires by weight.

Composition (% by weight)	Automobile tire	Truck tire
Natural rubber	14	27
Synthetic rubber	27	14
Carbon black	28	28
Steel	14–15	14–15
Fabric, filler, accelerators, antiozonants	16–17	16–17

Table 4.14 TDA physical characteristics.

TDA physical characteristics	
Specific gravity	1180
Fine aggregates size (mm)	0.29–0.55 (Albano et al., 2005)
Coarse aggregates size (mm)	5.50–10.8 (Raghvan et al., 1998)
Weight	Lighter by 30%–50%
Drains 10 times better than well graded soil	
Provides 8 times better insulation than gravel	

4.9 Municipal solid waste (MSW) combustor ash

Municipal Waste Combustion (MWC) residue is generated at several stages in the process of combustion of municipal solid waste (MSW) for energy recovery. The material discharged from the bottom of the furnace, primarily the grate but also as grate siftings is generally referred to as Bottom ash or clinker. Bottom ash comprises the major portion of the residues generated, depending on the design of the combustion facility, operating conditions, and characteristics of the waste being incinerated.

Particulate matter carried forward from the furnace and removed from the flue gas prior to injection of sorbents, including volatiles condensed during flue gas cooling, is normally referred to as Fly ash or air pollution control (APC) residue.

So far, the possibilities of using these ashes as a total or partial substitute for concrete aggregates have only been studied in the laboratory. When replacing part of the concrete aggregates, up to 25%, there are no significant variations in the mechanical behavior and durability of the concrete (García, 1998). Problems of workability and formation of expansive hydrates may occur due to the presence of chlorides, sulfates and heavy metals in the ashes.

4.9.1 Physical and chemical properties

MSW ash is a gray to black material containing high levels of heavy metals such as Cd and organic compounds such as dioxins, both of which are highly toxic. The nature and quality of this ash depends largely on the nature of the residue, the type of combustion and the nature of the air pollution control device. The physical properties of MSW ash are given in Table 4.15. Combustor ash is primarily a sandy material, most of which passes through a 4.75 mm (#4) sieve. Ash also contains a relatively high under-0.075 mm (#200 sieve) silt fraction. These ashes are highly absorptive compared to conventional aggregates (absorption value of <2%), with absorption rates from 5% to 17% in the case of fine particles and from about 4% to 10% for coarse particles.

Major elements in MSW ash are silica, calcium, and iron. Ash composition can also have a lime content from 24% to 27% as well as silicates and aluminosilicates, which make it potentially useful in the production of cement powder (Mangialardi et al., 1999). Chemical composition as reported by IAWG (1994) is detailed in Table 4.16.

Table 4.15 Physical properties of MSW ash.

Property		Bottom ash (hesner et al., 1986)	Fly ash (Koppelman et al., 1993)
Bulk specific gravity	Fine	–	1.86–2.03
	Coarse	2.11–2.23	1.96–2.24
Absorption (%)	Fine	12.0–17.0	4.8–14.8
	Coarse	4.1–4.7	3.6–10.0
Moisture (% dry wt)		29–66	17–76
Unit weight (kg/m^3)		960–1400	990–1170
LOI (%)		3.7	2.5–13.5
Maximum density (kg/m^3)		1260–1570	1260–1730

The most common mineral phases in bottom ash include quartz (silicates), plagioclase, ghelenite, piroxene, olivine, alite, belite (Ca and Na aluminosilicates), metal oxides, portlandite (hydroxides), anhydrite (sulfates), calcite and siderite (carbonates) as well as metal and alloys, as reported by Clozel-Leloup et al. (1999) and Zevenbergen and Comans (1994).

4.10 Inert/incinerated sewage sludge

According to AQUASTAT, the FAO's global information system on water and agriculture, approximately 330 km^3 of wastewater is produced globally each year, of which only 141 km^3 - less than 50% - is treated.Table 4.17 shows the sewage sludge generated in EU, USA and China (UBA, 2012; WB, 2016; WEF, 2011).

The most common technology for wastewater treatment is the activated sludge process, which generates large quantities of organic sludge called Sewage Treatment Plant Sludge. The quality of this sludge is not constant: it varies according to the design characteristics of each treatment plant, the type of wastewater treated, the industries that produce it, etc. Fig. 4.11 shows the destinations of the sludge produced annually in the European Union (Eurostat, 2016).

4.10.1 *Physical and chemical properties*

The main physical characteristics of sewage sludge are shown in Table 4.18 (Shinogi & Kanri, 2003; Valls et al., 2007). The quality of a sludge for building material depends on three main groups of pollutants that accumulate after the water treatment process: heavy metals, organic pollutants and pathogens (Helaine, 2000). Incineration reduces the volume of inert sludge by 70%–90% and makes it suitable to be added to ceramic masses or vitrified. Heavy metals such as zinc, nickel and magnesium

Table 4.16 Composition of MSW ash. Elemental composition for Bottom ash and Fly ash form mass burn incinerators.

	Elements												
(ppm)	Al	Ca	Cd	Chloride (% dry wt)	Cr	Cu	Fe	K	Mg	Mn	Ni	Pb	Sulfate	Zn
Bottom ash	$22 \cdot 10^3$ – $73 \cdot 10^3$	$37 \cdot 10^3$ – $120 \cdot 10^3$	0.3 – 71	800–4200	23 – 3200	190 – 8200	4100 – $150 \cdot 10^3$	750 – $16 \cdot 10^3$	400 – $26 \cdot 10^3$	83 – 2400	7 – 4300	98 – $14 \cdot 10^3$	1000 – 5000	610 – 7800
Fly ash	$49 \cdot 10^3$ – $90 \cdot 10^3$	$74 \cdot 10^3$ – $130 \cdot 10^3$	50 – 450	$29 \cdot 10^3$ – $210 \cdot 10^3$	140 – 1100	600 – 3200	$12 \cdot 10^3$ – $44 \cdot 10^3$	$22 \cdot 10^3$ – $62 \cdot 10^3$	$11 \cdot 10^3$ – $19 \cdot 10^3$	800 – 1900	60 – 260	5300 – $26 \cdot 10^3$	$11 \cdot 10^3$ – $45 \cdot 10^3$	9000 – $70 \cdot 10^3$

Table 4.17 Sewage sludge generated in EU, USA and China.

Quantities of sewage sludge generated	EU	USA	China
Million tons DS/year	11.25	7	6
Million m^3/year	58	35	30

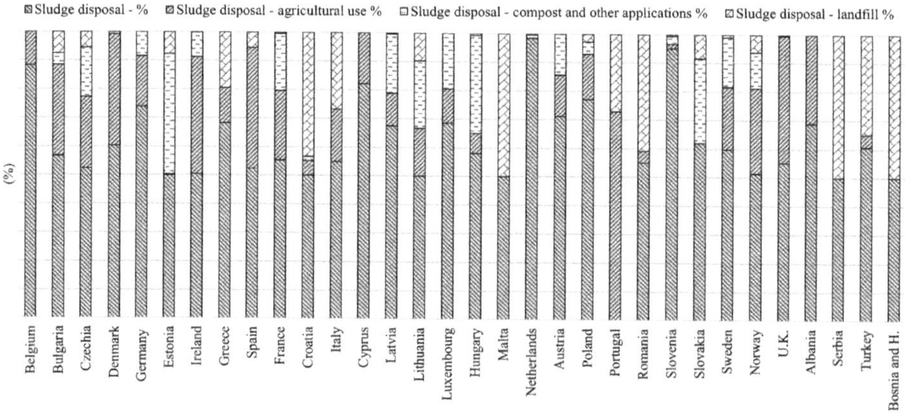

Figure 4.11 Distributions of annual sludge disposal/reuse Distributions of annual sludge disposal/reuse EU.

Table 4.18 Main physical characteristics of sewage sludge.

	Inert sludge	Incinerated sludge
Specific weight (g/cm^3)	2.21	2.44–2.96
Surface area	Generally below 100 m^2/g	
Humidity (%)	5	0.5–28
Density (T/m^3)	–	0.90–1.10
Loss on ignition (%)	–	3.60–5.20
Water absorption (%)	–	7.80–9.60
Porosity (%)	–	63.5–69.2
pH	–	8.97–9.03

become more concentrated with the volume reduction caused by incineration, while other metals such as mercury, chromium and lead are volatilized. The changes in the concentrations of these metals can be seen in Table 4.19 (FHWA, 2016; Guillem et al., 2010). Ash from sludge incineration contains mainly quartz and to a lesser extent mica, vermiculite and smectite (collapsed).

4.11 Urban plastic waste

Plastics are an integral and unavoidable part of modern society. The immense volume of solid plastic waste grows steadily every year (see Fig. 4.12).

Table 4.19 Chemical characterization of sewage sludge.

	Oxide composition (%wt)										
	PF	SiO_2	Al_2O_3	Fe_2O_3	MgO	CaO	Na_2O	K_2O	MnO	P_2O_5	SO_3
Inert sludge (Guillem et al., 2010)	32	1.32	0	0.89	1.02	61	0.1	0.1	0.006	0.27	1.3
Incinerated sludge (FHWA)	1.4	14.4–57.7	4.6–22.1	2.6–24.4	0.8–2.2	8.9–36.9	–	–	–	3.9–15.4	0.01–3.4

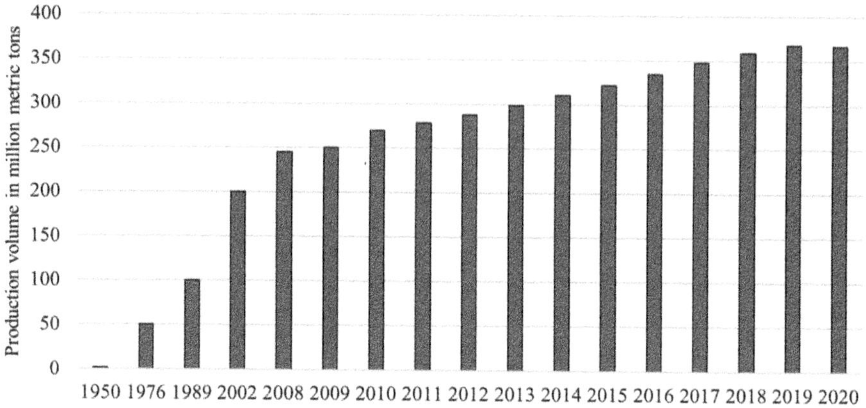

Figure 4.12 Production of plastics worldwide Annual production of plastics worldwide from 1950 to 2020.

As can be seen in Fig. 4.12, the world's annual consumption of plastic materials has increased from around five million tons in the 1950s to nearly 400 million tons in 2020. In addition, by 2050 the amount of plastic waste generated is expected to be around 50% more than in 2015, as shown in Fig. 4.13. Its low density, strength, user-friendly designs, fabrication capabilities, long life, light weight, and low cost are the factors behind such phenomenal growth. However, plastic waste feedstock is not uniform over time and it is estimated that only 7% of the total plastic waste generated is currently being recycled (Siddique, 2008).

4.11.1 Physical and chemical properties

Within the wide spectrum of waste plastics available for recycling, polypropylene represents the largest fraction, followed by PET. Lesser amounts of a variety of other plastics can also be found in the plastics waste stream as given in Table 4.20 (Kazemi Najafi, 2013).

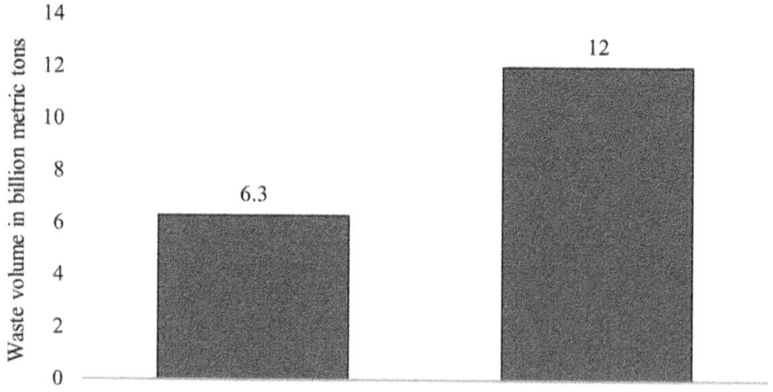

Figure 4.13 Plastic waste volume worldwide in 2015 and 2050 Plastic waste volume worldwide in 2015 and 2050.

Table 4.20 Types and quantities of plastics in municipal solid waste in the USA in 2010.

	Generated	Recovered
Type of plastic	10^3 tons	
Polyethyleneterephthalate (PET)	3980	560
High density polyethylene (HDPE)	5450	570
Low density polyethylene (LDPE)	7430	420
Polypropylene (PP)	7530	60
Polystyrene (PS)	2060	20
Other resins	3630	730

LDPE and PET are the plastics most frequently included in cement pastes and mortars. In order to be suitable for use as aggregates, they must be subjected to mechanical processing, such as shredding, and chemical processing, such as melting. The processing applied depends on the final purpose (Curlee, 1986). Some of the basic properties of these wastes are presented in Table 4.21 (Brydson, 1995; Carraher, 2000; Mark, 1985; Zoorob & Suparma, 2000).

4.12 Waste glass

Glass is a transparent substance formed at high temperatures by fusion of various materials such as silica, soda ash and calcium carbonate, followed by a cooling process in which no crystallization occurs. When waste glass is crushed to sand-like particle sizes, similar to that of natural sand, it has properties suitable for use as an aggregate. Thus there is significant interest in the development of concrete with waste glass. Recycling waste glass as an aggregate has environmental and economic benefits. Any type of waste glass is suitable for this purpose, both colorless and colored (Park et al., 2004; Topçu & Canbaz, 2004) with a very low recycling rate in general and generally dumped into landfill sites. In 2018, the glass industry reported recycling around 27 million metric tons worldwide: 21% of the total glass production in that year.

Table 4.21 Plastic waste main properties.

Properties	LDPE	PET
Granulate shape	Pellet	Flakes
Size (mm)	5.0–2.36	–
Density (g/cm^3)	0.92	1.41
Softening point (°C)	120	115
Melting point (°C)	140	265

4.12.1 Physical and mechanical properties

Crushed residual glass (cullet) is characterized by its angular shape and elongated, flat particles. The quantity and size of these particles depends on the degree of processing (crushing). Extra crushing results in smaller, less angular particles, and reduced quantities of flat and elongated particles. Table 4.22 gives some typical physical properties of waste glass collected from Materials Recovery Facilities (MRF) (WSDTED, 1993). Waste glass usually has a specific gravity of approximately 2.5.

Typical mechanical properties of waste glasses are given in Table 4.23.

The low thermal conductivity of waste glass is an important factor since aggregates and aggregate mixtures with low thermal conductivity can help to decrease the depth of frost penetration (Dickson, 1973).

Table 4.22 Selected physical properties of waste glass.

Test	Glass samples		Test method
	Coarse	Fine	
Particle shape angularity	Angular	Angular	ASTM D2488
Flat (%)	20–30	1	
Flat/Elongated (%)	1–2	1	
Specific gravity	1.96–2.41	2.49–2.52	ASTM D854 ASTM C127
Permeability (cm/sec)	$\sim 2 \times 10^{-1}$	$\sim 6 \times 10^{-2}$	ASTM D2434

Table 4.23 Typical mechanical properties of waste glass.

Test	Results	Test method
Los Angeles abrasión (%)	3042	ASTM C131
Maximun dry density (kg/m^3)	1800–1900	ASTM D1557
Optimum moisture (%)	5.7–7.5	ASTM D1557
Angle of internal friction (deg)	51–53	ASTM D3080
California bearing ratio (%)		
15% glass	132	ASTM D1883
50% glass	42–125	
Hardness (Moh's scale)	5.5	
Apparent thermal conductivity (W/m-°k) (WSDTED, 1993)	0.260–0.315	ASTM C518

4.12.2 Chemical properties

The major constituent is silicon dioxide (SiO_2) used in sand form. Common glass contains about 70% SiO_2. Soda ash, anhydrous sodium carbonate, (Na_2CO_3) acts as a fluxing agent in the melt, lowering the melting point and the viscosity of the formed glass. Table 4.24 lists the typical chemical compositions of these glasses as reported by Martin (2006) of soda-lime, by Sasai et al. (2008) of lead glasses, and by Pfaender (1996) of borosilicate glass.

4.13 Foundry sand

Foundry sand is a by-product of the ferrous and nonferrous metal casting industry. In this sector, sand has been used for hundreds of years as a molding material due to its thermal conductivity. The sands are successfully recycled and reused several times in a foundry until they reach a limit and are disposed of and referred to as foundry sands. Foundry sand is high-quality silica sand with uniform physical characteristics, which is being used as a fine aggregate substitute in construction applications and as kiln feed in the manufacture of Portland cement.

4.13.1 Physical and mechanical properties

Based on ASTM C33-03 recommendations, foundry sands can be used as a substitute for fine aggregates in Portland cement pastes. The American Foundry Society (AFS) has studied foundry sand as a substitute for fine aggregate in concrete (FIRST, 2004). However, foundry sands are generally too fine to allow full substitution. More specifically, between 5% and 12% of the spent foundry sands are smaller than 0.075 mm. Therefore, to meet the specifications of the standard it is necessary to discard part of the fines or to blend the foundry sands with coarse sand.

Another characteristic of this material is its content of organic impurities, which can vary widely, reaching very high levels. This can affect its use in applications, such as Portland cement concrete aggregate, in which the control of these elements is

Table 4.24 Typical chemical composition of waste glass types.

Constituent (wt%)	Borosilicate glass (Pfaender, 1996)	Lead glass (Sasai et al., 2008)	Soda–lime glass (Martin, 2006)
SiO_2	70–80	35.3	70
Al_2O_3	2–8	–	–
CaO	–	–	10
Na_2O	4–8	0.5	15
K_2O	4–8	2.3	–
PbO	–	56.8	–
B_2O_3	7–25	–	–

Table 4.25 Physical properties of foundry sands.

Property	Result	Test method
Moisture content (%)	0.1–10.1	ASTM D2216
Coefficient of permeability (cm/sec)	10^{-3}–10^{-6}	AASTHO T2215/ ASTM D2434
Absorption (%)	0.45	ASTM C128
Specific gravity	2.39–2.55	ASTM D854
Clay lumps and friable particles (AFS, 1991)	144	ASTM C142/AASTHO T112
Bulk relative density (kg/m^3)	2589	ASTM C48/AASTHO T84
Plastic index	Nonplastic	ASTM D4318/ AASTHO T90

Table 4.26 Mechanical properties of foundry sands.

Property	Result	Test method
Microdeval abrasion loss (%)	<2	
Magnesium sulfate soundness loss (%)	5–15	ASTM C88
Friction angle (deg)	33–40	
California bearing ratio (%)	4–20	ASTM D1883

important (MOEE, 1993). The typical physical and mechanical properties of spent foundry sand are given in Table 4.25 (Javed & Lovell, 1994; Johnson, 1981) and Table 4.26 (Javed & Lovell, 1994; MNR, 1992).

4.13.2 Chemical properties

Chemical composition of foundry sands is directly related to the metal mold of the foundry, the fuel used and the type of binder used. Spent foundry sand consists primarily of silica sand, coated with a thin film of burnt carbon, residual binder (bentonite, sea coal, resins) and dust. The binder and the type of metal in the foundry can determine a pH range in foundry sands between four and 8 (Johnson, 1981). Cases of foundry sands corrosive to metal have been studied (MNR, 1992). The typical chemical composition of foundry sand is shown in Table 4.27 (AFS, 1991).

4.14 Summary

In order to enhance the application of recycled materials it is important to identify the different types of waste and by production according to its specific country and region. Additionally, regarding the construction engineering project goals, the knowledge

Table 4.27 Typical chemical composition of foundry sand.

Oxide composition (wt%)											
SiO_2	Al_2O_3	Fe_2O_3	CaO	MgO	SO_3	Na_2O	K_2O	TiO_2	Mn_2O_3	SrO	LOI
87.91	4.7	0.94	0.4	0.3	0.09	0.19	0.25	0.15	0.02	0.03	5.15

related to the physical, mechanical and chemical properties is imperative to select the suitable materials. Considering the circular economy concepts, recycled aggregates are promising alternative for both structural and no-structural purposes.

References

ACHE C 2 GT 2/5. (2006). *Monograph M-11 ACHE: Use of recycled aggregate for the manufacture of structural concrete (Spanish)*. Asociación Científico-Técnica de Hormigón Estructural.

ACI Committee 555. (2002). Removal and reuse of hardened concrete. Special Publication. In *ACI committee 555* (p. 26). American Concrete Institute.

Adegoloye, G., Beaucour, A. L., Ortola, S., & Noumowe, A. (2016). Mineralogical composition of EAF slag and stabilised AOD slag aggregates and dimensional stability of slag aggregate concretes. *Construction and Building Materials, 115*, 171–178. https://doi.org/10.1016/j.conbuildmat.2016.04.036

AFS. (1991). *Alternative utilization of foundry waste sand. Final report (phase I) (Illinois department of commerce and community affairs*. American Foundrymen's Society.

Albano, C., Camacho, N., Reyes, J., Feliu, J. L., & Hernández, M. (2005). Influence of scrap rubber addition to Portland I concrete composites: Destructive and non-destructive testing. *Composite Structures, 71*(3–4), 439–446. https://doi.org/10.1016/j.compstruct.2005.09.037

ASTM. (1993). *Standard specification for coal fly ash and raw or calcined natural pozzolan for use as a mineral admixture in concrete*. Annual Book of ASTM Standards. ASTM C 618.

ASTM D 6270-98. (2002). Standard practice for use of scrap tires in civil engineering applications. Annual book of ASTM standards: Environmental assessment; hazardous substances and oil spill responses. *Waste Management, 11*(04).

ASTM-C989. (2018). *Standard specification for slag cement for use in concrete mortars.*

Baciocchi, R., Costa, G., Di Bartolomeo, E., Polettini, A., & Pomi, R. (2010). Carbonation of stainless steel slag as a process for CO2 storage and slag valorization. *Waste and Biomass Valorization, 1*(4), 467–477. https://doi.org/10.1007/s12649-010-9047-1

Baradaran-Nasiri, A., & Nematzadeh, M. (2017). The effect of elevated temperatures on the mechanical properties of concrete with fine recycled refractory brick aggregate and aluminate cement. *Construction and Building Materials, 147*, 865–875. https://doi.org/10.1016/j.conbuildmat.2017.04.138

Barbudo, A., Agrela, F., Ayuso, J., Jiménez, J. R., & Poon, C. S. (2012). Statistical analysis of recycled aggregates derived from different sources for sub-base applications. *Construction and Building Materials, 28*(1), 129–138. https://doi.org/10.1016/j.conbuildmat.2011.07.035

Barra, M., Ramonich, E. V., & Munoz, M. A. (2001). Stabilization of soils with steel slag and cement for application in rural and low traffic roads. In *Proceedings of the beneficial use of recycled materials in transportation application* (pp. 423–432).

BN - 79. (2009). *Ash and slag from hard and brown coal-fired boilers. Classification, terms and designations (in Polish).*

Brydson, J. A. (1995). *Butter-worth-Heinemann.*

BS-8500. (2006). *Concrete – complementary British standard to BS EN 206-1 – Part 2: Specification for constituent materials and concrete* (Vol 52). UK: British Standards Institution.

BSI. (1983). *BS 1047:1983 Specification for air-cooled blast furnace slag aggregate for use in construction*. BSI.

Cabral, A. E. B., Schalch, V., Molin, D. C. C. D., & Ribeiro, J. L. D. (2012). Performance estimation for concretes made with recycled aggregates of construction and demolition waste of some Brazilian cities. *Materials Research, 15*(6), 1037–1046. https://doi.org/10.1590/S1516-14392012005000119

Carraher, C. E. (2000). *Polymer chemistry*.

CEDEX. (2014). Catalogue of residues that can be used in construction. *Centre of studies and experimentation in civil engineering (CEDEX). CEDEX*. http://www.cedexmateriales.es/2/catalogo-de-residuos/.

CEN. (2002). *EN-1260. Aggregates for concrete* (Vol 56). Comité Européen de Normalisation (CEN).

CEN. (2003). *EN 12457. Characterisation of waste—leaching—Compliance test for leaching of granular waste materials and sludges*. Comité Européen de Normalisation (CEN).

CEN. (2013). *EN 13242. Aggregates for unbound and hydraulically bound materials for use in civil engineering work and road construction*. Comité Européen de Normalisation (CEN).

Chen, W., Jin, R., Xu, Y., Wanatowski, D., Li, B., Yan, L., Pan, Z., & Yang, Y. (2019). Adopting recycled aggregates as sustainable construction materials: A review of the scientific literature. *Construction and Building Materials, 218*, 483–496. https://doi.org/10.1016/j.conbuildmat.2019.05.130

Chesner, W.H., Collins, R.J., & Fung, T. (1986). Characterisation of incinerator residues in New York City. Proceedings of the 1986 National Waste Processing Conference, ASME Solid Waste Processing Division, USA.

Clozel-Leloup, B., Bodenan, F., & Piantone, P. (1999). Bottom ash from municipal solid waste incineration. Mineralogy and distribution of metals. In *Proc. STAB & ENV (vol. 99). Proc. STAB & ENV 99 46*.

Costa, J. S., Martins, C. A., & Baldo, J. B. (2013). Masonry mortar containing mixed recycled aggregates from the traditional ceramic industry. *InterCeram: International Ceramic Review, 62*(1), 30–36. http://www.interceram-review.info/download.cfm?param1=INTERC&param2=2013&param3=1&param4=30&param5=36.

Curlee, T. R. (1986). *The economic feasibility of recycling: A case study of plastic wastes*.

Debieb, F., Courard, L., Kenai, S., & Degeimbre, R. (2010). Mechanical and durability properties of concrete using contaminated recycled aggregates. *Cement and Concrete Composites, 32*(6), 421–426. https://doi.org/10.1016/j.cemconcomp.2010.03.004

Dickson, P. (1973). *Albuquerque symposium on utilization of waste glass in secondary products*.

DIN 4226-100:2002-02. (2002). *Aggregates for concrete and mortar - part 100: Recycled aggregates*. German Institute for Standardisation (Deutsches Institut für Normung).

EHE-08. (2010). *Code on structural concrete*. Spanish Ministry of Development, Permanent Commission of Concrete.

Elçi, H. (2016). Utilisation of crushed floor and wall tile wastes as aggregate in concrete production. *Journal of Cleaner Production, 112*, 742–752. https://doi.org/10.1016/j.jclepro.2015.07.003

EPRI. (1987). *Classification of fly ash for use in cement and concrete. CS-5116*. Electric Power Research Institute.

EU. (2008). *DIRECTIVE 2008/98/EC of the European Parliament and of the Council on waste and repealing certain Directives*. http://data.europa.eu/eli/dir/2008/98/oj.

European Aggregates Association. (2020). *UEPG annual review 2019-2020*. https://uepg.eu/mediatheque/media/UEPG-AR20192020_V13_(03082020)_spreads.pdf.

European Tire & Rubber Manufacturers Association. (2018). *End of life tyres management – Europe*. European Tire & Rubber Manufacturers Association. https://www.etrma.org/wp-content/uploads/2020/09/Copy-of-ELT-Data-2018-002.pdf.

European Union. (2008). *Directive 2008/98/CE, of 19 November 2008, on waste and repealing certain Directives. L 312 22/11/2008*. Official Journal of the European Union. https://eur-lex.europa.eu/legal-content/EN/TXT/?uri=celex%3A32008L0098.

Eurostat. (2016). *Sludge disposal in European countries*. https://appsso.eurostat.ec.europa.eu/nui/show.do?dataset=env_ww_spd&lang=en.

Evangelista, L., & de Brito, J. (2007). Mechanical behaviour of concrete made with fine recycled concrete aggregates. *Cement and Concrete Composites, 29*(5), 397–401. https://doi.org/10.1016/j.cemconcomp.2006.12.004

Faleschini, F., Brunelli, K., Zanini, M. A., Dabalà, M., & Pellegrino, C. (2016). electric arc furnace slag as coarse recycled aggregate for concrete production. *Journal of Sustainable Metallurgy*, 44–50. https://doi.org/10.1007/s40831-015-0029-1

FHWA. (2016). *RD-97-148 user guidelines for waste and byproduct materials in pavement construction*. https://www.fhwa.dot.gov/publications/research/infrastructure/pavements/97148/index.cfm.

FIRST. (2004). *Foundry industry recycling starts today. Foundry sand facts for civil engineers*. Federal Highway Administration Environmental Protection Agency.

García, M. (1998). Feasibility of the use of incineration ashes in construction materials, construction and environment: XIV course of major construction studies. In *, 98. CCEMCO*. Instituto de Ciencias de la Construcción Eduardo Torroja.

GEAR Project. (2012). *Spanish guide of recycled aggregates from CDW*. GERD.

Goldstein Market Intelligence. (2020). *Global tire recycling industry analysis by rubber type, by product type, by end user and by geography & COVID-19 impact with market outlook 2017-2030*. https://www.goldsteinresearch.com/report/global-tire-recycling-industry-market-trends-analysis.

González-Taboada, I., González-Fonteboa, B., Martínez-Abella, F., & Carro-López, D. (2016a). Study of recycled concrete aggregate quality and its relationship with recycled concrete compressive strength using database analysis. *Materiales de Construccion, 66*(323). https://doi.org/10.3989/mc.2016.06415

Grand View Research. (2020). *Aggregates market size, share & trends analysis report by type (gravel, crushed stone, sand), by application (concrete, road base & coverings), by region (APAC, MEA), and segment forecasts, 2020 – 2027. GVR-3-68038-458-1*. Grand View Research. https://www.grandviewresearch.com/industry-analysis/aggregates-market.

Grubeša, I. N., Barisic, I., Fucic, A., & Bansode, S. S. (2016). *Characteristics and uses of steel slag in building construction*.

Guillem, M., Fos, C., Blanco-Varela, M., & Husillos, N. (2010). Utilización de lodos de EDAR tratados con cal como materia prima alternativa en la fabricación de clínker. *Cemento Hormigón, 936*, 66–76.

Hansen, T. C. (1992). Recycling of demolished concrete and masonry. In *Report of technical committee 37 – DRC, demolition and reuse of concrete (RILEM)*.

Helaine, D. (2000). *Les boues de curage des rèseaux dassaissement; caracteristiques et techniques de traitement/calorisation*. TMS.

IAWG. (1994). *An international perspective on municipal waste incineration residue characterization and management. Int. Conf. Environmental Implications of Construction Materials and Technology Developments, Maastricht*. The International Ash Working Group.

Javed, S., & Lovell, C. (1994). *Use of foundry sand in highway construction. Joint Highway Report No. C-36-50N*. Department of Civil Engineering, Purdue University.

Johnson, C. K. (1981). Phenols in foundry waste sand. *Modern Casting, 71*(1), 48–49.

Joshi, R. C. (1970). *Experimental production of synthetic fly ash from kaolinite.*

de Juan, M. S., & Gutiérrez, P. A. (2009). Study on the influence of attached mortar content on the properties of recycled concrete aggregate. *Construction and Building Materials, 23*(2), 872–877. https://doi.org/10.1016/j.conbuildmat.2008.04.012

Kasai, Y. (1994). Guidelines and the present state of the reuse of demolished concrete in Japan". Demolition and reuse of concrete and masonry. In *Proceedings of the third international RILEM symposium* (pp. 93–104).

Kazemi Najafi, S. (2013). Use of recycled plastics in wood plastic composites - a review. *Waste Management, 33*(9), 1898–1905. https://doi.org/10.1016/j.wasman.2013.05.017

Khatib, Z. K., & Bayomy, F. M. (1999). Rubberized Portland cement concrete. *Journal of Materials in Civil Engineering, 11*(3), 206–213. https://doi.org/10.1061/(ASCE)0899-1561(1999)11:3(206)

Kikuchi, M., Yasunaga, A., & Ehara, E. (1993). The total evaluation of recycled aggregate and recycled concrete. In *Proc. Third int. RILEM symp. Demolition reuse concr. Mason* (pp. 425–436).

Koppelman, L.E., and Tanenbaum, E.G. (1993) The potential for beneficial use of ashes from waste-to-energy facilities: volume 4, engineering properties New York State Energy Research and Development Authority, USA.

Kosmatka, S., & Wilson, M. (2016). *Design and control of concrete mixtures* (16th ed.). The Portland Cement Association.

Luz, A. P., Tomba Martinez, A. G., López, F., Bonadia, P., & Pandolfelli, V. C. (2018). Slag foaming practice in the steelmaking process. *Ceramics International, 44*(8), 8727–8741. https://doi.org/10.1016/j.ceramint.2018.02.186

Maghool, F., Arulrajah, A., Horpibulsuk, S., & Mohajerani, A. (2020). Engineering and leachate characteristics of granulated blast-furnace slag as a construction material. *Journal of Materials in Civil Engineering, 32*(7). https://doi.org/10.1061/(ASCE)MT.1943-5533.0003212

Mangialardi, T., Paolini, A. E., Polettini, A., & Sirini, P. (1999). Optimization of the solidification/stabilization process of MSW fly ash in cementitious matrices. *Journal of Hazardous Materials, 70*(1–2), 53–70. https://doi.org/10.1016/S0304-3894(99)00132-6

Manso, J., & Marquínez, J. (2006). Investigación de nuevos usos de las escorias de horno eléctrico de arco (EAF). La oportunidad de los hormigones. *Hormigón Y Acero, 241*, 51–57.

Manz, O. E. (1999). Coal fly ash: a retrospective and future loo. *Fuel, 78*, 133–136. https://doi.org/10.1016/S0016-2361(98)00148-3

Mark, H. F. (1985). *Encyclopedia of polymer science and engineering.*

Martin, J. W. (2006) (pp. 133–158). Elsevier BV. https://doi.org/10.1533/9781845691608.2.133

Mehta, P. K. (1988). Standard specifications for mineral admixtures – an overview. *ACI SP, 91*, 637–658.

MNR. (1992). *Report prepared by John emery geotechnical engineering limited for aggregate and petroleum resources section, Ontario Ministry of Natural Resources, Queen's printer for Ontario.*

MOEE. (1993). *Report prepared by John emery geotechnical engineering limited for Ontario Ministry of the Environment and Energy and the Canadian Foundry Association, Queen's Printer for Ontario.*

Monosi, S., Ruello, M. L., & Sani, D. (2016). Electric arc furnace slag as natural aggregate replacement in concrete production. *Cement and Concrete Composites, 66*, 66–72. https://doi.org/10.1016/j.cemconcomp.2015.10.004

Nagataki, S., Gokce, A., Saeki, T., & Hisada, M. (2004). Assessment of recycling process induced damage sensitivity of recycled concrete aggregates. *Cement and Concrete Research, 34*(6), 965–971. https://doi.org/10.1016/j.cemconres.2003.11.008

NBR-15.116. (2005). *Recycled aggregate of solid residue of building constructions – requirements and methodologies. Brazilian Norm.*

O'Brien, J. (2016). Global aggregates industry under the spotlight. *Modern Quarrying*, 14–15.

Özkan, Ö., Yüksel, I., & Muratoğlu, Ö. (2007). Strength properties of concrete incorporating coal bottom ash and granulated blast furnace slag. *Waste Management, 27*(2), 161–167. https://doi.org/10.1016/j.wasman.2006.01.006

Pal, S. C., Mukherjee, A., & Pathak, S. R. (2003). Investigation of hydraulic activity of ground granulated blast furnace slag in concrete. *Cement and Concrete Research, 33*(9), 1481–1486. https://doi.org/10.1016/S0008-8846(03)00062-0

Park, S. B., Lee, B. C., & Kim, J. H. (2004). Studies on mechanical properties of concrete containing waste glass aggregate. *Cement and Concrete Research, 34*(12), 2181–2189. https://doi.org/10.1016/j.cemconres.2004.02.006

Pellegrino, C., & Gaddo, V. (2009). Mechanical and durability characteristics of concrete containing EAF slag as aggregate. *Cement and Concrete Composites, 31*(9), 663–671. https://doi.org/10.1016/j.cemconcomp.2009.05.006

Persistence Market Research. (2015). *Construction aggregates market. Global market study on construction aggregates: Supported by infrastructure investments to reflect over 5% CAGR by 2021.* Persistence Market Research. https://www.persistencemarketresearch.com/market-research/construction-aggregates-market.asp.

Pfaender, H. G. (1996). *Schott guide to glass.*

Raghavan, D., Huynh, H., & Ferraris, C. F. (1998). Workability, mechanical properties, and chemical stability of a recycled tyre rubber-filled cementitious composite. *Journal of Materials Science, 33*(7), 1745–1752. https://doi.org/10.1023/A:1004372414475

RILEM. (1992). *Recycled aggregates and recycled aggregate concrete. Recycling of demolished concrete and masonry. RILEM Report 6. (First). E&FN Spon 2-6 Boundary Row.*

RILEM TC 121-DRG. (1994a). Specifications for concrete with recycled aggregates. *Materials and Structures, 27*, 557–559. https://doi.org/10.1007/BF02473217

Rodríguez, G., Medina, C., Alegre, F. J., Asensio, E., & Sánchez de Rojas, M. I. (2015). Assessment of construction and demolition waste plant management in Spain: In pursuit of sustainability and eco-efficiency. *Journal of Cleaner Production, 90*, 16–24. https://doi.org/10.1016/j.jclepro.2014.11.067

Rubber Manufacturer's association.(2000).

Sasai, R., Kubo, H., Kamiya, M., & Itoh, H. (2008). Development of an eco-friendly material recycling process for spent lead glass using a mechanochemical process and Na2EDTA reagent. *Environmental Science and Technology, 42*(11), 4159–4164. https://doi.org/10.1021/es0719576

Sekaran, A., Palaniswamy, M., & Balaraju, S. (2015). A study on suitability of EAF oxidizing slag in concrete: An eco-friendly and sustainable replacement for natural coarse aggregate. *The Scientific World Journal*, 1–8. https://doi.org/10.1155/2015/972567

Shi, C. (2004). Steel slag - its production, processing, characteristics, and cementitious properties. *Journal of Materials in Civil Engineering, 16*(3), 230–236. https://doi.org/10.1061/(ASCE)0899-1561(2004)16:3(230)

Shinogi, Y., & Kanri, Y. (2003). Pyrolysis of plant, animal and human waste: Physical and chemical characterization of the pyrolytic products. *Bioresource Technology, 90*(3), 241–247. https://doi.org/10.1016/s0960-8524(03)00147-0

Siddique, R. (2008). Waste materials and by-products in concrete. *Recycled/Waste Plastic, 94*, 93–120.

Siddique, R., & Kaur, D. (2012). Properties of concrete containing ground granulated blast furnace slag (GGBFS) at elevated temperatures. *Journal of Advanced Research, 3*(1), 45–51. https://doi.org/10.1016/j.jare.2011.03.004

Silva, R. V., De Brito, J., & Dhir, R. K. (2014). Properties and composition of recycled aggregates from construction and demolition waste suitable for concrete production. *Construction and Building Materials, 65*, 201–217. https://doi.org/10.1016/j.conbuildmat.2014.04.117

Skaf, M., Manso, J. M., Aragón, Á., Fuente-Alonso, J. A., & Ortega-López, V. (2017). EAF slag in asphalt mixes: A brief review of its possible re-use. *Resources, Conservation and Recycling, 120*, 176–185. https://doi.org/10.1016/j.resconrec.2016.12.009

Topçu, I. B., & Canbaz, M. (2004). Properties of concrete containing waste glass. *Cement and Concrete Research, 34*(2), 267–274. https://doi.org/10.1016/j.cemconres.2003.07.003

Tsakiridis, P.E., Papadimitriou, G.D., Tsivilis, S., Koroneos, C. (2008) Utilization of steel slag for Portland cement clinker production, *J. Hazard. Mater. 152(*2), 805–811, ISSN 0304-3894, https://doi.org/10.1016/j.jhazmat.2007.07.093.

UBA. (2012). *Umwelt Bundesamt*. https://www.umweltbundesamt.de.

Valls, S., Kuchinow, V., Vázquez, E., & Yague, A. (2007). Utilización de lodo seco de depuradora de aguas residuales como adición en prefabricados. In *III Jornadas Técnicas de estión de Sistemas de Saneamiento de Aguas Residuales*.

Vincke, J., & Rousseau, E. (1994). Recycling of construction and demolition waste in Belgium: Actual situation and future evolution." Demolition and reuse of concrete and masonry. In *Proceedings of the third international RILEM symposium* (pp. 57–69).

WB. (2016). *World bank*. https://blogs.worldbank.org/water.

WEF. (2011). *The world economic forum*. https://www.weforum.org/.

WSDTED. (1993). *Glass feedstock evaluation project*. Washington State Department of Trade and Economic Development.

Yagisita, F., Sano, M., & Yamada, M. (1993). Behaviour of reinforced concrete beams containing recycled coarse aggregate." Demolition and reuse of concrete and masonry. In *Proceedings of the 3rd international RILEM symposium* (pp. 331–342).

Zevenbergen, C., & Comans, R. (1994). Geochemical factors controlling the mobilization of major elements during weathering of MSWI bottom ash. Environmental Aspects of Construction with Waste Materials. In *Science*. Elsevier.

Zoorob, S. E., & Suparma, L. B. (2000). Laboratory design and investigation of the properties of continuously graded Asphaltic concrete containing recycled plastics aggregate replacement (Plastiphalt). *Cement and Concrete Composites, 22*(4), 233–242. https://doi.org/10.1016/S0958-9465(00)00026-3

Quality improvement of recycled aggregate

5

Mahfooz Soomro [1], *Vivian W.Y. Tam* [1], *Mitsuhiro Shigeishi* [2], *Ni Nyoman Kencanawati* [3], *Takao Namihira* [4] *and Katarzyna Kalinowska* [5]
[1]School of Engineering, Design and Built Environment, Western Sydney University, Penrith, NSW, Australia; [2]Department of Civil and Environmental Engineering and Architecture, Kumamoto University, Japan; [3]Civil Engineering, University of Mataram, Mataram, Indonesia; [4]Institute of Industrial Nano Materials, Kumamoto University, Japan; [5]Faculty of Civil and Environmental Engineering, Bialystock University of Technology, Poland

5.1 Introduction

By: Mahfooz Soomro and Vivian W.Y. Tam.

Construction industry is a major consumer of resources. Increasing activities in infrastructure development, renovation, urbanization and industrialization around the globe are the key drivers, which increase the demand for virgin aggregate consumption in construction projects and it is growing year on year.

Concrete is widely used as a basic material for construction and infrastructure. About 500 million tons of concrete were produced in Japan around 1990s. In recent years, around 35 million tons of demolished concrete are being generated every year. Due to the regulations in Japan, not a little concrete is illegally dumped or mixed with construction soil that is not properly treated as waste. At present, 95% of the concrete is recycled using cascade recycling and subsequently reused as a low-quality road subbase. Few advanced processing to reproduce high-quality recycled aggregate from demolished concrete are carried out at present. Soon, a substantial amount of concrete will reach its end of life from the construction undertaken during the economic growth of 1960 and 1970s, and the generation of demolished concrete is expected to rapidly increase and may create extremely serious environmental problems (Shima et al., 2005b).

The demolition of concrete structures produces a large amount of waste and most of it end-up in the landfills. It will lead to a potential problem of filling up the landfills soon because more of the old and dilapidated concrete structures need to be demolished (Azúa et al., 2019). The waste is also generated from the construction process. The construction sector in Europe generates 50% of the waste originating from the construction and demolition waste (CDW) of all the total waste. Until now, concrete and mortar are the most popular materials in the construction world. In addition, for every ton of material mined, on average more than 85% is wasted (Villoria-Sáez et al., 2020). CDW has become one of the major problems in the construction industry as it directly impacts the environment. It is estimated that around 35% of CDW ends up

Recycled Concrete. https://doi.org/10.1016/B978-0-323-85210-4.00004-7

in landfills globally; therefore, an effective CDW management process is required to minimize the negative impacts on the environment (Kabirifar et al., 2020). This issue has been a concern in the development of a national policy of many countries worldwide to prevent large landfills by regulatory measures to enhance recycling and recovery (Villoria-Sáez et al., 2020). Especially in developing countries, more actual efforts are required even if there is a national waste regulation; however, the implementation of regulations on the local level are deficient and sparse (Meidiana & Gamse, 2010).

Another critical issue regarding the use of concrete is the depletion of fresh natural aggregate, which is one of the primary materials in concrete manufacturing. However, the truth is that there are still plentiful natural resources available for the supply of aggregate to the construction industry; meanwhile, attention should be paid to the long-term negative environmental impacts caused by the extraction of nonrenewable natural resources. In addition, the process includes extensive deforestation, topsoil abrasion, air and water pollution (Pacheco-Torgal et al., 2020). Materials are taken and produced from a considerable distance from where the project site is built. In addition to the extraction time, transporting materials from the production site to the project site also harms the environment, because transportation requires fuel and contributes pollutants to the atmosphere. The natural aggregates require an exceedingly long time to form, and if the level of uptake is remarkably high, the level of availability of natural aggregate will be lower and depleted (Calkins, 2008; Ismail et al., 2013).

Furthermore, concrete as a construction material affects global warming by adding CO_2 to the atmosphere. All these processes not only cause material extinction, environmental damage, emissions of CO_2 to the air but also will have an impact on human health, indirectly. Thus, alternative materials that are more environmentally friendly and sustainable are urgently needed at this time (Calkins et al., 2008; Ismail et al., 2013; Kibert et al., 2000).

Using recycled aggregate as replacement of natural aggregate is an example of efforts to solve the problem (Assi et al., 2018; Chen et al., 2019; Kenai et al., 2018). For this purpose, several criteria have been studied and researched (Martínez-Lage et al., 2020). Recycling materials from demolished building waste are an alternative to reduce the exploitation of depleted natural aggregates, and also overcome environmental problems created by the landfills and this solution provides economic benefits.

Rapid infrastructure development demands huge volumes of varying types of construction materials. Recycled aggregate can be obtained quickly from CDW and post-disaster building debris. Some countries already have regulations in using the recycled aggregate as construction material. Aggregate recycling centers around major cities have been established. To reduce transportation costs, mobile crushing and screening plants (track-mounted) are also used in aggregate production by construction waste recycling (Kenai et al., 2018; Miraldo et al., 2021; Tam et al., 2013).

The global virgin aggregate consumption almost doubled from 21 billion tons in 2007 to 40 billion tons in 2014. The market demand reached to 43.3 billion tons in 2016 and is expected to reach 62.9 billion tons by 2024 (Jim O'Brien et al., 2019).

Based on the estimates of 2017, countries with the highest consumption of virgin aggregate are China (38.16%), India (9.54%), Asian region (excluding China and India) (15.26%), Japan (0.67%), Oceania (0.76), Australia (0.38%), North America

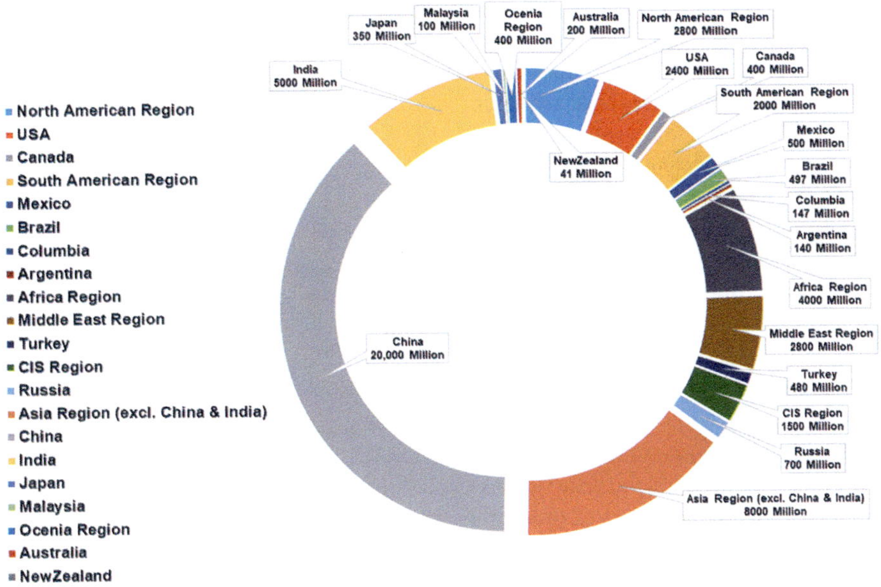

Figure 5.1 Global consumption of natural aggregate (Tam et al., 2018).

region (5.34%), USA (4.58%), Canada (0.76%), South America region (3.82%), Mexico (0.95%), Brazil (0.948%), Africa region (7.63%), Middle East region (5.34%), Turkey (0.92%), CIS (2.86%) and Russia (1.33%) as depicted in Fig. 5.1 (Tam et al., 2018).

The depletion of natural resources at such a vast scale and the growing awareness of sustainable waste management practices in the developed and developing countries, environmental organizations are giving an ever-increasing heed to conserve natural resources by increasing the consumption of recycle aggregate in nonstructural, precast and structural concrete in the construction projects.

Numerous studies have been conducted, focusing primarily on improving the quality of recycled concrete aggregate (RCA) to replace natural aggregate in conventional concrete and promote the sustainable development of construction industry. It looks a viable alternative in the production of structural and nonstructural concrete. It reduces depletion and helps in conserving the natural resources, avoids waste disposal in landfills and preserves the environment (Noguchi et al., 2015; Shaban et al., 2019; Tam et al., 2016).

Recycled aggregate, produced only by crushing waste concrete to coarse or fine aggregate, cannot have the same quality as natural aggregate because of high water absorption and low density of recycled aggregate. This is due to the presence of old residual mortar on the surface of the recycled aggregate.

The examination of the microstructure and interfacial transition zone (ITZ) of recycled aggregate has revealed that due to the existence of old mortar, recycled aggregate shows a higher porosity and lower solid profile compared to the natural aggregate. This leads to lower mechanical properties of recycled aggregate concrete (Djerbi et al.,

2018). Thus, the surface of the recycled aggregate must be clean and clear from the old mortar so that its quality is increased; and as a result, the concrete produced with recycled aggregate can also show an increase in its quality. Numerous studies on improving the quality of recycled aggregate are being carried out worldwide.

To separate the old mortar from the aggregate surface, several studies using different methods have been carried out. Pulsed power technology has been applied to produce high-quality aggregates (Bluhm et al., 2006; Kencanawati et al., 2013; Narahara et al., 2007). The Marx generator was used as a source of pulsed power. Several parameters such as the number of discharges, the amount of voltage, and the discharge medium were studied to produce the optimum aggregate quality (Narahara et al., 2007). An effort to remove old mortar, by presoaking it in chemical solution was carried out by (Tam et al., 2007). Recycled aggregate was presoaked in solutions of HCL, H_2SO_4, and H_3PO_4 with a concentration of 0.1 mol, for 24 hours. After that, it was kept immersed in distilled water for 24 hours, to remove the traces of acid from the aggregate. An improvement was observed in the quality of both, recycled aggregate and the concrete made from treated aggregate. Presoaking in the chemical solutions can remove residual mortar without making the recycled aggregate hazardous.

With the purpose of reducing the amount of the residual mortar on the surface of recycled aggregate, an autogeneous cleaning process was introduced. The concrete produced from this kind of recycled aggregate reached the compressive strength with a slight reduction of 8.9% compared to the strength of conventional natural aggregate concrete (Pepe et al., 2014).

The improved quality of recycled aggregate concrete has been reviewed by two approaches: (a) treatment methods to improve the recycled aggregate quality by removal of attached mortar and (b) treatment methods to improve the recycled aggregate quality by strengthening and fortification of the old residual mortar. The first method was implemented by removing the attached mortar, and the second method was carried out by adding pozzolan or emulsion polymer to fill the microcracks and by surface coating (Mistri et al., 2020) or by improving the mixing process (Tam et al., 2005).

A combination of thermal-mechanical-chemical processes has been proposed to produce a higher quality of recycled coarse aggregate (RCA) (Kencanawati et al., 2015, 2017). The treatment process is continuously carried out sequentially to obtain a better quality of RCA. This process is quite capable of producing RCA with better quality than the RCA obtained by the conventional methods (Xiao et al., 2005). The process consists of heating followed by grinding and the last process of soaking with chemical solutions. This treatment process does not require high-power equipment. Therefore, it is suitable for application in developing countries.

The concept of complete recycle of construction and demolition (C&D) waste by producing high quality recycled aggregate with either the removal of attached mortar fraction or by reinforcing the residual porous mortar by filling its pores and the void spaces around its ITZ. After treatment, the improved quality aggregate can be utilized in conventional ready-mix concrete, in precast concrete and in structural concrete. The mortar removed from aggregate and ground to powder or as fine aggregate can be used either as cement substitution material or as moderate strength cement after thermal activation or as precursor raw material for alkali-activated concrete, as soil

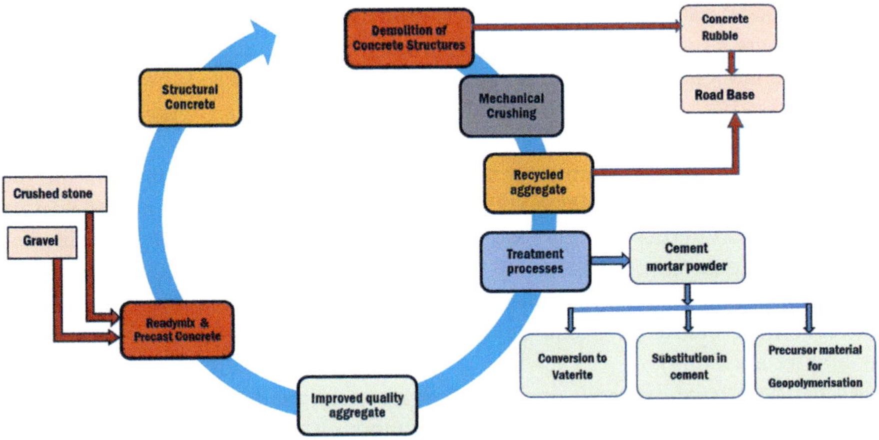

Figure 5.2 Schematics of concrete recycling system.
(Modified Shima, H., Tateyashiki, H., Matsuhashi, R., & Yoshida, Y. (2005a). An advanced concrete recycling technology and its applicability assessment through input-output analysis. Journal of Advanced Concrete Technology, 3(1), 53e67. https://doi.org/10.3151/jact.3.53).

stabilizer/soil nutrient (plant fertilizer) or to produce concrete admixture (Vaterite) a $CaCO_3$ polymorph (Chen et al., 2017) (Fig. 5.2).

The recycled concrete aggregate derived from C&D waste is composed of aggregate partially covered with residual porous cement mortar, inhomogeneous, lower in density and with porous and weak ITZ between mortar and aggregate. The volume of residual mortar around the recycled concrete aggregate varies from 25% to 60% depending on the size of aggregate (Corinaldesi et al., 2010). Some studies have reported about 20% cement mortar attached around the aggregate having 20–30 mm grain size (Nassar et al., 2012).

The type of crushing method has considerable influence on the shape and texture of RCA and the development of tiny cracks in the mortar and ITZ. It is widely accepted that in RCA, the presence of porous mortar and weaker ITZ, which is about 40–50 μm in thickness between mortar and natural aggregate plays important role in reducing the bond strength between RCA and fresh concrete produced from it (Xiao et al., 2013). The ITZ is composed of, much of calcium hydroxide (CH), little of calcium-silicate hydrate (C–S–H) (with similar composition as in the bulk phase), partially with Ettringite (Aft) and mono-sulfoaluminate (Afm) phases (similar as in bulk phase) and has a relatively high capillary porosity (Kjellson et al., 1998; Scrivener et al., 2004).

The angular shape, rough texture and the porosity of RCA compared to natural aggregate (NA) together with its high water absorption, adversely affects its quality, hence it is not favored for use in ready mix, precast or structural concrete (Evangelista et al., 2015; Hongzhu et al., 2010). Since the adhered mortar is porous, the water absorption of recycled concrete aggregate is considerably higher and can range from 3% to 10% as compared to $\leq$1% to 5% for natural aggregate (Gomez-Soberon et al., 2002; Narud et al., 1983).

Compared to natural aggregate, the tendency of RCA for higher water absorption results in lower workability in fresh concrete and produces higher shrinkage and higher creep in the hardened concrete. This results in lower strength and lower durability of the concrete. These impediments limit the wider use of recycled aggregate in civil engineering applications (Etxeberia et al., 2006).

Because of various impediments mentioned above, the improvement in the quality of RCA has become imperative. The quality of RCA must be improved to the extent that its properties compare well with those of NA and comply with the specified requirements for bulk density, water absorption, compressive and flexural strengths, regulated by various countries. It has been observed that the RCA with mortar content below 44% can be used in the production of structural concrete (de Juan et al., 2009).

Researchers have explored various (separate or in combination/hybrid) treatment approaches to produce high quality RCA by adopting two main pathways;

(i) Removal of adhered old cement mortar from the aggregate grains
(ii) Fortification of attached cement mortar to make it stronger, less porous and hydrophobic.

In addition, the natural coarse aggregate (NCA) have been gradually exhausted, which is depleting the natural resources and filling up the landfills, creating environmental issues. Therefore, the development of new recycling technology for waste concrete rubble has become of paramount importance in Japan. One of the conventional treatments to produce higher quality recycled aggregate from the waste concrete rubble is based on the rubbing method and some research and development have been carried out in Japan (Ogawa et al., 2010). "Taiheiyo Cement Corporation" has developed TRASS, the Taiheiyo Recycled Aggregate Solution System, which incorporates a dry screw grinding system with twin cones that can produce quality aggregate from concrete chunks without removing the cement paste and thus preventing damage to the aggregate. With this treatment, aggregate can be produced of required size specifications. The machines have low vibration and noise emissions compared with rod mills or impact crushers and have energy-saving small motor loads. Furthermore, the system is sectional and hence can be transported easily (The Cement Sustainability Initiative: Recycling Concrete, 2009). Another system, the "Aurora Max" is a horizontal rod mill rubbing machine, which was developed by Kurimoto Ltd. to produce products such as fine aggregate for concrete (Fumoto & Yamada, 2002). Also, Mitsubishi Materials Corporation commercialized a high-quality aggregate product, "Diagate," which is produced by the heat and rubbing method (Shima et al., 2003). However, the rubbing or the heating and rubbing method also have quality restrictions on the reproduced material since the applied high stresses result in extensive degradation of not only the mortar but aggregate as well. In addition, the recovery rate of high-quality aggregate is particularly low, and much of the original waste concrete ends-up in the form of fine particles.

In this chapter, only the treatments applied to remove the adhered mortar on RCA have been presented, due to the limited number of pages allowed for each chapter.

5.2 Quality improvement of recycled aggregate using thermal-mechanical-chemical process

By: Ni Nyoman Kencanawati and Mitsuhiro Shigeishi.

5.2.1 *RCA production and treatment*

According to (Purkiss et al., 2007), when concrete material is exposed to the temperature of 200°C, the excess water, which lies in microstructures, mostly in the boundary between cement paste and aggregate explodes due to pore hydro-pressure. The explosion weakens the bond between two materials and leads to mortar delamination. This mechanism is adopted to weaken the bond between the aggregate surface and mortar. Then, the grinding process is needed for separating the attached mortar. The mortar separation is achieved more effectively by rubbing the aggregate surfaces with each other or between aggregate surfaces and iron balls. Soaking the product in an acid solution helps the cleaning of old mortar continuously (Tam et al., 2007). Fig. 5.3 explains the treatment processes of thermal (heating), mechanical (grinding), and chemical (acid solution soaking) to reduce the amount of old mortar on the aggregate surface.

5.2.1.1 *Thermal-mechanical process*

The first combination is thermal-mechanical by heating and grinding process. Lumps of concrete waste were crushed and reduced to the size of around 10–20 cm in diameter. These lumps were put in an oven and heated at a temperature of 100°C for 24 h. Then the lumps were put in a grinding machine for 500 revolutions. The rubbing process was done by the rotation of the machine and the impact of iron balls inside the machine. This type of treatment is regarded as heating-grinding to produce high grade recycled coarse aggregate (HG RCA).

5.2.1.2 *Thermal-mechanical-chemical process*

The second treatment method is a combination of heating, grinding, and acid soaking process. The first of two processes are the same as mentioned above, and treatment is

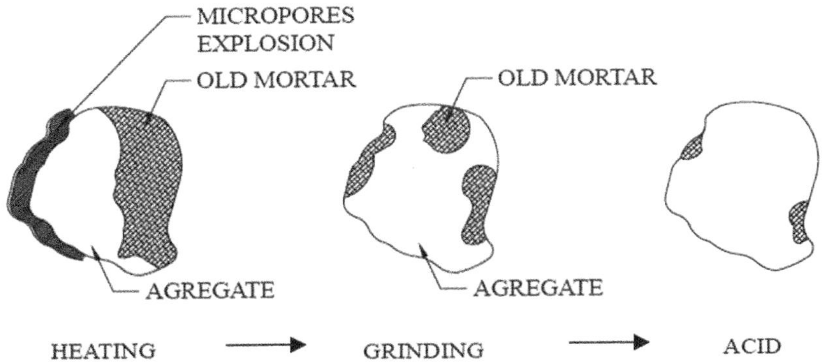

Figure 5.3 RCA production and treatment process.

further continued by soaking the aggregate in an acid solvent, sulfuric acid (H_2SO_4), for 24 h. Sulfuric acid helps release the attached mortar around the aggregate surface by converting it into a water-soluble coagulant. This treatment type is regarded as heating-grinding-acid soaking to produce high quality recycled coarse aggregate (HGA RCA).

5.2.1.3 Physical properties of recycled aggregate

Based on visual examination, the presentations illustrated in Fig. 5.4 (Kencanawati et al., 2017), both HG RCA and HGA RCA surfaces are clean from the old residual mortar. Although the old mortar has separated from the aggregate surface, Fig. 5.4 show that the appearance of HGA RCA is considerably cleaner from cement paste than the appearance of HG RCA. Soaking in chemical solutions causes detachment of cement paste more effectively, hence HGA RCA displays cleaner surface.

To meet the regulatory requirements, the quality of aggregate is further investigated for density, water absorption, fineness modulus, and sieve analysis. Almost equivalent properties are obtained compared to natural coarse aggregate, indicating the improved quality of treated recycled aggregate. Fig. 5.5 shows the comparison of the physical properties of recycled coarse aggregate with natural coarse aggregate.

5.2.1.4 Mechanical properties of RCA concrete

Study was conducted to assess the comparative performance of RCA concrete containing the two types of treated recycled coarse aggregate (HG and HGA) to the specimens produced from natural coarse aggregate. The water-cement ratio was kept at 48% and the specimens were of cylindrical shape. The mixture proportions used are shown in Table 5.1 (Kencanawati et al., 2017).

The mechanical properties of compression, flexural load and splitting-tensile of the RCA concrete were obtained along with those of natural aggregate concrete (NAC),

Figure 5.4 Recycled coarse aggregate.

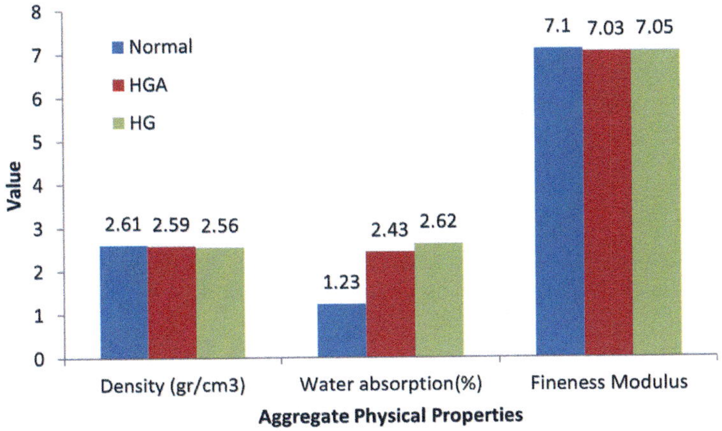

Figure 5.5 Comparison of the physical properties of natural and treated aggregate.

Table 5.1 Concrete mixture proportion.

Concrete type	Concrete components			
	Cement (kg/m^3)	**Water (kg/m^3)**	**Sand (kg/m^3)**	**Gravel (kg/m^3)**
Normal coarse aggregate concrete	427	205	675	1013
HG recycled coarse aggregate concrete	427	205	667	1001
HGA recycled coarse aggregate concrete	427	205	671	1007

which are presented in Table 5.2 (Kencanawati et al., 2017). In general, the mechanical properties of conventional concrete are slightly higher than the concrete specimens containing treated coarse aggregate. Furthermore, HGA RCA concrete displays better mechanical properties than those of HG RCA concrete. The difference in the mechanical properties of the two treated RCA is caused due to their difference in quality. The continuous treatment process by immersing in acid solution further enhances the mechanical properties of HGA concrete. However, the improvement in mechanical properties is not significantly different between the two types; hence both, HG RCA or HGA RCA have the potential to be utilized in producing concrete.

Table 5.2 Mechanical properties of concrete.

Mechanical properties (average)	NCA concrete	HGA concrete	HG concrete
Compressive strength (MPa)	41.77	39.79	39.32
Modulus of elasticity (MPa)	30,280	29,643	29,472
Splitting-tensile strength (MPa)	4.57	4.49	4.31
Flexural strength (MPa)	6.29	5.73	5.58

Surface hardness of RCA concrete

To ensure the surface hardness of concrete using RCA, the surface hardness test is carried out using the rebound hammer test (Malhotra & Carino, 2004). The surface hardness of the concrete is symbolized by the rebound number listed on the hammer after the test is carried out. As comparison, the surface hardness of the concrete using natural aggregate is also carried out. A total of 150 test points were conducted with the hammer test.

As illustrated in Fig. 5.6, in terms of the average value of the rebound number, NAC reaches the highest value, followed by HGA RCA concrete, while the lower value is of HG RCA concrete. The average rebound number of HGA RCA concrete is 8% lower than the rebound number of NAC. At the same time, the rebound number of HG RCA concrete is 12% lower compared to the rebound number of NAC. Between HGA RCA concrete and HG RCA concrete, the difference in rebound numbers is only 4%. It may be emphasized that the surface hardness properties obtained from rebound numbers are not only related to strength but also to the durability properties associated with the abrasion resistance (Bungey & Millard, 1995).

Ultrasonic pulse velocity of RCA concrete

Ultrasonic pulse velocity (UPV) is measured on both, HGA RCA concrete and HG RCA concrete samples. For comparison, the UPV of NAC was also measured. The average value assessed from several tests, the highest UPV is seen in natural aggregate concrete of 4468.2 m/s. Comparatively, the value of HGA RCA concrete is lower by about 3%, while, for HG RCA concrete, this value is 8.5% lower than NAC. The velocity between HGA RCA and HG RCA concrete is not much different, around 5%, where HGA RCA concrete shows higher velocity than HG RCA concrete. Fig. 5.7 presents an overview of the ultrasonic pulse velocity in NAC and RCA concrete specimens.

The velocity of ultrasonic waves can indicate the level of density in the concrete material. In concrete of good quality, the value of the wave propagation speed shows a more excellent value because of the denser characteristics of the concrete. On the

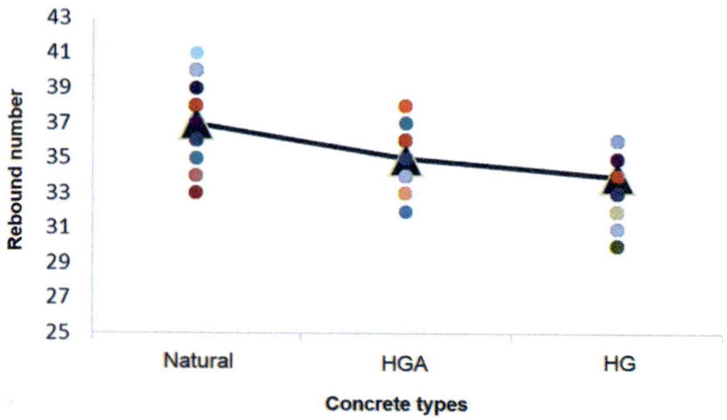

Figure 5.6 Average rebound number of NCA and RCA concrete.

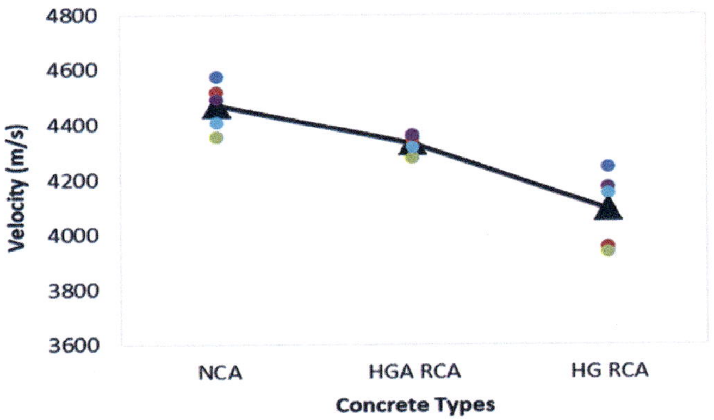

Figure 5.7 Ultrasonic pulsed velocity of concrete.

other hand, low-strength concrete is usually porous so the resulting wave velocity is lower (Jain et al., 2013; Kencanawati et al., 2018). The HGA RCA concrete velocity shows a slightly higher value compared to HG RCA concrete because fewer pores exist due to the better removal of old mortar, providing a better bonding of the new cement paste with the interfacial transition zone (ITZ) of HGA RCA.

5.2.2 Pulsed power technology application

By: Takao Namihira and Mitsuhiro Shigeishi.

On the other hand, the destruction of concrete structures by lightning strikes have been reported for a long time. It is clear from these strikes that lightning has sufficient energy to destroy concrete. Based on this idea, the "Pulsed Power," which can also be called manmade lightning, has been used for smashing and recycling concrete, shattering of solid material (Bluhm et al., 2000) and drilling of wells, as well. (Timoshkin et al., 2004). This new recycling technique uses pulsed power discharges into waste concrete lumps immersed in water, which separates residual mortar from aggregate due to lightning effect and reproduces high quality coarse aggregate (Inoue et al., 2009, 2012). This method is based on two principles: (a) the formation of discharge pass in the concrete and (b) the formation of shockwave due to heat expansion of the discharge pass, during the electrical breakdown of concrete underwater.

Fig. 5.8 shows (a) the schematic diagram and (b) the still image of a typical pulsed power source for the discharge treatment of concrete lumps. The generator of "Pulsed Power" is based on Marx circuit, which can easily deliver higher voltage pulses. Fig. 5.8c indicates typical waveforms of the applied voltage and the discharge current into the treatment chamber for concrete lumps. The peaks of the voltage and the current reached up to 400 kV and 20 kA, respectively. The electrical power is enough to induce the electrical discharge breakdown of concrete lumps immersed in water.

The typical reactor for discharge treatment to smash the big concrete chunks to smaller pieces is shown in Fig. 5.9a. The reactor consists of a point and a hemisphere-mesh electrode. For the discharge treatment, the positive voltage from

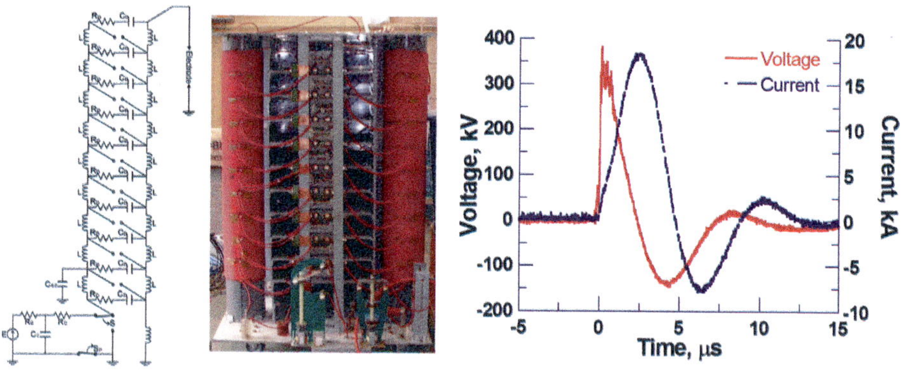

(a) (b) (c)

Figure 5.8 Typical pulsed power generator used in the experiments.

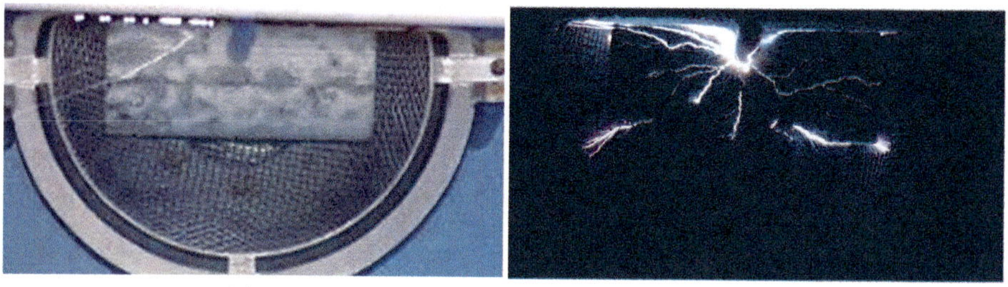

(a) (b)

Figure 5.9 (a) Treatment chamber using pulsed power and appearance of concrete in the discharge chamber (b) electrical discharge paths formed into the concrete lump.

the generator is applied to the point electrode while the hemisphere electrode is grounded. The concrete chunks/blocks are put on the hemisphere electrode for the discharge treatment. When the positive pulse voltage is applied to the concrete lumps/blocks, firstly the discharge occurs in air gaps in the concrete because the dielectric breakdown strength of air is lower than the concrete. The dielectric breakdown strength of the concrete becomes lower by the discharge occurrences in air gaps and the discharge paths are constructed in the concrete. Fig. 5.9b shows electrical discharge paths formed in the concrete lump. During the repetitive discharge treatment, the small pieces of treated concrete drop off from the hemisphere-mesh electrode.

Fig. 5.10 shows an X-ray CT image of the treated concrete by a single discharge application. In the grayscale of the figure, white and black color indicate the aggregate and the cement paste. From Fig. 5.8, it can be confirmed that the electrical discharge avoided the aggregate and traveled through the cement paste. This is because the breakdown voltage of the cement paste is lower than that of the aggregate. Therefore, it can be said that the discharge destroys the cement paste selectively. After full development of discharge paths between the point and the hemisphere electrodes, the large current flows in the discharge paths, and the discharge paths are rapidly thermalized

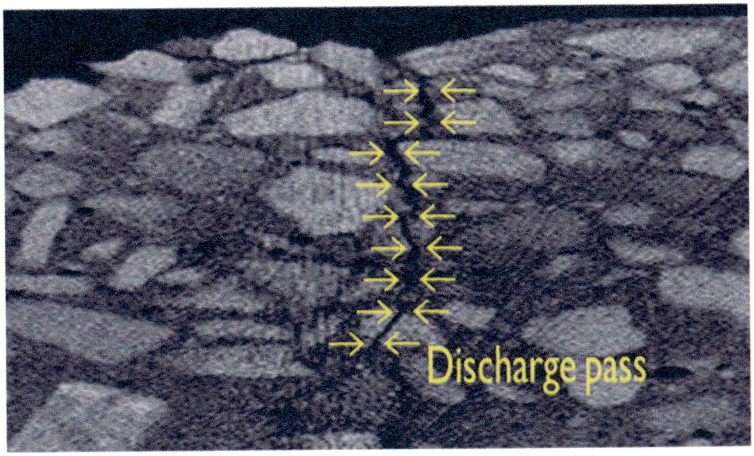

Figure 5.10 X-ray CT image of the current path by the electric breakdown in the treated concrete.

and expanded by the joule heating in microseconds range. The volume expansion of the discharge paths in a quite short time period induce the shockwave in concrete and the shockwaves propagate through the concrete. During the propagation of the shockwaves in concrete, it splits into the transmitted and the reflected shockwaves due to the difference of acoustic impedances of the aggregate and cement paste. As the result of the separation of the shockwaves, the tensile stress is generated at the boundary between the aggregate and the cement paste. The tensile stress is suitable for the removal of the cement paste from the aggregate.

Fig. 5.11 shows the appearance of pulse discharge treated recycled coarse aggregates, which have been collected as the residue on the hemisphere-mesh electrode after the repetitive discharge treatment.

From Fig. 5.11a, the progress of separation of aggregate and cement paste with increasing the number of the discharge applications and after 100 times discharge applications ("shots"), most of the residue remaining on the mesh electrode are higher quality recycled aggregate with almost all of the mortar removed. Fig. 5.11b shows a piece of the coarse aggregate after 100 pulse discharge applications. We can see that almost no cement paste remains on the surface of the recycled aggregate. It means that the recycled aggregates produced by pulsed discharge treatment have high quality. Fig. 5.11c shows the fine recycled aggregate, which dropped down through the mesh electrode during the discharge treatment. Even though, the visibility of Fig. 5.11c is low, the recycled fine aggregates consist of medium quality fine aggregate and the powdered cement paste mass.

5.2.2.1 RCA by pulsed power

Fig. 5.12 indicates the dependence of the quality of recycled coarse aggregate i.e., the water absorption ratio and the density in oven-dry condition, on the number of

Figure 5.11 Photographs of pulse discharge treated concrete.

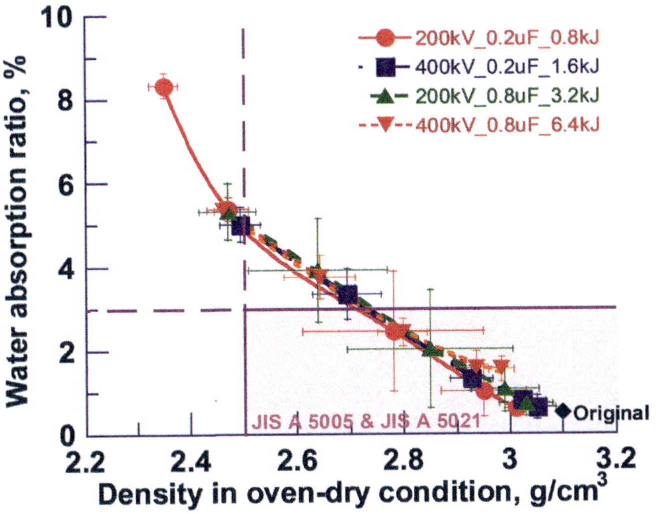

Figure 5.12 Quality of recycled aggregate treated by pulsed power discharging method.

discharge applications in the pulsed power treatment. In the experiments, four pulsed power generators were used, each one was operated on different energy per pulse from 0.8 to 6.4 kJ/pulse. Fig. 5.12 also shows that the quality of the recycled coarse aggregate improves and reaches the high-quality level, specified by Japanese Industrial Standards. From the results, it can be concluded that the performance of the pulsed power treatment for waste concrete have no dependence on the output energy from the generator in the range from 0.8 to 6.4 kJ/pulse.

Kumamoto University in Japan has developed a prototype plant called the "Pulsed Power Based Aggregate Recycler" (PULSER Mk. I) to process big concrete lumps. The processor is composed of the pulsed power source, the scale hopper and the discharge reactor, which can treat concrete repetitively till the high quality aggregate meeting the specifications of Japanese Industrial Standards is produced.

The performance of the *PULSER MK.1* is depicted in Fig. 5.13 it shows that 98% of the concrete lumps are disintegrated into coarse and fine recycled aggregate and just 2% of the concrete lumps converted to mortar powder. These facts demonstrate that the treatment by pulsed power discharge is one of the promising methods for reducing the concrete waste and producing high quality recycled aggregate.

The comparative performance between the pulsed power discharging method and the heating-rubbing method is tabulated in Table 5.3. The pulsed power discharging method shows higher performance in producing recycled aggregate with 59% lower energy consumption. Moreover, the fine powder discharged in pulse power treatment is just 2% compared to 52% discharged by heating and rubbing method.

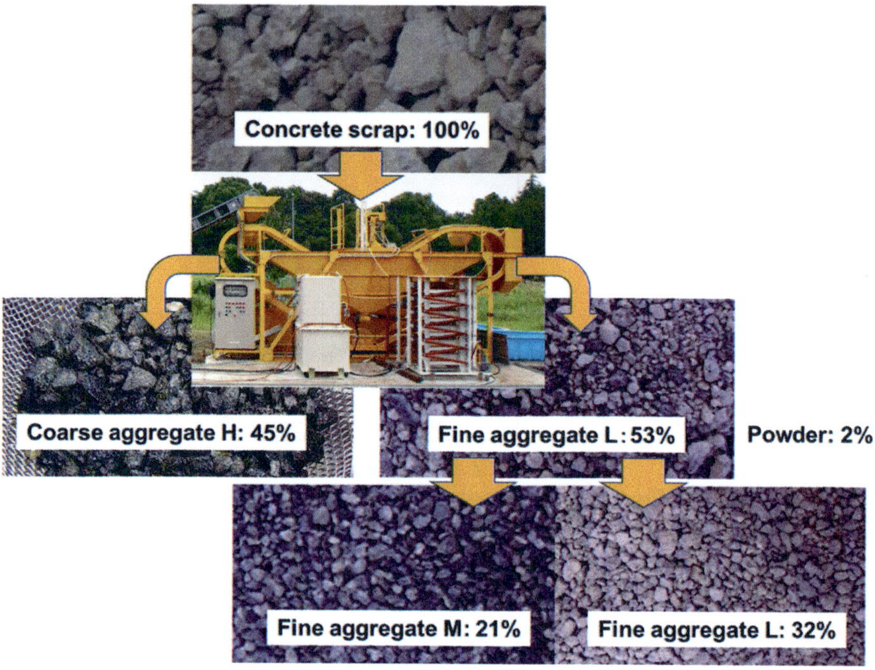

Figure 5.13 Performance of the prototype processor with the aid of pulsed power discharging method.

Table 5.3 Performance Comparison between the pulsed power discharging method and the heating-rubbing method.

Processing for 1 ton of concrete			By pulsed power discharging method	By heating and rubbing method
Energy consumption (MJ/ton)			**279**	**680**
Reclaimed products contents (%)	Coarse aggregate	(High-quality*)	45	34
	Fine aggregate	(High-quality*)	–	14
		(Medium-quality*)	32	–
		(Low-quality*)	21	–
	Fine powder	(By-product)	2	52

5.3 Alternative uses of removed residual mortar

By: Mahfooz Soomro and Vivian W. Y. Tam.

5.3.1 *Removed cement mortar conversion to "vaterite"*

A viable way to improve the quality of RCA is to remove the porous residual mortar from RCA. Cement mortar is composed of the hydration products (C—S—H gel, portlandite, AFm, and Aft phases). These can be dissolved by soaking the RCA in acid solutions and the removal efficiency of mortar from RCA depends on the factors such as mortar porosity, the nature of RCA, the choice of acid solution, the molar concentration of acids and the conditions of treatment e.g., volume of the material to be processed, duration of treatment etc.

Various studies (Akbarnezhad et al., 2011; Al-Bayati et al., 2016; Ismail and Ramli, 2013; Tam and Le, 2007; Wang et al., 2017) have been carried out using different acids, hydrochloric acid (HCl), sulfuric acid (H_2SO_4), phosphoric acid (H_2PO_4) and acetic acid (CH_3COOH) having different molar concentrations for the residual mortar removal. However, on one hand, the use of strong acids (hydrochloric acid and sulfuric acid) for mortar removal can introduce detrimental SO_4 and Cl^- ions to treated RCA, which may cause durability issues in concrete containing treated RCA and on the other, they can pose a threat to the safety of the industrial workers. Also, large amounts of water are needed to soak and rinse off the SO_4 and Cl^- ions from treated RCA. Moreover, disposal of this wash water as well as the residual treatment solution will create new environmental problems.

Acetic acid is free from these impediments and is considered safer, cleaner and cheaper compared to other acids. It has shown superior performance in removing the attached mortar on RCA. Acetic acid reacts with the hydration products such as portlandite, C—S—H gel and minor amounts of $CaCO_3$ (developed due to the environmental carbonation of portlandite in RCA after long exposure to air) and dissolves them, weakening the bonds in the ITZ, thereby, allowing the delamination and easy removal of adhered mortar by mechanical rubbing.

$$Ca(OH)_{2(s)} + 2CH_3COOH_{(liq)} \rightarrow Ca^{2+}{}_{(aq)} + 2CH_3COO_{(aq)} + 2H_2O_{(aq)} \quad (5.1)$$

$$CaO—SiO_2—H_2O + CH_3COOH_{(liq)} \rightarrow Ca^{2+}{}_{(aq)} + CH_3COO_{(aq)} + SiO_{2(s)} + H_2O \quad (5.2)$$

$$CaCO_3 + CH_3COOH_{(liq)} \rightarrow Ca^{2+}{}_{(aq)} + CH_3COO_{(aq)} + CO_{2(g)} + H_2O_{(aq)} \quad (5.3)$$

After acetic acid treatment, the waste slurry obtained containing calcium acetate can be converted by CO_2 sequestration to form "Vaterite", a polymorph of $CaCO_3$ (Fig. 5.14).

$$Ca^{2+}{}_{(aq)} + CH_3COO_{(aq)} + CO_{2(g)} + H_2O_{(aq)} \rightarrow CaCO_3 + CH_3COOH_{(liq)} \quad (5.4)$$

Calcium rich waste solution (Eq. 5.3) is rich in Ca^{2+} and acetate ions. When CO_2 is bubbled through the solution, precipitated $CaCO_3$ is produced. By properly controlling

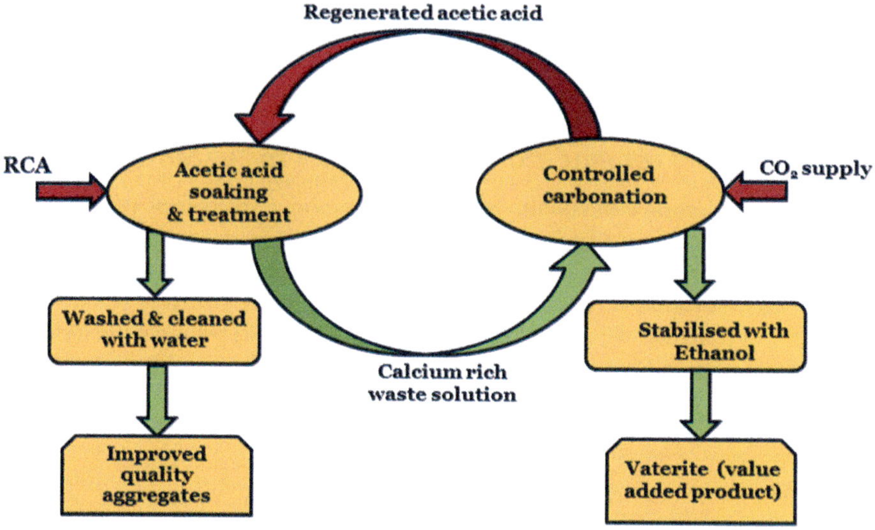

Figure 5.14 Schematic of RCA treatment and conversion of Ca + rich waste solution to Vaterite, a value–added product (Chen et al., 2017).

Table 5.4 Purity of Vaterite at different ethanol concentrations (Chen et al., 2017).

Ethanol concentration (%)	Purity of vaterite (%)
0	32.36
5	40.00
10	50.33
15	71.50
20	96.49

the reaction conditions, "Vaterite" a metastable polymorph of $CaCO_3$ is produced (Chen et al., 2017). Due to its phase transformation attributes, metastable polymorph is stabilized with ethanol. After stabilization, its purity was found to be 96.49%, while the dependence of its purity is based on the concentration of ethanol. Table 5.4 shows the dependence of ethanol purity on stabilized Vaterite.

Producing Vaterite from Ca^{2+} rich waste solution can permanently store CO_2 and reduce CO_2 emissions of concrete in life cycle assessment. In addition, acetic acid can be recovered by CO_2 sequestration because carbonic acid is stronger than acetic acid (Wang et al., 2017). The process Vaterite formation through CO_2 sequestration is very promising, eco-friendly, sustainable and cost-effective treatment of waste mortar. It is valuable product and is used in chemical industries such as paper production, paints, rubber and plastic manufacture and manufacture of reaction catalyst etc.

5.3.2 As partial substitution in portland cement

Acceptance for the use of alternative binders to partially replace Portland cement has grown in recent years. The low activity of waste cementitious powder from recycled concrete aggregate is the main reason that limits its application in cementitious materials. The un-treated waste powder mainly consists of inert hydrated products, quartz or calcite, with thermal activation, new active compounds are formed, and the utilization of thermal-activated waste powder promotes the hydration reaction of new cementitious materials. A number of studies reported in the literature, have been carried out to investigate the use of thermally activated residual mortar fines (RMF) as partial cement replacement to produce cement mortars, concrete products and material for wall rendering.

Shui et al. (2008) studied the effect of temperature on the phase transformation of cement mortars and found that a sequence of physical and chemical processes takes place. The main phenomena is the evaporation of water from cement hydrated phases, i.e., dehydration of Ettringite at 105°C, dehydration of C—S—H gel between 200 and 400°C, the morphological changes in C—S—H gel takes place mainly above 600°C. Portlandite dehydrates between 530 and 560°C (Alonso et al., 2004).

All these decompositions take place with mass loss comprising essentially of water and CO_2. However, one of the processes that take place during dehydration at about 580°C is the phase transformation, which takes place without mass loss but only with a change in structure i.e., the transformation of crystalline silica from its ά form to β form (Castellote et al., 2004).

Florea et al. (2014) demonstrated that RMF when activated at 800°C can be used up to 20% as cement replacement. To maintain workability, 20% flyash or superplasticizer can be added in the mix, without significant loss in strengths of mortars.

Shui et al. (2009) thermally activated residual mortar powder at temperatures (300—900°C) for 2.5 h, it was then cooled to room temperature. The cementitious properties of each RMF sample were evaluated by the required water of standard consistency, setting time, degree of rehydration, compressive strength development and microstructure analysis. The water requirement for consistency showed linear relationship with the incremental heating temperature while the compressive strength of RMF activated at 800°C exhibited maximum compressive strength of 11.7, 14.0 and 19.0 MPa cured for 3-, 7- and 28-days, respectively. The authors observed that RMF activity was influenced by the activation temperature.

Kalinowska et al. (2019) studied the possible use of RMF as cementitious material for partial substitution in cement after thermal activation. The thermal activation was based on two parameters (i) the RMF calcination temperatures varied from 280 to 712°C and (ii) the time of thermal activation was varied from 30 to 90 min. The authors observed that the most optimum temperature was 650°C and optimum heating time 60 min, for maximum activation of RMF. The RMF applicability as a reactive SCM was assessed based on the characteristics such as Strength Activity Index (SAI), compressive and flexural strength and water absorbability. The mortars produced with 25% substitution of activated RMF in CEM-I 42.5R, showed even higher compressive strength, 107% compared to the control specimens after 28 days of curing.

Wu et al. (2021) used different activation temperatures ranging from 600 to 1200°C to activate waste mortar powder. To assess the degree of effectiveness of thermal activation of waste mortar powder at the above respective temperatures, they assessed the mechanical properties and durability performance of cement paste substituted with 30% of activated waste mortar powder, and to determine the optimal temperature of thermal activation. The authors observed that (a) the cement paste containing 30% waste mortar powder activated at 600–800°C showed superior properties such as higher mechanical strength and transport properties due to improved microstructure (b) taking into consideration the energy consumption and thermal activation efficiency, they suggest 600–800°C as the optimum activation temperature.

5.3.3 As raw material for geopolymerization

Recycling of removed residual mortar fines (WMF) using alkali activation method has become an attractive topic for researchers in civil and materials engineering. Alkali activated cements/mortars/concrete develop high strengths and durability when silicate or alumino-silicate wastes are used as precursor materials.

Geopolymer nomenclature is used largely to describe low Ca^+ alkali-activated alumino-silicate binders, therefore WMF can be used as precursor raw material by itself or in combination with SCMs e.g., coal flyash, blast-furnace slag, calcined clays and natural pozzolan (Provis & van Deventer, 2013). Other materials, which have pozzolanic or hydraulic character such as ferrous/nonferrous air or water cooled finely ground slags (Komnitsas et al., 2007; Pontikes et al., 2013), flyash, which do not meet the criteria of specified standards (Škvarla et al., 2011), coal bottom ash (Donatello et al., 2014), kaolin purification sludge (Longhi et al., 2016), agro-waste ashes (Bernal et al., 2016), calcined clays (McIntosh et al., 2015), red mud (Hertel et al., 2020, p. 120610) (Ascensão et al., 2017; Singh et al., 2018) or natural pozzolan (Ahmad et al., 2022; González-García et al., 2022; ilkentapar et al., 2022) can also be used to produce alkali-activated mortars/concrete. The hybrid combinations have demonstrated to give good results and are highly beneficial to both, resource recovery and the environment.

Aranda et al. (2021) investigated the substitution of waste concrete powder (WCP) with its particle size reduced to 37 μm after grinding it in the ball mill for 12 h. This fine RCP was substituted at 10%, 20%, 30%, 40% and 50% to produce geopolymer mortars containing binder to sand ratio of 1:3 and water to binder ratio of 0.6 according to ASTM C109. As activator, sodium hydroxide in pellet form with 98% purity at a concentration of 4M (Moles) was used, the water to binder ratio was kept at 0.6 for all the mixes. The mortar cubes were demolded after 24 h and cured at 70°C for 72 h in the oven. The compressive strength after 28 days showed 14.7% higher values for 10% WCP substitution and 6.8% higher values compared to control sample. The higher compressive strength indicates the densification of geopolymer mortar cubes due to the formation of additional C–S–H gel with improved ITZ, compactness and density of mortars/concrete.

Gong et al. (2014) studied the mix ratio of 70/30 of RMF/powdered slag, to produce geopolymer maintaining sodium silicate/NaOH ratio of 1.4. Laboratory prepared cement mortars were ground to Blaine specific surface area ~380 m^2/kg. The alkali-activated specimens cured at 20°C with relative humidity >90%, showed compressive strengths of 18.4 and 45.5 MPa and flexural strength of 4.4 and 7.5 MPa for three- and 28-days curing, respectively.

To utilize waste concrete powder (WCP) based geopolymer for high specification applications Chen et al. (2018), produced geopolymer material by partially replacing WCP with flyash (type F) at 25%, 50% and 75% replacement ratios, using higher molar concentrations of 10.0, 14.0 and 18.0 M of NaOH and maintaining the sodium silicate/NaOH ratio at 1.5, and the liquid/solid ratio at 0.4. They observed that with 75% flyash substitution for WCP at molar concentration of 18 M, compressive strength achieved was 37.5 MPa at 7 days when the samples were thermally cured at 50°C for 24 h and then left in the open atmosphere for testing at 3-, 7-, 14- and 28-days.

5.4 Recycling of concrete rubble using thermo-mechanical treatment

By: Katarzyna-Kalinowska-Wichrowska.

Concrete is currently used as a building material in almost every new construction. Estimates indicate that up to 25 billion tons of concrete are produced around the world each year. The consequence of this is 510 million tonnes of construction waste is generated in Europe, around 325 million tonnes in the US, and around 77 million tonnes in Japan (Ferrari et al., 2014).

As the consumption of concrete increases, the demand for cement and aggregate increases. The need to protect the Earth's resources, mainly natural aggregates, and the high energy consumption of the cement production impose on the EU Member States, the need to reuse waste concrete, i.e., recycling. It is estimated that the concrete industry consumes about 11 billion tons of natural aggregate per year, and the cement industry is responsible for about 5%–8% of anthropogenic CO_2 emissions in the world (Jin & Chen, 2015).

Recycling is currently one of the effective methods of managing concrete rubble. The huge consumption of concrete in the world and the fact that its production consumes huge amounts of nonrenewable resources encourage a responsible approach in the search for the methods of effective recycling.

Mechanical methods of recycling the rubble concrete are the most frequently used and they consist of single or multiple crushing. They are mainly focused on obtaining aggregate, which is usually used for road foundations. This is specifically due to the high water absorption of the aggregate, caused by the presence of residual porous mortar on the aggregate surface, which adversely affects the properties of both, the aggregate itself and the concrete.

The emerging advanced recycling methods are based on improving the properties of crushed secondary coarse aggregates, e.g., by surface impregnation of the aggregate with oil or other chemicals (Tsujino et al., 2007). The other methods employed are to remove the residual cement mortar from the surface of recycled aggregate grains by mechanical abrasion in screw ball mills or eccentrically rotating cylinders, or by gravitational classification based on their density. Method of treating recycled aggregate proposed by Tam et al. (2007), consists of presoaking the recycled aggregate in acids, respectively: HCl, H_2SO_4, and H_3PO_4, to reduce the amount of old cement mortar. These methods to varying degrees improve the properties of recycled aggregate, specially by reducing its water absorption capacity. It should be emphasized that all of the above methods focus only on the recovery of improved quality coarse aggregates, excluding the utilization of the fine fraction (i.e., old cement mortar).

In connection with the benefits of comprehensive methods (complete processing of rubble concrete waste) and to improve the properties of the obtained coarse and fine fractions, it was considered necessary to carry - out research in the field of recycling technology using thermo-mechanical treatment of concrete rubble. This line of research was carried out by the research team at the Bialystok University of Technology, Poland.

5.4.1 A comprehensive recycling technology for structural concrete

A comprehensive thermo-mechanical technology (Boltryk et al., 2018) for recycling of concrete rubble is under development. Concrete rubble is transported via a belt conveyor to a grinding device (jaw crusher) for further fractionation. After crushing to a maximum size of 4 cm, the so-called "secondary grains", are subjected to thermal and mechanical treatment. Thermal treatment takes place for a sufficiently long time, at a high, experimentally selected temperature, supplemented by a mechanical process. During full processing, the aggregate-cement mortar interfacial transition zone (ITZ) is broken. The high temperature inside the rotating drum eliminates the presence of light impurities, e.g., paper, plastic, or wood. The step-wise process is depicted in Fig. 5.15.

Products obtained after thermo-mechanical treatment are screened to separate the coarse aggregates from the fines. The treated coarse aggregate meeting the specified Standards can be used for concrete as a substitute for natural aggregate, while fine fractions can be used as an ingredient (raw material) in the cement manufacturing process or as pozzolan additive for cement substitution or an active ingredient for autoclaved products.

The aim of the research carried-out by Kalinowska-Wichrowska et al. (2020) was to select the optimal parameters for the treatment of concrete rubble (such as time and temperature) of the treatment and processing of rubble, which would enable to obtain both, valuable aggregate as well as recycled mortar, which could further be used in the cement production, cement substitution and construction processes.

The laboratory tests conducted by the authors focused mainly on the selection of heating time and temperature of the concrete rubble combined with a mechanical proces. The best parameters of the thermo-mechanical process were determined by trials: heating temperature of 650°C and heating time of 60 min were found to be

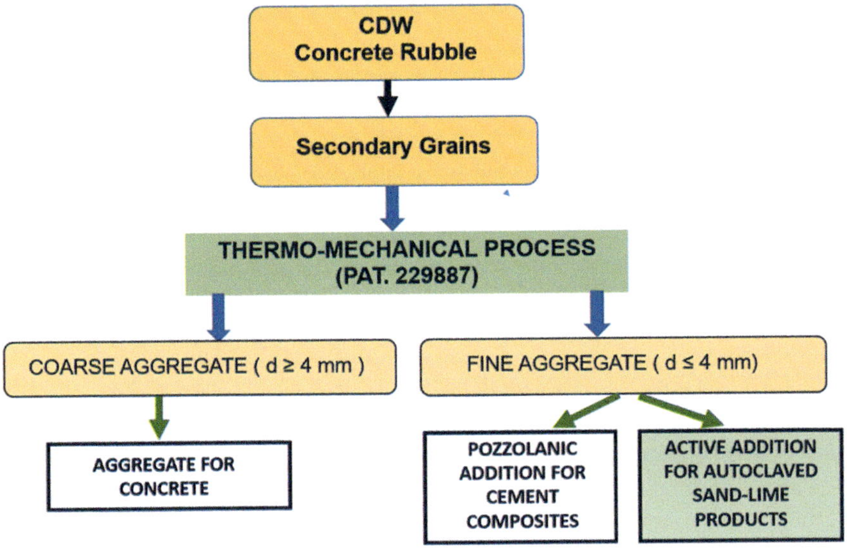

Figure 5.15 The process of thermo-mechanical treatment of concrete rubble.

optimum. The full process was based on eight stages, which in laboratory conditions were carried out in the following stages:

Stage I—the concrete rubble was crushed in a jaw crusher to <40 mm in size.

Stage II—the material obtained in this way was placed in a thermal furnace (11 kW maximum power and 1250°C maximum temperature) and heated according to the parameters mentioned above (after this stage partial separation of the cement mortar from the aggregate was observed).

Stage III—after removing it from the furnace, the heat treated material was placed in a Los Angeles drum and subjected to the mechanical treatment at 500 revolutions of the drum containing five steel balls to finally separate the cement mortar from the surface of coarse aggregate grains.

Stage IV—the cooled material was sieved through a 4 mm sieve to separate fine fractions (<4 mm) from the coarse fractions (≥4 mm). The coarse fractions (≥4 mm) were additionally divided into two fractions of 4−8 mm and 8−16 mm.

The method described above for separating the residual mortar, will simplify existing recycling techniques for producing improved quality aggregate for use in structural concrete, improved efficiency of production lines, significantly improve the environmental conditions, and, above all, allow to obtain the high quality recycled aggregate. The application of the treatment can reduce the cost of manufacturing of the new precast and reinforced concrete structures. Moreover, the disintegrated and activated cement mortar can be used in the production of blended cements, autoclaved concrete, conventional concrete as well as lime and sand products.

The designed device [PAT. 229,887] combines thermal and mechanical treatment of the rubble and allows the recovery of valuable recycled aggregate and recycled mortar having pozzolanic properties.

The method and operation of the proposed device consist of continuous feeding of the tubular mill with a suitable angle of inclination of the fragmented concrete rubble from the upper part to the lower part. The temperature in the rotating device is not below 650°C, which is essential for dehydration of the cement in the recycled mortar. Four shelves are arranged along the inner perimeter of the furnace in the roasting chamber, which allows for quick and even heating of the concrete debris as a result of the rotation of the device around its axis. The cyclical gravity lifting and dropping of heated concrete rubble allow the initial separation of the mortar from the aggregate as a result of its collision with the mantle and shelves of the device. Due to the inclination of the device, the material gradually moves toward the grinding and crushing chamber of the mortar, equipped with a special system of rollers. As a result of the rotation of the entire device, the rollers also rotate with a reciprocating movement, which causes complete cleaning of the aggregate from the mortar and its partial fragmentation. After leaving the device, the material is transferred to a set of sieves using conveyors and separated into different size fractions.

The above method of separating the hardened cement mortar from coarse aggregate and grinding this mortar and the device for using this method will simplify the existing technologies for recycling reinforced concrete structures, increase the efficiency of production lines, significantly improve environmental protection conditions, and above all, allow to obtain very high-quality recycled aggregate with properties better than those of natural aggregate. The treated recycled aggregate can reduce the production costs of new reinforced concrete structures. Also, crushed and activated cement mortar having pozzolanic properties can be used in the production of autoclaved products, cement, and as cement substitution material (SCM) in conventional concrete.

5.4.2 The properties of recycling products received according to the thermo-mechanical method

As it results from the research, the fine fraction (<4 mm) constitutes over 64% of concrete rubble and therefore it is necessary to properly manage it. Table 5.5 presents the properties of the obtained recycled aggregate and natural aggregate, and Table 5.6 presents the chemical composition and properties of the fine recycled fraction obtained in the thermo-mechanical treatment process.

The properties of recycled aggregates obtained according to the described method presented in Table 5.4 are similar to the parameters of natural aggregates. They are characterized by a slightly lower density, acceptable water absorption, and a similar crushing index. Relics of the dehydration process, formed as a result of high-temperature roasting of rubble, which are present on the surface of the aggregate grains, partially react with the new cement paste, which improves the compressive and bending strength of concrete (Kalinowska-Wichrowska et al., 2020). Recycling mortar (Table 5.5) has a high specific surface even without the remixing process, and thanks to the content of reactive pozzolanic oxides (total content of $SiO_2 + Al_2O_3 + Fe_2O_3 > 70\%$) meets the requirements of PN-EN 450-1: 2012 in the context of activity assessment) pozzolana) in combination with cement binder in new composites can be used as an additive to concrete, which is also validated by the obtained test results (Kalinowska-Wichrowska et al., 2019).

Table 5.5 The properties of natural and recycled coarse aggregate.

Properties	Natural aggregate (gravel)		Recycled coarse aggregate after treatment according to PAT.229,887	
	4–8 mm	8–16 mm	4–8 mm	8–16 mm
Bulk density (g/cm^3)	2.64	2.64	2.53	2.59
Dry bulk density (g/cm^3)	2.62	2.62	2.50	2.52
Saturated bulk density (g/cm^3)	2.66	2.66	2.57	2.58
Water absorption (%)	1.3	1.0	2.5	1.80
Crushing indicator (%)	10.4	11.4	13.6	11.8

Table 5.6 The properties of recycled cement mortar after thermo-mechanical treatment.

Component	Amount, %	Surface area (blaine) (cm^2/g)	Bulk density, (g/cm^3)
SiO_2	75.6	3600	2.73
Fe_2O_3	3.10		
Al_2O_3	5.9		
Mn_2O_3	0.05		
CaO	17.40		
Calcite	9.2		
Free CaO	<0.03		
Reactive CaO	5.28		

5.4.3 Properties of composites with treated cement mortar used as supplementary cementitious material (SCM)

5.4.3.1 Characteristics of recycled cement mortar (RCM)

Strength activity index (SAI)

The highest compressive strength was obtained for the composite containing 25% RCM calcined at 650°C temperature (SAI = 1.07), shown in Fig. 5.16. The increase in SAI value for these samples was 79%, compared to the noncalcined RCM (Series 12, SAI = 0.59), while it was 66% compared to RCM calcined at 350°C.

Thermogravimetry (TG) and differential thermal analysis (DTA)

The quite high content of calcium carbonates in both tested specimens is noteworthy. It results from the thermal treatment of samples to 650°C, which does not cause the decomposition of $CaCO_3$. This phenomenon occurs at a temperature above 750°C, as indicated by peaks associated with mass losses in Figs. 5.17 and 5.18. The higher $CaCO_3$ content in calcined RCM can be explained by the presence of aggregate in the specimen tested. In the RCM specimen after heat treatment, there was no peak for portlandite, which confirmed a sufficiently well-selected RCM treatment temperature

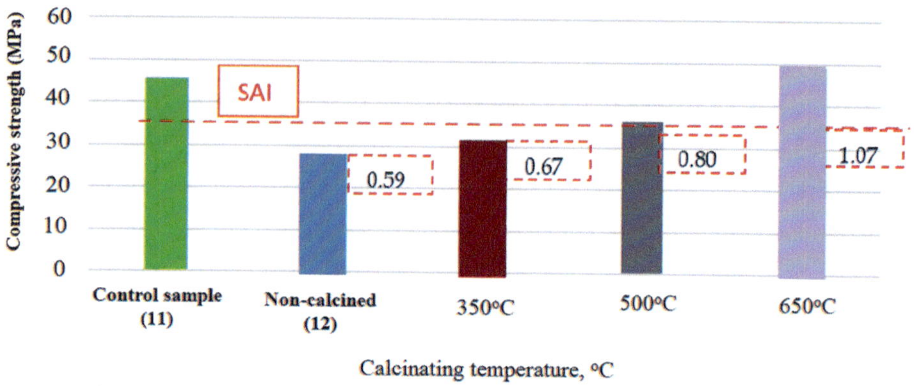

Figure 5.16 Compressive strength of the composites with and without RCM after 28 days of curing with the strength activity index (SAI) (Kalinowska-Wichrowska et al., 2019).

(Fig. 5.18). Earlier heating of concrete rubble resulted in the disintegration of calcium hydroxide into calcium oxide and water, as evidenced by its lack in the tested RCM.

5.4.4 Properties of hardened cement composites

5.4.4.1 Compressive Strength

The average compressive strength results obtained for the test series are shown in Fig. 5.19.

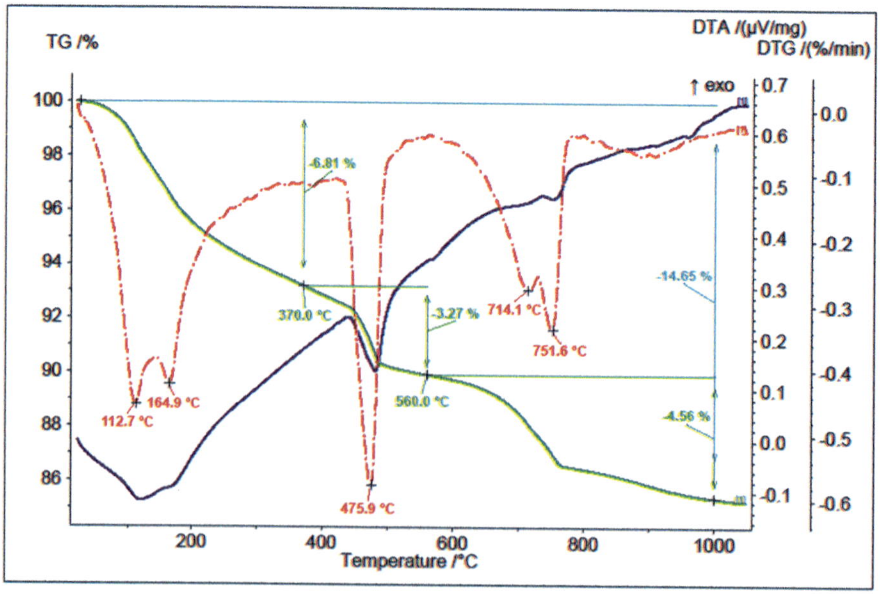

Figure 5.17 Thermal changes of RCM without thermal and mechanical treatment (Kalinowska-Wichrowska et al., 2019).

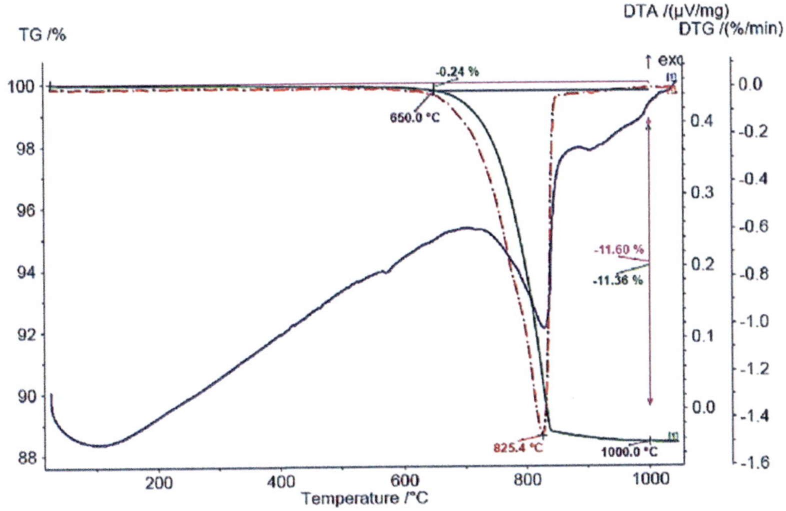

Figure 5.18 Thermal changes in RCM after thermal and mechanical treatment at temperature of 650°C (Kalinowska-Wichrowska et al., 2019).

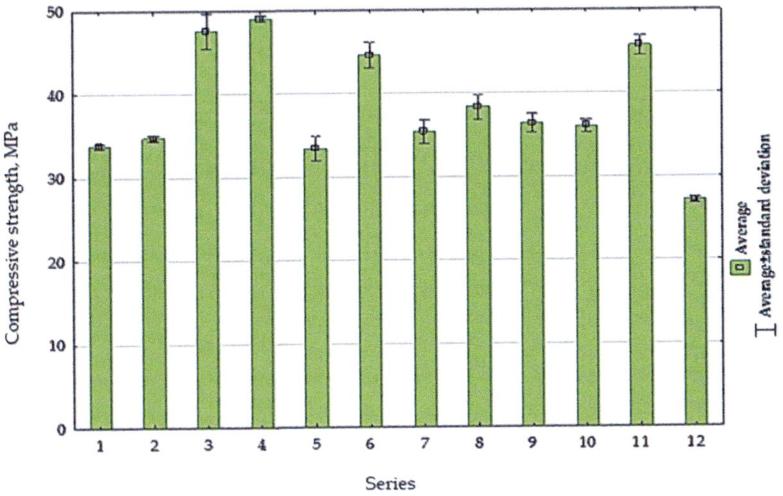

Figure 5.19 The compressive strength of composites after 28 days (Kalinowska-Wichrowska et al., 2019).

The obtained compressive strength test results indicate that the increase in calcination temperature from around 300 to 712°C caused an increase in compressive strength of about 32%. The highest strength values, exceeding those obtained for the control series, were recorded for composites with RCM calcined at 650°C temperature. In case of using the RCM calcined at lower temperatures (up to 350°C), as well as in the presence of noncalcined addition, the compressive strength results of the composite did not exceed 35 and 30 MPa, respectively.

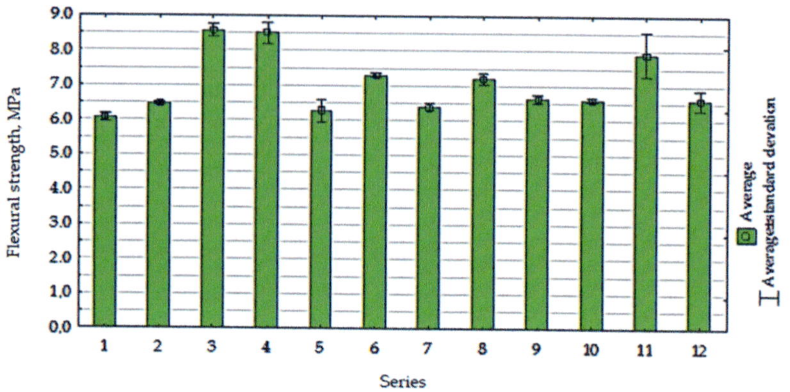

Figure 5.20 The Flexural strength of composites with RCM after 28 days (Kalinowska-Wichrowska et al., 2019).

5.4.4.2 Flexural strength

The analysis of test results showed that the most favorable results of flexural strength of cement composites containing RCM were obtained in cases of using the secondary material after calcination at temperature higher than 650°C (series 3, 4, 6), and the results were similar to those obtained in the control series. The average flexural strength results are given in Fig. 5.20.

5.5 Summary

Both HG and HGA RCA have almost similar quality to natural coarse aggregate in terms of density, water absorption, and sieve analysis. The mechanical properties, including the compressive strength, modulus of elasticity, splitting-tensile strength, and flexural strength of HGA RCA concrete have shown slightly higher values than those of HG RCA concrete. Similarly, HGA RCA concrete's surface hardness and ultrasonic wave velocity are slightly higher than those of HG RCA concrete. The chemical treatment during HGA RCA production has removed the old cement paste more effectively from the aggregate surface than that of HG RCA; therefore, the HGA RCA has shown much better performance than HG RCA. More delamination and removal of the attached mortar increases the bonding of the ITZ between the new cement paste and the RCA surface. However, the properties between HGA RCA concrete and HG RCA concrete are not significantly different, even compared to those of NCA concrete. Thus, both types of treated RCA can be potentially used in construction.

The huge popularity of concrete in today's world leads to the search for effective methods allowing for such processing of concrete rubble that will allow the recovery of valuable natural resources i.e. coarse aggregates together with separated cement mortar as much as possible. The proposed method of comprehensive recycling is based on the use of thermal and mechanical treatment of concrete rubble. The best parameters for the treatment of concrete rubble were selected, which enable its comprehensive processing into high-quality recycling products: recycled aggregate with improved parameters and partially active recycled mortar. These products, thanks to their properties (which has been confirmed by tests), can be used in the construction industry: recycling aggregate can be used even as a 100% replacement for the necessary coarse natural aggregate, and the recycled mortar, depending on the degree of addition—e.g., as a pozzolanic additive in cement composites. The proposed technology is completely waste-free and fits perfectly with the idea of sustainable development and the circular economy.

References

Ahmad, S., et al. (2022). Properties of natural pozzolan-based geopolymer concrete: Effects of natural pozzolan content, type of alkaline activator, and silicate/alkali ratio. *European Journal of Environmental and Civil Engineering, 0*(0), 1–18. https://doi.org/10.1080/19648189.2022.2043941

Akbarnezhad, A., et al. (2011). Microwave-assisted beneficiation of recycled concrete aggregates. *Construction and Building Materials, 25*(8), 3469–3479. https://doi.org/10.1016/j.conbuildmat.2011.03.038

Al-Bayati, H. K. A., et al. (2016). Evaluation of various treatment methods for enhancing the physical and morphological properties of coarse recycled concrete aggregate. *Construction and Building Materials, 112*, 284–298. https://doi.org/10.1016/j.conbuildmat.2016.02.176

Alonso, C., & Fernandez, L. (2004). Dehydration and rehydration processes of cement paste exposed to high temperature environments. *Journal of Materials Science, 39*(9), 3015–3024. https://doi.org/10.1023/B:JMSC.0000025827.65956.18

Aranda, C. A., Zavaleta, S. A., & Vega, A. (2021). Alkaline activation of mortars made from solid construction waste. *Journal of Sciences and Engineering (SEJ), 5*(2), 79–84.

Ascensão, G., et al. (2017). Red mud-based geopolymers with tailored alkali diffusion properties and pH buffering ability. *Journal of Cleaner Production, 148*, 23–30. https://doi.org/10.1016/j.jclepro.2017.01.150

Assi, L., Carter, K., Deaver, E., Anay, R., & Ziehl, P. (2018). Sustainable concrete: Building a greener future. *Journal of Cleaner Production, 198*, 1641–1651. https://doi.org/10.1016/j.jclepro.2018.07.123

Azúa, G., González, M., Arroyo, P., & Kurama, Y. (2019). Recycled coarse aggregates from precast plant and building demolitions: Environmental and economic modeling through stochastic simulations. *Journal of Cleaner Production, 210*, 1425–1434. https://doi.org/10.1016/j.jclepro.2018.11.049

Bernal, S. A., et al. (2016). Management and valorisation of wastes through use in producing alkali-activated cement materials. *Journal of Chemical Technology and Biotechnology, 91*(9), 2365–2388. https://doi.org/10.1002/jctb.4927

Bluhm, H. (2006). *Pulsed power systems*.

Bluhm, H., Frey, W., Giese, H., Hoppe, P., Schultheiss, C., & Strassner, R. (2000). Application of pulsed HV discharges to material fragmentation and recycling. *IEEE Transactions on Dielectrics and Electrical Insulation, 7*(5), 625–636. https://doi.org/10.1109/94.879358

Bołtryk, M., Kalinowska-Wichrowska, K., & Pawluczuk, E. (2018). *Method for separation of set cement mortar from coarse aggregate and for crushing that mortar, and the device for the application of this method, PAT.229887*. Available online:).

Bungey, J. H., & Millard, S. G. (1995). *Testing of concrete in structures* (3rd ed.). Informa UK Limited: CRC Press. https://doi.org/10.4324/9780203487839

Calkins, M. (2008). *Materials for sustainable sites: A complete guide to the evaluation, selection, and use of sustainable construction materials.*

Castellote, M., et al. (2004). Composition and microstructural changes of cement pastes upon heating, as studied by neutron diffraction. *Cement and Concrete Research, 34*(9), 1633–1644. https://doi.org/10.1016/S0008-8846(03)00229-1

Chen, P., et al. (2017). Producing vaterite by CO_2 sequestration in the waste solution of chemical treatment of recycled concrete aggregates. *Journal of Cleaner Production, 149*, 735–742. https://doi.org/10.1016/j.jclepro.2017.02.148

Chen, W., Jin, R., Xu, Y., Wanatowski, D., Li, B., Yan, L., Pan, Z., & Yang, Y. (2019). Adopting recycled aggregates as sustainable construction materials: A review of the scientific literature. *Construction and Building Materials, 218*, 483–496. https://doi.org/10.1016/j.conbuildmat.2019.05.130

Chen, W., & Zhu, Z. (2018). Utilization of fly ash to enhance ground waste concrete-based geopolymer. In *Advances in materials science and engineering*. Hindawi. https://doi.org/10.1155/2018/4793917, 2018.

Corinaldesi, V. (2010). Mechanical and elastic behaviour of concretes made of recycled-concrete coarse aggregates. *Construction and Building Materials, 24*(9), 1616–1620. https://doi.org/10.1016/j.conbuildmat.2010.02.031

Djerbi, A. (2018). Effect of recycled coarse aggregate on the new interfacial transition zone concrete. *Construction and Building Materials, 190*, 1023–1033. https://doi.org/10.1016/j.conbuildmat.2018.09.180

Donatello, S., et al. (2014). The early age hydration reactions of a hybrid cement containing a very high content of coal bottom ash. *Journal of the American Ceramic Society, 97*(3), 929–937. https://doi.org/10.1111/jace.12751

Etxeberia, M., Vazquez, E., & Marı, A. (2006). Microstructure analysis of hardened recycled aggregate concrete. *Magazine of Concrete Research, 58*(10), 683–690.

Evangelista, L., et al. (2015). Physical, chemical and mineralogical properties of fine recycled aggregates made from concrete waste. *Construction and Building Materials, 86*, 178–188. https://doi.org/10.1016/j.conbuildmat.2015.03.112

Ferrari, G., Miyamoto, M., & Ferrari, A. (2014). New sustainable technology for recycling returned concrete. *Construction and Building Materials, 67*, 353–359. https://doi.org/10.1016/j.conbuildmat.2014.01.008. PART C.

Florea, M. V. A., Ning, Z., & Brouwers, H. J. H. (2014). Activation of liberated concrete fines and their application in mortars. *Construction and Building Materials, 50*, 1–12. https://doi.org/10.1016/j.conbuildmat.2013.09.012

Fumoto, T., & Yamada, M. (2002). *Influence of the quality of recycled fine aggregate on properties of concrete. 4* pp. 97–103). Memoirs of the Faculty of Engineering Osaka City University. https://dlisv03.media.osaka-cu.ac.jp/il/meta_pub/G0000438repository_KJ00000026454

Gomez-Soberon, J. M. V. (2002). Porosity of recycled concrete with substitution of recycled concrete aggregate an experimental study. *Cement and Concrete Research, 32*, 1301–1311.

Gong, Y. F., et al. (2014). Investigation on alkali activated recycled cement mortar powder cementitious material. *Materials Research Innovations, 18*, S2784–S2787. https://doi.org/10.1179/1432891714Z.000000000477

González-García, D. M., et al. (2022). Evolution of a natural pozzolan-based geopolymer alkalized in the presence of sodium or potassium silicate/hydroxide solution. *Construction and Building Materials, 321*(January). https://doi.org/10.1016/j.conbuildmat.2021.126305

Hertel, T., & Pontikes, Y. (2020). Geopolymers, inorganic polymers, alkali-activated materials and hybrid binders from bauxite residue (red mud)—putting things in perspective. *Journal of Cleaner Production, 258*. https://doi.org/10.1016/j.jclepro.2020.120610. p. 120610.

Hongzhu, Qian, & Hideo Kasami, T. T. (2010). Experimental study on effects of adhered mortar content on quality of recycled coarse aggregate and properties of concrete. In *2nd international conference on waste engineering and management—ICWEM 2010* (pp. 1059–1062). RILEM Publications.

Ilkentapar, S., & Ozsoy, A. (2022). Investigation of mechanical properties, high-temperature resistance and microstructural properties of diatomite-containing geopolymer mortars. *Arabian Journal of Geosciences, 15*, 502–514.

Inoue, S., Araki, J., Aoki, T., Maeda, S., Iizasa, S., Takaki, M., Wang, D., Namihira, T., Shigeishi, M., Ohtsu, M., & Akiyama, H. (2009). Coarse aggregate recycling by pulsed discharge inside of concrete. *Acta Physica Polonica A, 115*(6), 1107–1109. https://doi.org/10.12693/APhysPolA.115.1107

Inoue, S., Iizasa, S., Wang, D., Namihira, T., Shigeishi, M., Ohtsu, M., & Akiyama, H. (2012). Concrete recycling by pulsed power discharge inside concrete. *International Journal of Plasma Environmental Science and Technology, 6*(2), 183–188.

Ismail, S., Hoe, K. W., & Ramli, M. (2013). Sustainable aggregates: The potential and challenge for natural resources conservation. *Procedia—Social and Behavioral Sciences*, 100–109. https://doi.org/10.1016/j.sbspro.2013.07.183

Jain, A., Kathuria, A., Kumar, A., Verma, Y., & Murari, K. (2013). Combined use of non-destructive tests for assessment of strength of concrete in structure. *Procedia Engineering, 54*, 241–251. https://doi.org/10.1016/j.proeng.2013.03.022

Jim O'Brien. (2019). *Global aggregate growth examined. Aggregateresearch.com*. Available at: http://www.aggregateresearch.com/news/global-aggregate-growth-examined-by-gain-convenor-jim-obrien/

Jin, R., & Chen, Q. (2015). Investigation of concrete recycling in the U.S. Construction industry. *Procedia Engineering, 118*, 894–901. https://doi.org/10.1016/j.proeng.2015.08.528

de Juan, M. S., & Gutiérrez, P. A. (2009). Study on the influence of attached mortar content on the properties of recycled concrete aggregate. *Construction and Building Materials, 23*(2), 872–877. https://doi.org/10.1016/j.conbuildmat.2008.04.012

Kabirifar, K., Mojtahedi, M., Wang, C., & Tam, V. W. Y. (2020). Construction and demolition waste management contributing factors coupled with reduce, reuse, and recycle strategies for effective waste management: A review. *Journal of Cleaner Production, 263*. https://doi.org/10.1016/j.jclepro.2020.121265

Kalinowska-Wichrowska, K., Kosior-Kazberuk, M., & Pawluczuk, E. (2019). The properties of composites with recycled cement mortar used as a supplementary cementitious material. *Materials, 13*(1), 64. https://doi.org/10.3390/ma13010064

Kalinowska-Wichrowska, K., Pawluczuk, E., & Bołtryk, M. (2020). Waste-free technology for recycling concrete rubble. *Construction and Building Materials, 234*, 117407. https://doi.org/10.1016/j.conbuildmat.2019.117407. The Authors.

Kenai, S. (2018). Recycled aggregates. In *Waste and supplementary cementitious materials in concrete: Characterisation, properties and applications* (pp. 79–120). Elsevier. https://doi.org/10.1016/B978-0-08-102156-9.00003-1

Kencanawati, N. N., Akmaluddin, Merdana, I. N., Nuraida, N., Hadi, I. R., & Shigeishi, M. (2017). Improving of recycled aggregate quality by thermal-mechanical-chemical process. *Procedia Engineering, 171*, 640–644. https://doi.org/10.1016/j.proeng.2017.01.399

Kencanawati, N. N., Iizasa, S., & Shigeishi, M. (2013). Fracture process and reliability of concrete made from high grade recycled aggregate using acoustic emission technique under compression. *Materials and Structures/Materiaux et Constructions, 46*(9), 1441–1448. https://doi.org/10.1617/s11527-012-9986-z

Kencanawati, Ni, N., Akmaluddin, A., Anshari, B., Paedullah, A. G., & Shigeishi, M. (2018). The study of ultrasonic pulse velocity on plain and reinforced damaged concrete. In *MATEC web of conferences, 195*. https://doi.org/10.1051/matecconf/201819502026

Kencanawati, Ni, N., Fajrin, J., Anshari, B., & Shigeishi, M. (2015). Evaluation of high grade recycled coarse aggregate concrete quality using non-destructive testing technique. *Applied Mechanics and Materials, 776*, 53–58. https://doi.org/10.4028/www.scientific.net/amm.776.53

Kibert, C. J., Sendzimir, J., & Guy, B. (2000). Construction ecology and metabolism: Natural system analogues for a sustainable built environment. *Construction Management and Economics, 18*(8), 903–916. https://doi.org/10.1080/014461900446867

Kjellson, K. O., Wallevik, O. H., & Fjallberg, L. (1998). Microstructure and microchemistry of the paste-aggregate interfacial transition zone of high-performance concrete. *Advances in Cement Research, 10*(1), 33–40.

Komnitsas, K., Zaharaki, D., & Perdikatsis, V. (2007). Geopolymerisation of low calcium ferronickel slags. *Journal of Materials Science, 42*(9), 3073–3082. https://doi.org/10.1007/s10853-006-0529-2

Longhi, M. A., et al. (2016). Valorisation of a kaolin mining waste for the production of geopolymers. *Journal of Cleaner Production, 115*, 265–272. https://doi.org/10.1016/j.jclepro.2015.12.011

Malhotra, V., & Carino, N. (2004). *CRC handbook on nondestructive testing of concrete*.

Martínez-Lage, I., Vázquez-Burgo, P., & Velay-Lizancos, M. (2020). Sustainability evaluation of concretes with mixed recycled aggregate based on holistic approach: Technical, economic and environmental analysis. *Waste Management, 104*, 9–19. https://doi.org/10.1016/j.wasman.2019.12.044

McIntosh, A., et al. (2015). Selection and characterisation of geological materials for use as geopolymer precursors. *Advances in Applied Ceramics, 114*(7), 378–385. https://doi.org/10.1179/1743676115Y.0000000055

Meidiana, C., & Gamse, T. (2010). Development of waste management practices in Indonesia. *European Journal of Scientific Research, 40*(2), 199–210. http://www.eurojournals.com/ejsr_40_2_04.pdf

Miraldo, S., Lopes, S., Pacheco-Torgal, F., & Lopes, A. (2021). Advantages and shortcomings of the utilization of recycled wastes as aggregates in structural concretes. *Construction and Building Materials, 298*, 123729. https://doi.org/10.1016/j.conbuildmat.2021.123729

Mistri, A., Bhattacharyya, S. K., Dhami, N., Mukherjee, A., & Barai, S. V. (2020). A review on different treatment methods for enhancing the properties of recycled aggregates for sustainable construction materials. *Construction and Building Materials, 233*. https://doi.org/10.1016/j.conbuildmat.2019.117894

Narahara, S., Namihira, T., Nakashima, K., Inoue, S., Iizasa, S., Maeda, S., Shigeishi, M., Ohtsu, M., & Akiyama, H. (2007). Evaluation of concrete made from recycled coarse aggregates by pulsed power discharge. In , *1. 2007 16th IEEE international pulsed power conference* (pp. 748–751). https://doi.org/10.1109/PPPS.2007.4651948

Narud, T. C. H., & H. (1983). Strength of recycled concrete made from crushed concrete coarse aggregate. *Concrete International, 5*(1).

Nassar, R., & Soroushian, P. (2012). Strength and durability of recycled aggregate concrete containing milled glass as partial replacement for cement. *Construction and building materials, 29*, 368–377. https://doi.org/10.1016/j.conbuildmat.2011.10.061

Noguchi, T., Park, W. J., & Kitagaki, R. (2015). Risk evaluation for recycled aggregate according to deleterious impurity content considering deconstruction scenarios and production methods. *Resources, conservation and recycling, 104*, 405–416. https://doi.org/10.1016/j.resconrec.2015.08.002

Ogawa, H., Torii, K., & Ishikura, T. (2010). Trend on decommissioning technology of nuclear power plants in 21 st century (7); Decommissioning technologies-concrete recycling technologies. *Atomos, 52*(3), 158–162. https://doi.org/10.3327/jaesjb.52.3_158

Pacheco-Torgal, F., Pacheco-Torgal, F., Ding, Y., Colangelo, F., Tuladhar, R., & Koutamanis, A. (2020). 1—introduction to advances in construction and demolition waste. In *Woodhead publishing series in civil and structural engineering* (pp. 1–10). Woodhead Publishing. https://doi.org/10.1016/B978-0-12-819055-5.00001-2

Pepe, M., Toledo Filho, R. D., Koenders, E. A. B., & Martinelli, E. (2014). Alternative processing procedures for recycled aggregates in structural concrete. *Construction and Building Materials, 69*, 124–132. https://doi.org/10.1016/j.conbuildmat.2014.06.084

Pontikes, Y., et al. (2013). Slags with a high Al and Fe content as precursors for inorganic polymers. *Applied Clay Science, 73*(1), 93–102. https://doi.org/10.1016/j.clay.2012.09.020

Provis, J. L., & van Deventer, J. S. J. (2013). *Alkali activated materials: State-of-the-Art report. RILEM TC 224-AAM*. Dordrecht, Netherlands: Springer/RILEM.

Purkiss, J. (2007). *Fire safety engineering*. Informa UK Limited. https://doi.org/10.1201/b12845-1

Scrivener, K., Crumbie, A., & Laugese, P. (2004). The interfacial transition zone (ITZ) between cement paste and aggregate. *Interface Science, 12*, 411–421.

Shaban, W. M., Yang, J., Su, H., Hung Mo, K., Li, L., & Xie, J. (2019). Quality improvement techniques for recycled concrete aggregate: a review. *Journal of Advanced Concrete Technology, 17*, 151–167. https://doi.org/10.3151/jact.17.4.151

Shima, H., et al. (2005b). An advanced concrete recycling technology and its applicability assessment through input-output analysis. *Journal of Advanced Concrete Technology, 3*(1), 53–67. https://doi.org/10.3151/jact.3.53

Shima, H., Matsuhashi, R., Yoshida, Y., & Tateyashiki, H. (2003). Life cycle analysis of high quality recycled aggregate produced by heating and rubbing method. *IEEJ Transactions on Electronics (Part of Information and Systems), 123*(10), 1680–1687. https://doi.org/10.1541/ieejeiss.123.1680

Shima, H., Tateyashiki, H., Matsuhashi, R., & Yoshida, Y. (2005a). An advanced concrete recycling technology and its applicability assessment through input-output analysis. *Journal of Advanced Concrete Technology, 3*(1), 53–67. https://doi.org/10.3151/jact.3.53

Shui, Z., Xuan, D., Wan, H., & Cao, B. (2008). Rehydration reactivity of recycled mortar from concrete waste experienced to thermal treatment. *Construction and Building Materials, 22*(8), 1723–1729. https://doi.org/10.1016/j.conbuildmat.2007.05.012

Shui, Z., et al. (2009). Cementitious characteristics of hydrated cement paste subjected to various dehydration temperatures. *Construction and Building Materials, 23*(1), 531–537. https://doi.org/10.1016/j.conbuildmat.2007.10.016

Singh, S., Aswath, M. U., & Ranganath, R. V. (2018). Effect of mechanical activation of red mud on the strength of geopolymer binder. *Construction and Building Materials, 177*, 91–101. https://doi.org/10.1016/j.conbuildmat.2018.05.096

Škvarla, J., et al. (2011). The potential use of fly ash with a high content of unburned carbon in geopolymers. *Acta Geodynamica et Geomaterialia, 8*(2), 123–132.

Tam, V. W. Y., Butera, A., & Le, K. N. (2016). Carbon-conditioned recycled aggregate in concrete production. *Journal of cleaner production, 133*, 672–680. https://doi.org/10.1016/j.jclepro.2016.06.007

Tam, V. W. Y., Gao, X. F., & Tam, C. M. (2005). Microstructural analysis of recycled aggregate concrete produced from two-stage mixing approach. *Cement and Concrete Research, 35*(6), 1195–1203. https://doi.org/10.1016/j.cemconres.2004.10.025

Tam, V. W. Y., Lo, C. Y., & Xiao, J. (2013). Bringing recycled aggregate to its full potential. *Proceedings of Institution of Civil Engineers: Waste and Resource Management, 166*(3), 128–136. https://doi.org/10.1680/warm.12.00020

Tam, V. W. Y., Soomro, M., & Evangelista, A. C. J. (2018). A review of recycled aggregate in concrete applications (2000–2017). *Construction and Building Materials, 172*. https://doi.org/10.1016/j.conbuildmat.2018.03.240

Tam, V. W. Y., Tam, C. M., & Le, K. N. (2007). Removal of cement mortar remains from recycled aggregate using pre-soaking approaches. *Resources, Conservation and Recycling, 50*(1), 82–101. https://doi.org/10.1016/j.resconrec.2006.05.012

The Cement Sustainability Initiative: Recycling Concrete. (2009). *World business council for sustainable development*. WBCSD.

Timoshkin, I. V., Mackersie, J. W., & MacGregor, S. J. (2004). Plasma channel miniature hole drilling technology. *IEEE Transactions on Plasma Science, 32*(5 I), 2055–2061. https://doi.org/10.1109/TPS.2004.835489

Tsujino, M., et al. (2007). Application of conventionally recycled coarse aggregate to concrete structure by surface modification treatment. *Journal of Advanced Concrete Technology, 5*(1), 13–25.

Villoria-Sáez, P., Porras-Amores, C., & del Río Merino, M. (2020). *Estimation of construction and demolition waste*. Elsevier BV. https://doi.org/10.1016/b978-0-12-819055-5.00002-4

Wang, L., et al. (2017). An environmentally friendly method to improve the quality of recycled concrete aggregates. *Construction and Building Materials, 144*, 432–441. https://doi.org/10.1016/j.conbuildmat.2017.03.191

Wu, H., et al. (2021). Utilizing thermal activation treatment to improve the properties of waste cementitious powder and its newmade cementitious materials. *Journal of Cleaner Production, 322*(August), 129074. https://doi.org/10.1016/j.jclepro.2021.129074

Xiao, J., et al. (2013). Properties of interfacial transition zones in recycled aggregate concrete tested by nanoindentation. *Cement and Concrete Composites, 37*(1), 276–292. https://doi.org/10.1016/j.cemconcomp.2013.01.006

Xiao, J., Li, J., & Zhang, C. (2005). Mechanical properties of recycled aggregate concrete under uniaxial loading. *Cement and Concrete Research, 35*(6), 1187–1194. https://doi.org/10.1016/j.cemconres.2004.09.020

Recycled concrete for structural applications

6

João Pacheco[1], *Jorge de Brito*[2], *Carlos Chastre*[3] *and Luís Evangelista*[4]
[1]CERIS, C5Lab - Sustainable Construction Materials Association, Edifício Central Park, Linda a Velha, Portugal; [2]CERIS, Department of Civil Engineering, Architecture and Georesources, Instituto Superior Técnico, Universidade de Lisboa, Lisbon, Portugal; [3]CERIS, Department of Civil Engineering, Faculdade de Ciências e Tecnologia, Universidade Nova de Lisboa, Caparica, Portugal; [4]CERIS, Department of Civil Engineering, Instituto Superior de Engenharia de Lisboa, Rua Conselheiro Emídio Navarro, Lisbon, Portugal

6.1 Coarse recycled aggregates for concrete

Recycled aggregates are produced from processing construction and demolition waste. Construction and demolition waste is preliminarily sorted, in order to remove unintended deleterious materials (such as asbestos and gypsum-based materials), and recycled aggregates are then produced through a processing stage that includes several steps that remove the finer particles (mostly composed of soils and clay), lightweight materials, and other materials present in the construction and demolition waste being processed, since they are not suitable for use as recycled aggregates for concrete. Crushing and sieving are also part of this process (Pacheco & de Brito, 2021a, 2021b). Different standards and national specifications (EN-12620+A1, 2008; LNEC-E471, 2009; JIS-5022, 2012; JIS-5023, 2012; JIS-5021, 2011) define requirements and recommendations for the use of recycled aggregates on concrete. In general terms, most of these documents either stipulate or recommend that recycled aggregates for structural concrete comply with the following conditions:

- Recycled aggregates should be composed mainly of concrete waste and the content of asbestos, metals, floating materials, gypsum materials, wood and plastics should be minimum;
- The water absorption should comply with a maximum value (e.g., 7% (LNEC-E471, 2009)) and density should be above a minimum value (e.g., 2200 kg/m^3 (LNEC-E471, 2009));
- The fines content is limited (e.g., 4% (LNEC-E471, 2009));
- Chemical contamination should be small. As in the case of natural aggregates (NA), the content of chlorides, sulfates and other substances (e.g., sugars) should not impair the fresh and hardened behavior of concrete;
- Only the coarse fraction of the recycled aggregate particles may be used.

Recycled aggregates produced from concrete waste will likely comply with these requirements. Fig. 6.1 presents a recycled aggregate particle produced from concrete waste and an NA. As seen in the figure, while the NA is composed of only stone, the recycled aggregate is composed of stone (which is, in fact, an NA) and attached mortar.

Recycled Concrete. https://doi.org/10.1016/B978-0-323-85210-4.00011-4

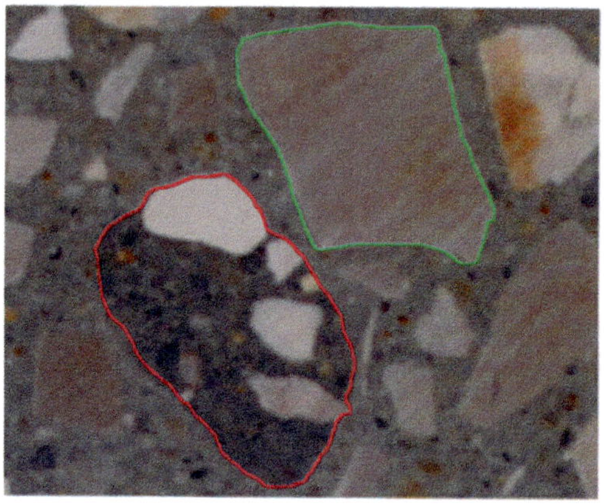

Figure 6.1 Natural aggregate (contour in green) and recycled aggregate produced from concrete waste (contour in red).

One of the aforementioned conditions is that only the coarse fraction of the recycled aggregates may be used. This is imposed in some standards since the coarse fraction of recycled aggregates tends to have better properties than the fine fraction (Sato et al., 2007). In the specific case of recycled aggregates produced from concrete waste, this is due to the coarser particles having lower content of attached mortar than finer ones (Artoni et al., 2016; Sato et al., 2007) because crushing tends to disaggregate the attached mortar before breaking stone (this results in finer particles being mainly composed of mortar (Artoni et al., 2016)). Table 6.1 presents the relationship between

Table 6.1 Influence of particle size on the attached mortar content of recycled aggregates produced from concrete waste.

RA source	Publication	Parameter			
Source 1	Abbas et al. (2008)	Particle size (mm)	4.75–9.5	9.5–12.7	12.7–19.0
		Attached mortar content	43%	39%	42%
Source 2	Abbas et al. (2008)	Particle size (mm)	4.75–9.5	9.5–12.7	12.7–19.0
		Attached mortar content	30%	21%	21%
Source 3	Sanchez de Juan and Alaejos Gutierrez (2009)	Particle size (mm)	4–8	8–16	–
		Attached mortar content	44%	34%	–

particle size and content of attached mortar of two publications on this topic (Abbas et al., 2008; Sanchez de Juan & Alaejos Gutierrez, 2009).

This chapter assumes that all common requirements presented in this section are complied with. Moreover, all mentions of recycled aggregates are meant to be understood as coarse recycled aggregate produced from concrete waste (henceforth, RA) and recycled aggregate concrete (RAC) is defined as concrete made with total or partial incorporation of this type of recycled aggregate.

6.2 Coarse recycled aggregate produced from concrete waste

The differences between the properties of NA and RA are relevant since aggregates usually account for 70% of the volume of concrete (Alexander & Mindess, 2005). The influence of an aggregate on concrete ranges from water demand and workability, to stiffness, strength and durability (Alexander & Mindess, 2005; Mehta & Monteiro, 2006). In general terms, aggregates should be relatively round and strong and their chemical composition should not favor chemical reactions (e.g., alkali-silica reactions) with cement (Alexander & Mindess, 2005; Neville & Brooks, 2010).

While NA are composed of stone and are a homogenous material, RA are made of stone and attached mortar. The attached mortar is rougher than stone, porous, and typically weaker. Additionally, the shape of RA is more elongated than that of NA (de Brito et al., 2016; Pedro et al., 2014, pp. 1−14) because most construction and demolition waste plants are focused on waste reception and sorting (rather than on the production of high-quality RA) and their production process may be simple (Pacheco & de Brito, 2021a, 2021b). This means that, while NA may be produced with up to three crushing stages and combining different equipment (e.g., primary jaw crushers followed by hammer mill or cone crushers) to ensure that the particles are well-shaped, RA may be produced with just primary jaw or impact crushing.

These aspects directly affect the properties of RA. Typical properties of RA are:

- 24-hour water absorption of about 5%;
- Saturated-surface-dry density of 2350 kg/m^3 and oven-dry density of 2250 kg/m^3;
- The attached mortar content is variable, depends on particle size, types and number of crushers used, and on the composition of the concrete waste. As presented in Table 6.1, the content of attached mortar is usually in the region of 20%−50% of the total mass of the RA;
- The content of fines is mostly dependent on the production process used (namely the crushing and sieving equipment used) and on the humidity conditions of RA at the time of sieving.

The values just reported are reference values based on data from several publications (Abbas et al., 2008, 2009; Agrela et al., 2011; Butler, 2012; Henry et al., 2011; Pepe et al., 2016; Sanchez de Juan & Alaejos Gutierrez, 2009; Silva et al., 2014). Since the properties of RA depend on the composition of the concrete waste used and grading, the actual properties of an RA derived from a specific source are somewhat uncertain. For instance, the 24-hour water absorption is usually in the range of 3.5%–6.0% and the saturated-surface-dry density is typically of 2300 kg/m^3 to 2500 kg/m^3 (Agrela et al., 2011; Silva et al., 2014). Furthermore, some of the trends reported may not be applicable to particular cases. An example is the production of RA with maximum aggregate diameter (D_{max}) that is larger than that of the concrete waste. In this case, the largest particles of the RA may have high attached mortar content (Albuquerque et al., 2021).

Concerning strength, RA are weaker than NA. This is well-understood from the single particle compression tests presented in Pepe et al. (2018), in which the RA were found to be weaker than NA. Also, a metanalysis presented in Pacheco et al. (2021) compared the aggregate crushing value of limestone NA (the most common type of NA used worldwide) with that of RA and found that the mean aggregate crushing value of RA is 1.7 times that of limestone NA.

The very large water absorption of RA means that, unless RA are saturated at the time of mixing, a significant portion of the mixing water will be absorbed by them. Because of this, the mix design of RAC is often presented in terms of the effective water and the effective water/binder ratio. These parameters are calculated by subtracting the water absorbed by RA from the total water. This requires not only testing the water absorbing of RA, but also measuring the water content of RA at the time of mixing.

The properties of RA can be improved with beneficiation methods, which are based on the reasoning that the attached mortar is responsible for the decrease of the quality of RA in comparison to NA. So, these methods are based on either the removal of attached mortar (e.g., through thermal heating (Shima et al., 2005), microwave heating (Akbarnezhad et al., 2011) or acid presoaking (Tam et al., 2007)), or on the treatment of the surface of the particles of the RA (Tsujino et al., 2007), which offsets the detrimental effects caused by the porosity of the attached mortar. Beneficiation methods are not a standard solution for the production of RA due to difficulties in industrial implementation, cost, and environmental impacts. Therefore, they are not treated in this chapter.

6.3 Mechanical behavior of recycled aggregate concrete

The mechanical behavior of RAC differs from that of natural aggregate concrete (NAC) and this influences structural behavior. There are two reasons for the incorporation of RA resulting in a decrease of the mechanical and durability properties of concrete and they are both directly related to the properties of RA presented in the last section:

- RA are weaker, less stiff, and more porous than NA and this has direct detrimental effects on the strength, modulus of elasticity and durability of RAC;
- The roughness and elongated shape of RA means that, for the same content of binder, the effective water content and the effective water/binder content of RAC tends to be larger than that of NAC and this will decrease the hardened-state mechanical and durability properties of concrete.

Roughness and shape influence fresh-state behavior because elongated particles roll and slide with more difficulty than round particles and because rough aggregates have higher surface area, thus increasing friction between aggregates (Alexander & Mindess, 2005; Mehta & Monteiro, 2006).

An alternative way to offset the influence of the incorporation of RA on the fresh-state behavior of concrete is to increase the content of binder. While higher effective water/binder ratios decrease the properties of the binder paste, of the interfacial transition zone between aggregates and binder paste (ITZ), and ultimately of concrete, increases of binder content are usually a worse option and should be avoided because they lead to larger cost and environmental impacts since they result in higher clinker content.

The effect of RA on the most common mechanical properties of concrete is now briefly summarized.

The compressive strength of concrete (f_c) decreases when RA are incorporated. For total incorporation of RA, this decrease is in the region of 5%–15% (Etxeberria et al., 2007; Kou & Poon, 2012; Malešev et al., 2010; Rao et al., 2011; Sato et al., 2007) and is mostly caused by the increase of the water/binder ratio. Moreover, the relationship between incorporation ratio of RA and f_c is approximately linear (Soares et al., 2014). Concerning stress-strain behavior, there is no consensus on the influence of the incorporation of RA. The main reasons for the incorporation of RA influencing the stress-strain relationship are that RA are more deformable than stone (so, they are more compatible with the new mortar of concrete and this is expected to lead to higher elastic strains (Xiao et al., 2005)) and that the crack development of RAC differs from that of NAC (with larger occurrences of transaggregate fracture surfaces, influencing inelastic strains (Guo et al., 2017; Xiao et al., 2012)). Examples of contradictory findings are those of Xiao et al. (2005) and Li et al. (2012), in which it was found that the incorporation of RA leads to larger elastic strains and increased brittleness of the softening branch of the stress-strain relationship, and the finding of González-Fonteboa et al. (2011) that the strain at peak strength and the maximum strain increased when RA were incorporated. No consensus exists because the effect of RA on the stress-strain relationship will depend on the properties of the RA used, on previous damage of the RA (b0th during the lifetime of the source concrete and due to the crushing of the source concrete into RA), and on the compositions of the concrete tested.

The stiffness of coarse aggregates is a determining factor for Young's modulus, which is largely influenced by the incorporation of RA due to the deformability of the attached mortar (Xiao et al., 2013). The majority of research agrees that the decrease of Young's modulus caused by the incorporation of RA is linear with their ratio (González-Fonteboa, Martínez-Abella, et al., 2011; Pacheco et al., 2019a, 2019b, 2019c, 2019d; Soares et al., 2014)—this linear relationship is justified by the fact that the Young's modulus is defined for elastic behavior. Therefore, the contribution of NA and RA toward the stiffness of concrete is proportional to their content. Since the Young's modulus of concrete depends strongly on the type of NA used (Alexander, 1996; Noguchi et al., 2009), Fig. 6.2, produced with data presented in Pacheco et al. (2019a, 2019b, 2019c, 2019d), compares the f_c versus Young's modulus

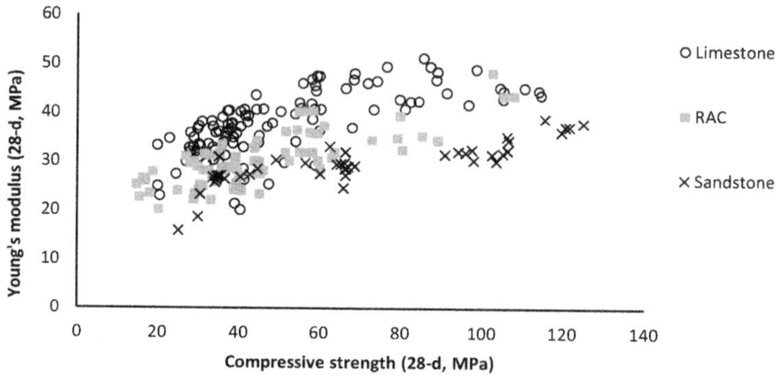

Figure 6.2 Influence of the coarse aggregate on the compressive strength versus Young's modulus relationship.

relationship of NAC made with limestone, NAC made with sandstone, and RAC made with full incorporation of RA. For concrete with the same f_c, RA results in lower Young's modulus than limestone, but in higher Young's modulus than sandstone.

The database collected in Pacheco et al. (2019a, 2019b, 2019c, 2019d) included only cylindrical specimens for both f_c and Young's modulus testing. In comparison with limestone NA, full incorporation of RA was found to result in a mean decrease of Young's modulus of 22%. This means that the Young's modulus is one of the properties that is the most affected by the use of RA.

Similar studies found that the incorporation of RA increases shrinkage (Silva et al., 2015a, 2015b; Tošić et al., 2018) and creep (Silva et al., 2015a, 2015b; Tošić et al., 2019). When the shrinkage and creep models of *fib* Model Code 2010 (Anonymous, 2010) are compared with experimental data of shrinkage (Tošić et al., 2018) and creep (Tošić et al., 2019), average increases of 39% and 32% are found for shrinkage and creep, respectively, in comparison with NAC of the same f_c. The main reason for this finding is the same as in the case of the Young's modulus: these properties depend strongly on the stiffness of the aggregates and RA are more deformable than NA.

Findings on the tensile strength of RAC are not as consensual, since this property has been found to either: (i) marginally increase or be unaffected (Butler et al., 2013; Etxeberria et al., 2007; Malešev et al., 2010) by the incorporation of RA; as well as to (ii) decrease when RA are used (Kheder & Al-Windaw, 2005; Ozbakkaloglu et al., 2018; Pacheco et al., 2019a, 2019b, 2019c, 2019d). These contradictory findings are caused by two aspects:

- The quality of the RA is a determining factor for this property and high quality RA may increase the tensile strength (Soares et al., 2014), while poor quality RA decrease it (González-Fonteboa, Martínez-Abella, et al., 2011);
- The tensile strength depends more on the quality of the ITZ than the f_c and Young's modulus (Mehta & Monteiro, 2006). The incorporation of RA has contradictory effects on the ITZ (on the one hand, the effective water/binder ratio increases and the ITZ becomes worse; on the other hand, the roughness of RA and "nailing" of the attached mortar to the new paste (Ferreira et al., 2011) improve the ITZ).

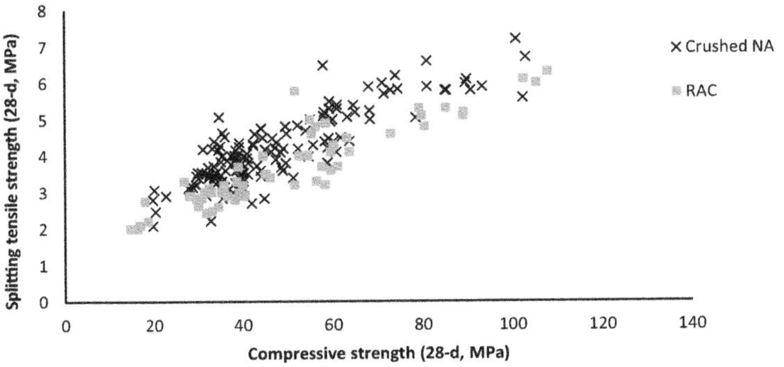

Figure 6.3 Influence of recycled aggregates on the compressive strength versus tensile strength relationship.

Fig. 6.3 was produced from data collected in Pacheco et al. (2019a, 2019b, 2019c, 2019d). Both f_c and tensile strength were tested in cylindrical specimens. The figure shows that the incorporation of RA tends to decrease the tensile strength of concrete. Through meta-analysis, it is argued in Pacheco et al. (2019a, 2019b, 2019c, 2019d) that total incorporation of RA results in an average 15% decrease of the tensile strength in comparison with NAC produced with crushed NA and with the same f_c.

Fracture energy (G_F), which evaluates the ease of crack propagation (Darwin et al., 2001), decreases when RA are used. This is understood by the statistics presented in Table 6.2, which show the ratio between experimental values of fracture energy and the constitutive model of Model Code 2010 (Anonymous, 2010). This constitutive model depends only on the mean cylindrical compressive strength (f_{cm}):

$$G_F = 73 \cdot f_{cm}^{0.18} \tag{6.1}$$

Fracture energy depends on testing procedure, test setup and specimen geometry (Mechtcherine & Müller, 1998). The NAC and RAC mixes analyzed in Table 6.2 (Arezoumandi, Smith, et al., 2015; Bordelon et al., 2009; Butler, 2012; Butler et al., 2013; Casuccio et al., 2008; García-González et al., 2016; Ghorbel et al., 2019; Ghorbel & Wardeh, 2017; Guo et al., 2017) were sourced from the same publications, ensuring an unbiased comparison.

The detrimental effect of RA on fracture energy is validated by mechanical considerations. Fracture energy depends on the pattern of crack propagation inside the

Table 6.2 Influence full incorporation of RA on the fracture energy.

Concrete	Number of mixes	Mean value ($G_{F,test}$/ $G_{F,MC2010}$)
NAC	18	0.92
RAC	31	0.72

concrete element (Peterson, 1981) and depends on the strength of the aggregates (fib, 2013). The strength of aggregates influences fracture energy because:

- Strong aggregates impose that the fracture surface is developed around them. In this case, the fracture surface is tortuous since cracks change orientation when they reach the aggregates;
- On the other hand, weak aggregates are associated with planar fracture surfaces characterized by broken aggregates.

Tortuous fracture surfaces are associated with higher fracture energies, since the effective surface of the aggregates (and the overall length of cracks) is larger for the same unit area of crack development. Experimental evidence shows not only that RA are weaker than NA (Pepe et al., 2018), but also that the fracture surfaces of RAC tend to break through aggregates more often than in the case of NAC (Guo et al., 2017).

As for fracture energy, the shear strength is detrimentally affected by weak aggregates. Therefore, shear strength decreases when RA are incorporated. This hypothesis is confirmed by analyzing push-off experiments on the shear strength of RAC. For full incorporation of RA:

- An 11% decrease was observed in (Xiao et al., 2016) when RA were incorporated by volume;
- When RA are incorporated by weight, the decrease of shear strength is expected to be in the region of 5%–14% (Waseem & Singh, 2016, 2018).

The main rationale for the decrease of shear strength is understood from the basic mechanism of shear stress transfer. Concrete withstands shear stresses through dowel action, friction, and aggregate interlock (Walraven, 1980). Dowel action and friction depend on reinforcement, but, as shown in Fig. 6.4, aggregate interlock is mobilized through the stresses transferred between aggregates and the failure surface. Therefore, aggregate interlock is affected by the strength of coarse aggregates and by crack width.

From this explanation and assuming that other relevant parameters (D_{max}, the content of aggregates, and the strength of the binder paste) are unaffected by the incorporation of RA, RAC will have a smaller shear strength than NAC. This is due to three reasons:

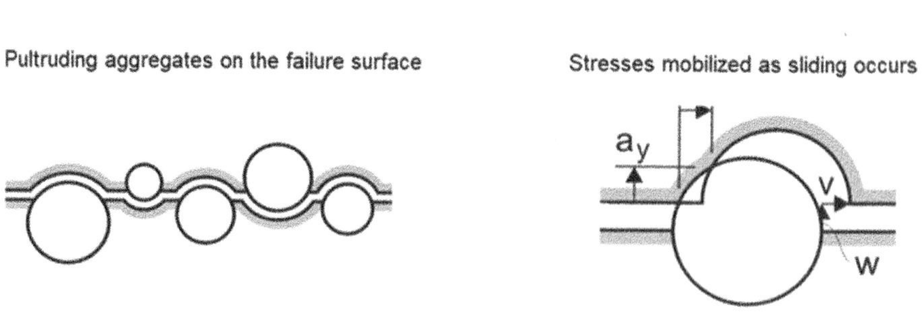

Figure 6.4 Aggregate interlock mechanism.
Adapted from Walraven, J. (1980). Aggregate interlock: A theoretical and experimental analysis, in Concrete structure's group. TU Delft: Delft, the Netherlands.

- As argued for fracture energy, RA are weaker and cracks will likely break through them;
- The crack surfaces of RAC are smoother than those of NAC because of the attached mortar (Fathifazl, Razaqpur, Isgor, et al., 2009) and this will reduce friction;
- Some authors observed that crack widths increase when RA are used (Ignjatović et al., 2017; Wardeh et al., 2018) and larger crack widths reduce aggregate interlock (Pacheco et al., 2020).

As understood from this section, the resistance models used for NAC design may not be as reliable when used for recycled aggregate concrete. This hypothesis is backed by the fact that the incorporation of RA influences each mechanical property of concrete differently and most of the ultimate limit state design provisions of codes are made in terms of f_c. Therefore, the influence of the incorporation of RA on other properties that affect mechanical behavior, such as tensile shear and shear strength, is not accounted for by the models used in codes.

6.4 Overview of the structural behavior of recycled aggregate concrete

The general consensus is that the structural behavior of RAC follows the same behavioral pattern found for NAC structural elements (Arezoumandi, Drury, et al., 2015; Choi et al., 2010; Ignjatović et al., 2017; Pacheco et al., 2015a, 2015b). Typically, authors state that the load-bearing capacity is only slightly affected but the effect of RA may be relevant for some resistance mechanisms (Pacheco et al., 2015a, 2015b; Pacheco et al., 2021). Deflections tend to be larger, due to the lower Young's modulus of RAC in comparison to NAC (Pacheco et al., 2015a, 2015b).

Some authors have proposed alternative production systems that mitigate the detrimental effect of RA, such as slabs produced with layered natural and recycled aggregates (Tošić et al., 2020; Xiao et al., 2015) and semi precast solutions where RAC is placed on an inner core and is surrounded by NAC (Xiao et al., 2014; Yang et al., 2017). All considerations presented in this chapter concern structural elements produced with conventional methods (that is, elements made with typical ready-mixed, precast, or onsite production) and alternative production methods are outside the scope of the chapter.

Most studies are made by producing NAC and RAC elements with the same concrete mix design, apart from slight changes (e.g., in the water/binder ratio). Then, these elements are tested up to failure and test parameters (e.g., resistance, cracking, and load vs. deflections relationships) are compared.

A general appraisal of research on the structural behavior of RAC with full incorporation of RA (Pacheco et al., 2015a, 2015b) found that the typical decrease of load-bearing capacity is in the region of 3%–7%, that deflections increase up to 10%, and that ductility, when evaluated by energy dissipation, is affected by less than 10%. Consequences for different types of structural elements are now stated:

- The cyclic behavior of RAC beam-column joints (Corinaldesi et al., 2011; Gonzalez & Moriconi, 2014; Xue et al., 2017) suits structural applications;
- The bending resistance of reinforced concrete beams is typically unaffected by the incorporation of RA (Arezoumandi, Smith, et al., 2015; Choi et al., 2012; Fathifazl, Razaqpur, Burkan Isgor, et al., 2009; Ignjatović et al., 2012);
- The shear resistance of RAC elements with shear reinforcement has been observed to be unaffected by the incorporation of RA in the majority of tests (Arezoumandi, Drury, et al., 2015; Choi et al., 2010; Fathifazl et al., 2010; González-Fonteboa & Martínez-Abella, 2007; Ignjatović et al., 2017);
- When shear reinforcement is absent, as is the case of slabs, most authors report decreases of the shear resistance (Choi & Yun, 2017; Fathifazl, Razaqpur, Isgor, et al., 2009);
- The normalized bond strength has been reported as either being unaffected/slightly increasing (Ajdukiewicz & Kliszczewicz, 2002; Breccolotti & Materazzi, 2013; Choi & Kang, 2008; Kim et al., 2012; Xiao & Falkner, 2007) by the incorporation of RA, or as decreasing (Butler et al., 2011, 2015; Seara-Paz et al., 2013; Wang, 2016);
- The seismic behavior of frame structures has been reported as being mostly unaffected by the use of RA, either when under cyclic loading (Wang & Xiao, 2013; Xiao et al., 2006) or when monotonic pushover tests are performed (Pacheco et al., 2015a, 2015b).

From these findings, it is concluded that the resistance of elements that are governed by the behavior of reinforcement (e.g., the bending resistance of properly-designed beams) is negligibly affected by the incorporation of RA. The same is valid for ductility, which is mostly dependent on reinforcement and detailing. However, when the resistance mechanism depends mostly on concrete (e.g., the shear resistance of elements without shear reinforcement), the incorporation of RA is expected to decrease the resistance of the structural member.

6.5 Model uncertainty of recycled aggregate concrete design

6.5.1 General concept

The last section summarized research on the structural behavior of reinforced RAC members. The findings of research programs on different resistance mechanisms in which both NAC and RAC specimens were produced and tested were stated in order to understand whether the incorporation of RA influences structural behavior. However, when it comes to quantifying the effect of the incorporation of RA on structural behavior, this is not the most technically sound option. Experiments on the structural behavior of reinforced concrete members are made for a specific concrete mix design and source of raw materials (including the NA and RA used), geometry, and reinforcement detailing. This means that specificities of the experiments skew the findings of each study. Two examples are provided:

- Concerning the concrete mix design of the mixes that will be compared (and not that of the concrete waste used to produce the RA), research in which the NA used is a high-strength aggregate (e.g., basalt) will lead to larger detrimental influence of the incorporation of RA in comparison to research in which other types of NA are used;
- Concerning reinforcement detailing and geometry, specimens whose resistance mechanism relies more on the properties of concrete will emphasize the effect of RA over other resistance mechanisms (e.g., lightly-reinforced elements subjected to bending will depend less on the properties of the compressed region of concrete - hence of RA, since the tensioned reinforcement will limit resistance).

To address this issue and ensure a fair analysis, this chapter evaluates the influence of the incorporation of RA through meta-analysis from several publications. The following resistance mechanisms are analyzed:

- Bond strength;
- Bending resistance;
- Shear resistance of elements without shear reinforcement;
- Shear resistance of elements with shear reinforcement.

This analysis is made through the model uncertainty (θ_R), which compares expectations of resistance with experimental results. The most common definition of θ_R for structural engineering purposes is used in this paper and is the following:

$$\theta_R = \frac{\text{Experimental resistance}}{\text{Predicted resistance}} \tag{6.2}$$

By definition, θ_R is evaluated for a resistance model. Since what is being evaluated is the suitability of the resistance model, the predicted resistance is calculated omitting partial factors from the resistance model and using mean values of material properties instead of characteristic values.

This chapter resorts to the resistance models presented in EN1992, Part 1 (EN-1992-1-1, 2008) except for the bond strength. This option is justified in subsection 6.5.3.

For each resistance model analyzed, all publications found by the authors that comply with strict validation criteria are gathered, a database for NAC and RAC is built, and the model uncertainty is calculated and used to assess the influence of RA. When θ_R is calculated for databases comprising several experiments, its statistics are used to assess the accuracy and prediction of resistance models:

- Mean values of θ_R close to 1.0 show that the model is accurate (experimental data is closely estimated by the resistance model), while mean values below 1.0 concern resistance models that underestimate resistance. Conversely, if the mean value of a resistance model is above 1.0, the resistance model is conservative;
- The precision of the resistance model is evaluated through the coefficient of variation (CoV) of θ_R. The larger the CoV, the less precise the prediction of the resistance model is. For most resistance models for reinforced concrete, the CoV of θ_R is between 5% and 10%. However, some resistance models are not as precise and their CoV may be larger (e.g., resistance models for shear resistance are expected to have CoV in the range of 10%–25% (Yousefpour & Bayrak, 2016).

Since θ_R is compiled with data coming from several experiments, the comparison of the θ_R of databases comprising NAC and RAC experiments is ideal to evaluate the influence of the incorporation of RA on structural behavior. This is true because, in comparison to evaluating NAC and RAC data coming from a single experiment, the evaluation of the θ_R of NAC and RAC minimizes the influence of parameters related to mix design and/or structural design that may prevent an unbiased assessment of the influence of RA when single experiments are analyzed.

The incorporation of RA is expected to influence the resistance of reinforced concrete (therefore, θ_R) because resistance mechanisms depend on several factors that are affected by the use of RA:

- RA are more heterogeneous than NA (Khoury et al., 2018; Pepe et al., 2018) and this increases the statistical size effects (Bažant & Yavari, 2005) of concrete (Pacheco et al., 2019a, 2019b, 2019c, 2019d). Resistance models are based on the cylindrical f_c, but higher statistical size effects result in larger different between the strength of specimens and the actual f_c of the structural element;
- Since RA are weaker than typical NA (Pepe et al., 2018), resistance mechanisms are affected by the incorporation of RA. The implications of this fact are understood in the discussion of shear strength and fracture energy presented in Section 6.3;
- As also argued in Section 6.3, the only parameter related to concrete of most resistance models is f_c, but each mechanical property is affected by the incorporation of RA to different extents.

6.5.2 Methodology

The methodology used to determine the θ_R presented in this chapter is based on the following considerations:

- Two databases were built, one for NAC and another for RAC with full incorporation of RA (RAC100). Databases with intermediate incorporation ratios of RA were not made because the number of experiments that would be included in such databases would be low;
- For an experiment to be included in a database, both NAC and RAC100 elements had to be tested. This option was taken to minimize the influence of mix design, geometry and detailing on the comparison of the θ_R of NAC and RAC100.

After gathering publications for a given resistance mechanism, a set of criteria is used to validate each publication. In general:

- The failure mode needs to be checked to make sure that the limiting resistance mechanism is that of the resistance model (for instance, beams reported as being tested for bending resistance were removed whenever failure was, in fact, due to shear or bond);
- All material properties included in the resistance model need to be tested (extrapolations from constitutive models or specified/characteristic properties are not used). None of studies state whether the tested values reported for the yield strength of the reinforcement (f_y) concern the proof strength, nonproportional extension or the yield stress as defined in EN 10002-1 (EN-10002-1, 2001). They are both assumed as equivalent and equal to f_y;
- An exception to this condition is the Young's modulus of the reinforcement. Whenever unreported, this property is assumed as 200 GPa;

- Geometry, detailing and concrete cover need to be stated and have to comply with the rules of EN1992, Part 1 (EN-1992-1-1, 2008);
- Test conditions have to be stated and conform with standards (material properties) and good practice (structural tests);
- Specific criteria for each resistance model are presented in the next subsections of this chapter.

After the databases are built, the θ_R of each test is calculated as shown in Eq. (6.1). Then, the mean and sample standard deviation are calculated (and the CoV is determined from both). These statistics are used to assess the influence of the full incorporation of RA on the accuracy and precision of the resistance models since this method results in statistically unbiased estimates of the mean and the CoV of the population that the database represents (Holicky et al., 2013).

No experiments were removed with outlier tests. This is common to most studies on θ_R and avoids erroneously removing experiments that correspond to the lower tail of probability distributions (Allen et al., 2005). Furthermore, it is assumed that hypothetical outliers due to poorly-executed experiments were already removed from the publications appraised by their corresponding authors.

The methodology just presented agrees with that presented in Annex C: "Design based on observations and experimental models" of ISO2394 (2014). Therefore, these θ_R may also be used for reliability analyses, which are outside the scope of this chapter.

A final check is made concerning the influence of the parameters of design on θ_R. This is made through linear correlations and scatterplots that check whether θ_R tends to affected as the values of the parameters of design change. The analysis of the coefficients of correlation is made following the criterion presented in Table 6.3 (Sykora et al., 2018). Since coefficients of correlation are only useful to check for linear correlations, scatterplots are used to visually check whether any nonlinear dependence exists (Benjamin & Cornell, 1970).

Typically, this type of analysis is made to ensure that a resistance model represents the mechanical phenomena well (Sýkora & Holicky, 2013). In the case of this chapter, this analysis will also serve to better understand the influence of the incorporation of RA on θ_R. As an example, if the f_c has negligible effect on the θ_R of the NAC database, but the same is not found for the RAC100 database, this strongly suggests that the strength of RAC (and, most probably, of RA) is limiting resistance (Pacheco et al., 2020, 2021).

Table 6.3 Classification of the coefficients of correlation (Sykora et al., 2018).

Classification of correlation	Weak	Medium	Strong
Absolute value of the coefficient of correlation	Below 0.30	Between 0.30 and 0.50	Above 0.50

6.5.3 Bond strength

Bond stress (f_b) is typically assumed as uniform along the full development length (l_b) of the reinforcement. Under this condition, and assuming that l_b is measured from an unbonded end of a bar, the stress mobilized in the bar through bond (f_{stm}) is given by:

$$f_{stm} = \frac{4 \times f_b \times l_b}{\phi} \tag{6.3}$$

In Eq. (6.2), ϕ represents the diameter of the reinforcement.

This subchapter analyzes the θ_R of the bond strength model of *fib* Bulletin 72 (fib, 2014). This option was taken because this bond strength model has prenormative character, is used in Model Code 2010 (Anonymous, 2010) and is incorporated on the draft of the next version of EN1992, Part 1 (CEN/TC-250/SC-2, 2020). Furthermore, both *fib* Bulletin 72 (fib, 2014) and background documents (PT-SC2-T1 D3BG, 2018) to the revision of EN1992 argue that the bond strength model of the current version of the code is empirical and not fully satisfactory.

The model developed in *fib* Bulletin 72 (fib, 2014) is semiempirical and is based on lap splice tests only. This model defines f_{stm} as:

$$f_{stm}(MPa) = \left[54\left(\frac{f_c}{25}\right)^{0.25}\left(\frac{25}{\phi}\right)^{0.2}\left(\frac{l_b}{\phi}\right)^{0.55}\right] \times \left[\left(\frac{c_{min}}{\phi}\right)^{0.25}\left(\frac{c_{max}}{c_{min}}\right)^{0.1} + k_m \times K_r\right] \tag{6.4}$$

In this model, f_c is in MPa must be in the range of 15−110 MPa; ϕ is in mm; $l_b \geq 10\ \phi$; c_{min} and c_{max} are from parameters related to concrete cover and $c_{min}/\ \phi$ must be between 0.5 and 3.5, while $25/\phi$ must be below 2.0. k_m and K_r are related to transverse reinforcement, which is absent in all tests appraised. Therefore, $k_m \times K_r = 0$.

Furthermore, f_{stm} must be limited to f_y and to $10\sqrt{f_{cm}} \times \left(l_b/\phi\right)$. The latter threshold accounts for strength limitations due to pullout failure.

As a matter of fact, Eq. (6.4) accounts for the fact that bond strength is limited by a combination of pullout failure and splitting failure (fib, 2014). Pullout failures occur through shearing of the concrete that surrounds the ribs of the reinforcement and occurs chiefly in highly-confined concrete elements (due to reinforcement or large concrete cover). Splitting failures occur due to the radial bursting stresses that develop from the concrete near the reinforcement as bond strength is mobilized. These stresses crack the concrete cover and are mostly dependent on the tensile strength of concrete (ACI-408, 2003). For typical cases of bond length design, the splitting tensile mechanism is more limiting than pullout. This is one of the reasons why the bond strength mechanisms of ACI-408 (2003) and *fib* Bulletin 72 (fib, 2014) were developed based on databases that include lap splice and beam end specimens only, since pullout strength tests are not as representative of actual reinforcement concrete design conditions (ACI-408, 2003; Cairns & Plizzari, 2003).

Eq. (6.4) shows that f_c is the only property of concrete included in the resistance model. However, the pullout strength mechanism is limited by the strength of aggregates (fib, 2009) and the splitting failure mechanism by the tensile strength. As seen in Section 6.3, RA are weaker than NA and Section 6.4 showed that, for the same f_c, the tensile strength of RAC is smaller than that of NAC. This means that the θ_R of the bond strength model of *fib* Bulletin 72 (fib, 2014) will presumably have a lower mean θ_R when used for RAC. This reduces the safety of RAC bond length design.

These hypotheses are backed by findings that higher aggregate crushing values are associated with lower bond strength (Butler et al., 2015) and with observations that, in bond strength tests, there are more instances of local crushing of RA around the ribs of reinforcement than what is reported for NA (Wang, 2016).

The database for bond strength was built from lap splice tests. Only straight deformed reinforcement bars are included. Specific criteria for database removal were:

- Any of the parameters required by Eq. (6.4) was lacking;
- Any of the conditions presented just below Eq. (6.4) was not complied with (e.g., c_{min}/ϕ was lower than 0.5 or higher than 3.5);
- The reinforcement yielded prior to bond failure. To check yielding, the mean tested value for f_y was used, unless this property was not tested. In this case, the characteristic value was used (it was assumed that f_y was at least as large as its characteristic value);
- Plain or corroded bars were not included in the database.

Notwithstanding the requirement that $(25/\phi)$ be limited to 2.0, lap splice tests on bars with $\phi = 12$ mm were included and Eq. (6.4) was computed with $(25/\phi) = 2.08$.

The number of tests that complied with these criteria is presented in Table 6.4. This table also shows the sample mean, standard deviation, and CoV of the θ_R of each dataset. The databases were built from references (Butler et al., 2015; Fathifazl et al., 2012; Gaurav & Singh, 2018; Hamad et al., 2018; Prince et al., 2017).

Several histograms were made to check the distribution of the parameters of design of the databases. The databases were found to be similar. Fig. 6.5 shows two of the histograms.

Table 6.4 shows that the mean θ_R decreases when RA are used by 8%. Furthermore, its CoV increases by 33%. Both effects have a detrimental influence for structural safety. The increase of the CoV is caused by the variability of the properties of the RA (Pacheco et al., 2019a, 2019b, 2019c, 2019d), while the decrease of the

Table 6.4 Statistics of θ_R, of bond strength tests.

Database	Number of tests	Mean	Standard deviation	CoV
NAC	40	1.03	0.15	15%
RAC100	56	0.95	0.19	20%

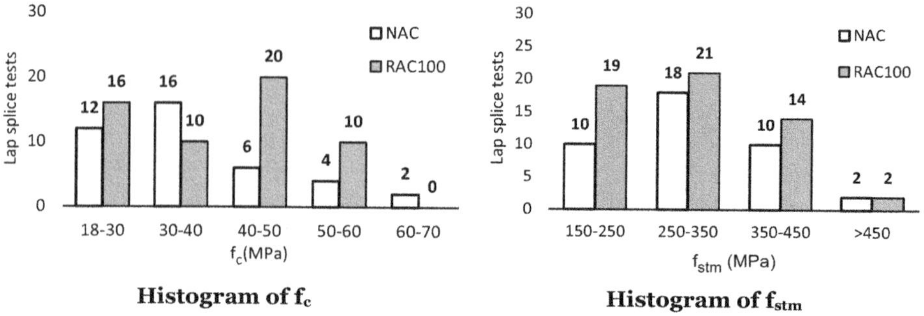

Figure 6.5 Histograms of the databases used for bond strength assessment.

mean θ_R is due to the aforementioned aspects: RA are weaker than NA and the tensile strength of RAC tends to be smaller than that of NAC with the same f_c.

Fig. 6.6 presents a scatterplot between f_c and θ_R. The θ_R of RAC100 decreases as f_c increases. However, this is not observed for NAC. From this finding, two conclusions are put forward:

- The hypothesis that θ_R decreases because RA are weaker than NA is validated. As f_c increases, the strength of the binder paste becomes higher and the strenght of the coarse aggregate starts to become a limiting factor (and this is found to be relevant in the case of RAC since RA are weak);
- For high-strength concrete, the decrease of θ_R due to the use of RA is more relevant. Therefore, it is advised that RAC applications prioritize low to medium strength concrete.

The detrimental effects of RA on the mean and CoV of θ_R should be accounted for and safety factors should be included in the design formulae of bond strength (or bond length) models of codes (Pacheco et al., 2021). Otherwise, bond length design is not as safe for RAC as it is for NAC.

Furthermore, additional experiments should be made to allow the determination of mechanically-based factor that would depend on the coarse aggregate used, including the distinction between types of NA. In order to allow a better understanding of the influence of the aggregate on the bond strength, research in which the aggregate crushing value and the bond strength are simultaneously tested for analogue concrete mixes with several types of coarse aggregate is recommended.

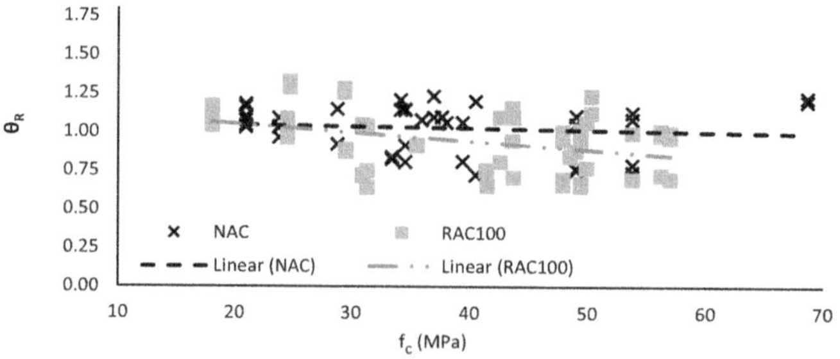

Figure 6.6 Scatterplot: fc versus θR of bond strength.

6.5.4 Bending resistance

The ultimate moment resistance (M_R) of reinforced concrete beams is typically calculated using the equivalent rectangular stress-block distribution in design made using EN1992, Part 1 (EN-1992-1-1, 2008). This stress-block is presented in Fig. 6.6 and neglects part of the compressed region of concrete. For large strains, the compressive stress is constant and equal to $\eta \times f_c$. The contribution of concrete under tension is disregarded.

The calculation of M_R is made for the stress-strain distribution presented in Fig. 6.7. The values of λ, η, and ε_{cud} depend on f_c and are presented in Eqs. (6.5)–(6.7):

$$\lambda = \begin{cases} 0.8 \; if \; f_c \leq 50MPa \\ 0.8 - \left(\frac{f_c - 50}{400}\right) if \; 50MPa < f_c \leq 100MPa \end{cases} \tag{6.5}$$

$$\eta = \begin{cases} 1.0 \; if \; f_c \leq 50MPa \\ 1.0 - \left(\frac{f_c - 50}{200}\right) if \; 50MPa < f_c \leq 100MPa \end{cases} \tag{6.6}$$

$$\varepsilon_{cp} = \begin{cases} 2.0‰ \; if \; f_c \leq 50MPa \\ 2.0‰ + 0.085(f_c - 50)^{0.53} \; if \; 50MPa < f_c \leq 100MPa \end{cases} \tag{6.7}$$

M_R is given by:

$$M_R = A_{sl} \times f_y \times \left[d - 0.5 \times \frac{A_{sl} \times f_y}{b \times \lambda \times (\eta \times f_c)}\right] \tag{6.8}$$

Where A_{sl} is the area of longitudinal reinforcement, d the effective depth of the reinforcement under tension, and b the width of structural element.

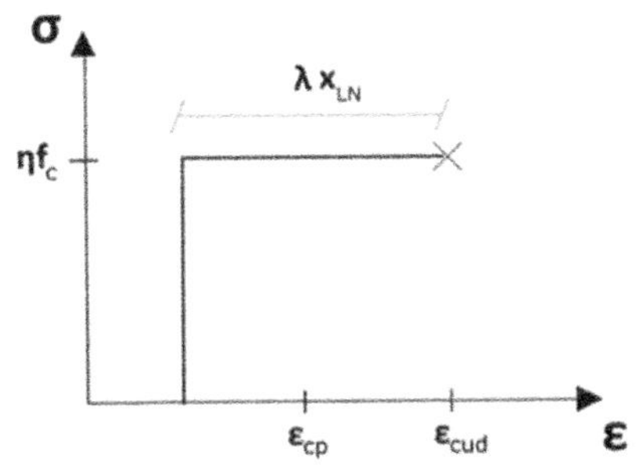

Figure 6.7 Parabolic rectangular stress-block.

Eq. (6.8) assumes that the stress-strain distribution of the reinforcement is elasto-plastic. This is the most common option of common reinforced concrete design. Furthermore, experimental data on the ultimate tensile strength of the reinforcement was not provided by most authors of the experiments used to build the database for this θ_R.

The behavior of RAC elements may differ from that of NAC elements made with concrete of the same f_c because of: (i) size effects (Pacheco et al., 2019a, 2019b, 2019c, 2019d); (ii) the influence of RA on the bond strength (presented in the previous chapter); (iii) possible effects of the incorporation of RA on the stress-strain relationship of concrete under compression. The stress-stain relationship of RAC has been scarcely studied and findings are not consensual, as presented in Section 6.4.

All tests collected were checked against the following criteria:

- Beams that did not fail in bending were removed from the database.;
- Beams whose reinforcement under tension did not yield before failure were also removed;
- Detailing requirements had to be complied with (e.g., the minimum and maximum reinforcement under tension ratios of EN1992, Part 1 (EN-1992-1-1, 2008)).

A conservative approach was used in the bond strength checks. All beams were checked against the bond strength model of EN1992, Part 1 (EN-1992-1-1, 2008). In the case of shear, the majority of the papers did not disclose sufficient information for shear resistance calculations. The approach taken was to check the claims of the authors and photographs of the experiments.

In the calculation of θ_R, the experimental ultimate moment was calculated accounting for the self-weight of each beam.

Table 6.5 presents the sample size, mean, standard deviation and CoV of each database. The references of the publications used are the following (Ajdukiewicz & Kliszczewicz, 2007; Arezoumandi, Smith, et al., 2015; Choi et al., 2012; Evangelista & de Brito, 2017; Fathifazl, Razaqpur, Burkan Isgor, et al., 2009; Ignjatović et al., 2012; Kang et al., 2014; Knaack & Kurama, 2015; Sato et al., 2007; Yagishita et al., 1993).

The statistics presented in Table 6.5 shows that the incorporation of RA has no influence on the θ_R for ultimate moment resistance. This is because the reinforcement under tension is a governing factor for the resistance of conventional reinforced concrete elements failing in bending. Table 6.5 also shows that the ultimate moment resistance is underestimated by Eq. (6.8). This was expected, since reinforcement hardening is neglected. Table 6.6 presents the θ_R of NAC design using EN1992, Part 1 (EN-1992-1-1, 2008) presented in publications of other authors (JCSS, 2001;

Table 6.5 Statistics of θ_R, for the ultimate moment resistance.

Database	Number of tests	Mean	Standard deviation	CoV
NAC	50	1.15	0.11	10%
RAC100	41	1.15	0.09	8%

Table 6.6 Appraisal of statistics of θ_R, for the ultimate moment resistance of NAC.

Statistic	JCSS (2001)	Vrouwenvelder and Siemes (1987)	Sykora et al. (2014)
Mean	1.20	1.10	1.08
CoV (%)	15	10	9

Vrouwenvelder & Siemes, 1987). The values proposed agree with those presented in Table 6.6 for NAC.

Histograms of the design parameters were produced and analyzed. Fig. 6.8 presents examples, including the ratio of the compression to tensile longitudinal reinforcement (β) and the experimental dimensionless ultimate moment (μ_{MUexp}), which was calculated using Eq. (6.9):

$$\mu_{MUexp} = \frac{M_{U,exp}}{b \times d \times f_{cm}} \tag{6.9}$$

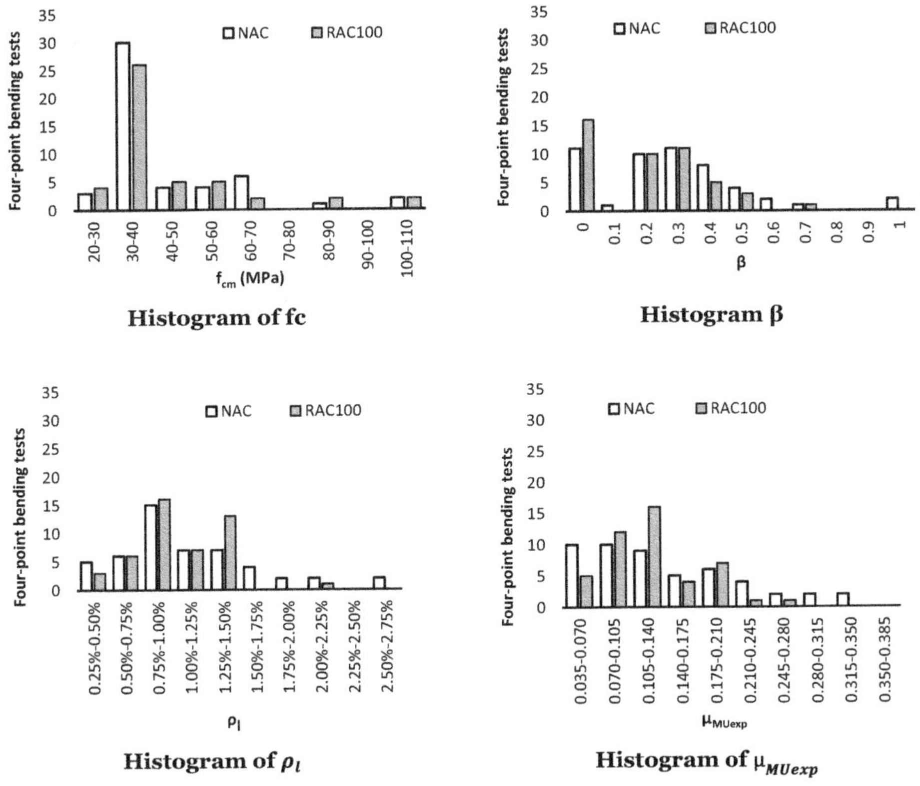

Figure 6.8 Histograms of the databases of the ultimate moment resistance.

Where $M_{U,exp}$ is the experimental ultimate moment resistance. In Fig. 6.8 and elsewhere, ρ_l is the geometric ratio of the longitudinal reinforcement under tension.

The analysis of the histograms shows that the parameters of design conform with common design of reinforced concrete beams and that the distribution of parameters of the NAC and RAC100 datasets is similar. The only noteworthy exception is β, which is smaller than usual. Nonetheless, low values of β were intended, since Eq. (6.8) neglects the contribution of the reinforcement under compression. Furthermore, the histogram of μ_{MUexp} also agrees with this option, since values of μ_{MUexp} above 0.250 are rare (therefore, the region of compressed concrete is not strongly stressed).

The largest coefficient of correlation was between θ_R and ρ_l and was of −0.30. According to Table 6.3, this corresponds to moderate correlation and has no practical relevance for structural design, especially when considered that any detrimental effect caused by large ρ_l is limited by the restriction of EN1992, Part 1 (EN-1992-1-1, 2008) that ρ_l must be below 4.0%. Additional noteworthy coefficients of correlation between the parameters of structural design and θ_R were not observed. Fig. 6.9 presents some of the scatterplots analyzed.

Furthermore, all trends observed between the parameters of design and θ_R were not affected by the incorporation of RA.

Since no relevant effect of the incorporation of RA is found on this θ_R, the authors recommend that common cases of reinforced concrete elements subjected to bending are designed with the common formulae of EN1992, Part 1 (EN-1992-1-1, 2008) without changes.

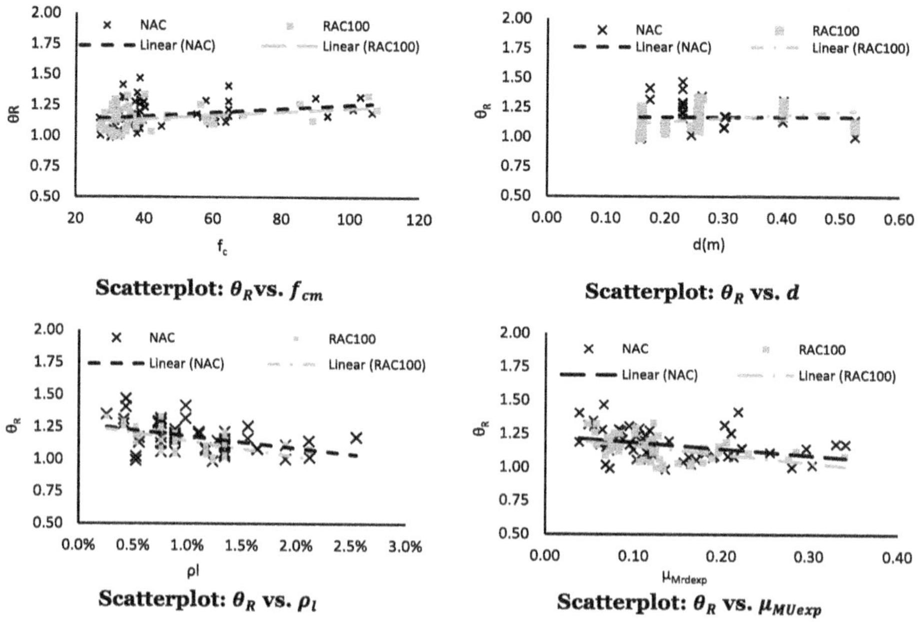

Figure 6.9 Scatterplots between parameters of design and the θR for ultimate moment resistance.

6.5.5 Shear resistance

The shear resistance (V_R) of reinforced concrete differs for elements with and without shear-specific reinforcement. In the case of elements without shear reinforcement, V_R is given by:

$$V_R = \left(0.18 \times k \times \sqrt[3]{100\rho_l \times f_c}\right) \times d \times b \tag{6.10}$$

Where $k = 1 + \sqrt{\frac{200}{d}} \leq 2$ accounts for size effects and ρ_l must be below 2%.

This model is empiric (PT-SC2-T1 D3BG, 2018) and most experimental data concerns simply supported beams under four-point bending, which the type of test of all beams included in the databases used to estimate this θ_R.

The V_R of elements with shear reinforced is calculated as presented in Eq. (6.11):

$$V_R = \frac{A_{sw}}{s} \times 0.9d \times f_{yw} \times \cot(\Omega) \leq V_{R,max} \tag{6.11}$$

Where A_{sw} is the area of shear reinforcement, s is the longitudinal distance between webs of the shear reinforcement, f_{yw} is the yield stress of the shear reinforcement, Ω may take any value between 21.8 and 35° and is the angle of the compression strut with the longitudinal axis of the element. $V_{R,max}$ is the resistance of the compression strut, calculated as:

$$V_{R,max} = \frac{b \times 0.9d}{\cot(\Omega) + \tan(\Omega)} \times \nu_1 \times f_c \tag{6.12}$$

EN1992, Part 1 (EN-1992-1-1, 2008) defines $\nu_1 = 0.6$ if $f_{ck} < 60\ MPa$, the case of the databases of this paper. $\Omega = 21.8^{\circ}$ was used in all cases since this results in the most conservative estimate of θ_R. An alternative would be to use the measured Ω of each test, but this information was not disclosed in most of the publications appraised. V_R was not limited by $V_{R,max}$ in any of the beams included in the databases.

The resistance of concrete without shear-specific reinforcement to shear load-effects is expected to be affected by the incorporation of RA due to reason put forward in Section 6.4: the incorporation of RA decreases shear strength, due to decreased aggregate interlock. Furthermore, this resistance model is particularly affected by structural size effects (PT-SC2-T1 D3BG, 2018; Collins & Kuchma, 1999; Lubell et al., 2004; Sigrist et al., 2013), since, as the size of structural elements increases, crack widths tend to increase and the potential for stress transfer through aggregate interlock will decrease. Past research (Ignjatović et al., 2017; Wardeh et al., 2018) has found that the incorporation of RA may lead to higher crack widths.

The V_R of the resistance model of elements with shear reinforcement does not include any parameter related to concrete. However, the resistance mechanism will also depend on concrete properties since part of the stress is transferred through concrete. Therefore, the θ_R of this resistance model is expected to also be affected by the incorporation of RA.

All tests concern beams under four-point loading tests. The criteria for database inclusion were the following:

- The ratio of shear span to the effective depth (a/d) has to be larger than 2.5. This option was taken because for a/d below this value, either flexural reinforcement (a/d below 1.0) or arching action (a/d between 1.0 and 2.5) take up a relevant role in this resistance mechanism. The same criterion was used in the development of the shear database that validated the V_R equations of Model Code 2010 (Sigrist et al., 2013);
- The beams were checked for bending and anchorage failure.

The validation of beams with shear reinforcement also included the following conditions:

- f_{yw} had to be disclosed;
- The maximum s presented in EN1992, Part 1 (EN-1992-1-1, 2008) had to comply with ($s <$ 0.75 d);
- The minimum geometric ratio of the shear reinforcement ($\rho_{w,min}$) stated in EN1992, Part 1 (EN-1992-1-1, 2008) had to comply with ($\rho_{w,min} = 0.08\sqrt{f_{ck}}/f_{ywk}$), where f_{ywk} is the characteristic yield stress of the shear reinforcement and f_{ck} is the characteristic compressive strength of concrete;

If any of the last two requirements is not complied with, EN1992, Part 1 (EN-1992-1-1, 2008) states that the structural element must be checked for shear resistance as if shear-specific reinforcement were absent. In this chapter, such elements were not included in the database of elements without shear reinforcement. This option was taken since the reinforcement is still present and would contribute to the shear resistance mechanism. Therefore, the θ_R of these elements would be larger than what is expected for elements without shear-specific reinforcement.

All elements had rectangular cross-section and none was subjected to axial stresses.

The results from a publication (Ignjatović et al., 2017) complied with all criteria stated but were still removed from the database. The tests presented in that publication had large θ_R and were not representative of conventional design conditions: the beams tested in this publication had $\rho_l = 4\%$ (the maximum admissible by EN1992, Part 1 (EN-1992-1-1, 2008)). This value of ρ_l resulted in large θ_R because elements with such geometric ratio of longitudinal reinforcement under tension not only have strong dowel action, but also limit crack widths (because this re1inforcement will control strains), thus increasing aggregate interlock.

Concerning elements without shear reinforcement, the number of beams that complied with these criteria is shown in Table 6.7. As in the other θ_R analyzed in this chapter, the sample mean, standard deviation and CoV of the θ_R are also presented. The publications used to produce the database are the following (Arezoumandi, Drury, et al., 2015; Choi et al., 2010; Choi & Yun, 2017; Etxeberria, 2004; González-Fonteboa & Martínez-Abella, 2007; Katkhuda & Shatarat, 2016; Kim et al., 2013; Rahal & Alrefaei, 2017; Sadati et al., 2016; Wardeh et al., 2018).

Table 6.7 Statistics of θ_R for shear resistance of elements without shear reinforcement.

Database	Number of tests	Mean	Standard deviation	CoV
NAC	35	1.03	0.11	11%
RAC100	26	0.95	0.11	12%

These results show that RA incorporation has negligible effect on the standard deviation and CoV of the θ_R of this resistance model. However, the mean θ_R decreases by 8% when RA are incorporated. This decrease is mostly caused by the effect of the incorporation of RA on shear strength/aggregate interlock. This finding is similar to that of the bond strength, whose decrease of the mean θ_R was also justified with physical reasons.

The results observed for NAC agree with those of other publications:

- In Sýkora and Holický (2012), a mean θ_R of 1.0 and a CoV of 13% are recommended;
- *fib* Bulletin 80 (fib, 2016) state that this θ_R may be modeled with a mean of 1.0 and a CoV of 15%.

As shown in the histograms of Fig. 6.10, the distributions of the parameters of design of the NAC and RAC100 databases are identical. Furthermore, the databases

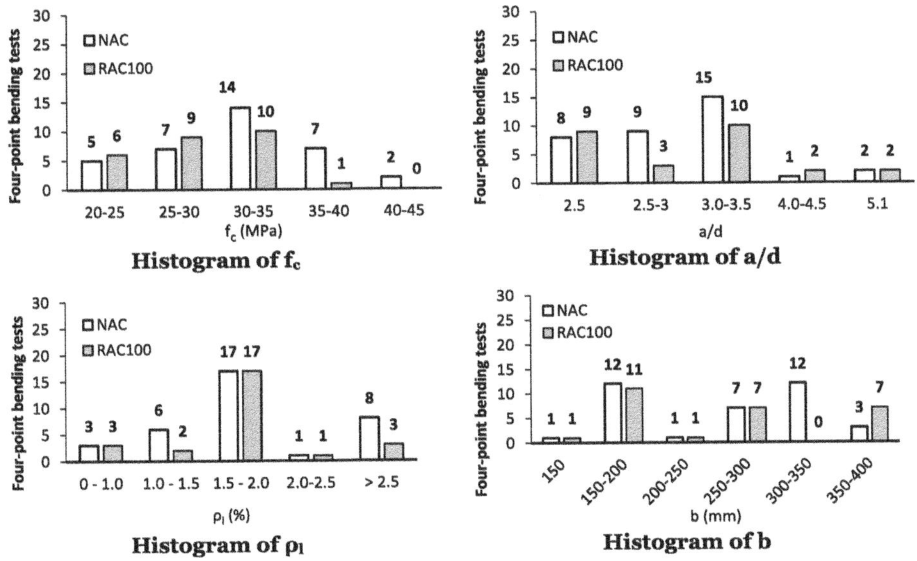

Figure 6.10 Histograms of the databases used for shear resistance of elements without shear reinforcement.

conform with common structural elements. D_{max} of the NAC and RAC100 datasets is similar: roughly 90% of the beams of either database has D_{max} of 25 mm, while D_{max} of the remaining beams is of either 19 or 20 mm.

All coefficients of correlation were below 30%, suggesting low correlation according to Table 6.3. Nonetheless, a consistent trend toward higher f_c leading to decreases of this θ_R was observed for the RAC100 database. This finding, which is shown in Fig. 6.11, strongly suggests that the shear resistance of RAC is indeed being limited by the fact that RA are weaker than NA. Increases of the f_c of concrete are associated with increases of the compressive strength of the binder paste, therefore the number of weak aggregates that are fractured when the shear surface is developed will also increase and this reduces aggregate interlock.

No noteworthy effect of a/d on θ_R was observed. This is a consequence of the criterion that only beams with $a/d > 2.5$ are included in the database. This criteria imposes that only one type of shear failure, characterized by localized tensile strains in the critical shear crack, occurs (Muttoni & Fernández Ruiz, 2010).

Since most of the beams of the two databases had the same D_{max} (25 mm), scatter-plots and correlations of θ_R with D_{max} were not made.

The number of beams that complied with the criteria for the database of shear resistance of elements with shear reinforcement is lower than in the other databases presented in this chapter. Table 6.8 presents the statistics of this θ_R, which was produced with tests coming from the following publications (Etxeberria, 2004; González-Fonteboa & Martínez-Abella, 2007; Rahal & Alrefaei, 2018).

By comparing the statistics presented in Table 6.8 with those presented in Table 6.7 for elements without shear reinforcement, it is found that this resistance model is more conservative and scattered than the latter. This finding is consensual (Sigrist et al., 2013; fib, 2016; Cladera & Marí, 2004) and is due to two reasons that are directly

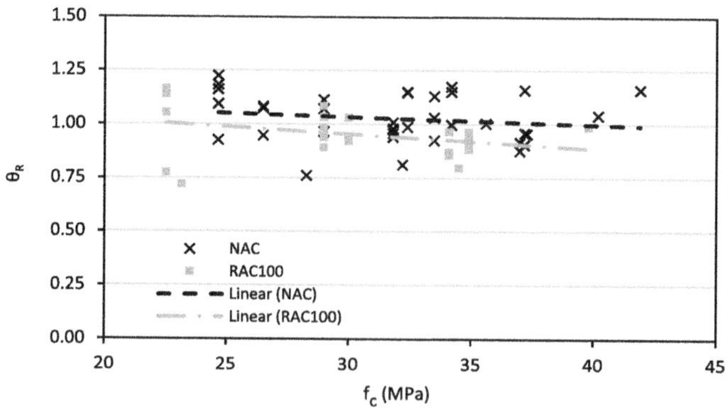

Figure 6.11 Scatterplot: θ_R versus f_{cm} of beams without shear reinforcement.

Table 6.8 Statistics of θ_R for shear resistance of elements with shear reinforcement.

Database	Number of tests	Mean	Standard deviation	CoV
NAC	10	1.10	0.25	23%
RAC100	8	1.03	0.23	22%

related to Eq. (6.11): (i) the hardening of the shear reinforcement is not considered; (ii) this resistance model only accounts for the contribution of shear reinforcement to the mechanical behavior, but concrete transfers part of the shear stress.

fib Bulletin 80 (fib, 2016) states that this θ_R may be modeled with a mean of 1.25 and a CoV of 25%, when used for elements with similar reinforcement to those of our database ($\rho_w \times f_{yw}$ in the range of 1.0–2.0 MPa).

The incorporation of RA is found to decrease the mean θ_R by 6%, while the standard deviation and CoV are negligibly affected. However, the size of the database and the fact that all tests came from just three publications mean that these findings should be complemented with additional research.

As shown in the histograms of Fig. 6.10, the distributions of the parameters of design of the NAC and RAC100 databases are identical. Furthermore, the databases conform with common structural elements.

The design parameters of the NAC and RAC100 databases are similarly distributed. In both datasets, six beams had $a/d = 3.0$ and the others had $a/d = 3.3$ (four beams for NAC and two for RAC100); in the case of NAC, 60% of the beams had D_{max} equal to 19 or 20 mm and 40% had D_{max} equal to 25 mm, while for RAC100, 75% had D_{max} equal to 19 or 20 mm and the D_{max} of the remaining beams was 25 mm. Table 6.9 shows the distribution of other parameters of design. In this table and elsewhere, ρ_w is the geometric ratio of shear reinforcement.

The analyses of coefficients of correlation between θ_R and the parameters of design is not presented since the number of tests on each database is low and the databases were built from just three publications. Any trend that would be detected could be related to specificities of test setups or quality of raw materials in the laboratories of each publication. This reinforces the assumption that more experiments are needed to validate the mean and CoV of the θ_R presented in Table 6.8.

Nonetheless, since $\rho_w \times f_{yw}$ is acknowledged as influencing the θ_R of resistance models for elements of shear with shear reinforcement (Holický et al., 2016), its scatterplot is presented in Fig. 6.12. The coefficients of correlation between θ_R and $\rho_w \times f_{yw}$ were of −60% (NAC) and −62% (RAC100). These results and the trend observed in Fig. 6.12 agree with what is consistently reported for NAC shear design (Sykora et al., 2018; fib, 2016; Holický et al., 2016).

Table 6.9 Comparison of the parameters of design of NAC and RAC100 for elements with shear reinforcement.

f_c (MPa)			ρ_w (%)			d (mm)			ρ_l (%)		
Range	NAC	RAC100	Range	NAC	RAC100	Range	NAC	RAC100	Value	NAC	RAC100
30–35	4	6	0.1–0.2	3	2	300–350	4	2	1.4	6	6
35–40	4	2	0.2–0.4	6	5	350–400	6	6	3.0	4	2
40–45	2	0	0.4–0.6	1	1						

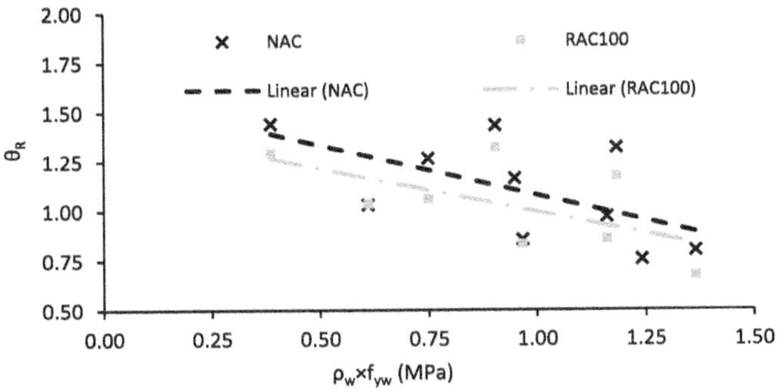

Figure 6.12 Scatterplot: θ_R versus $\rho_w \times f_{yw}$ of beams with shear reinforcement.

6.6 Concluding remarks

This chapter treated the structural behavior of concrete made with coarse recycled aggregates produced from concrete waste. This is relevant because, despite the abundant research arguing in favor of RAC as a structural material, clear design guidelines are lacking and comprehensive research on the understanding of the influence of recycled aggregates on structural behavior is needed for the development of such guidelines.

The production process of recycled aggregates and the requirements for their use in concrete were briefly presented as a means to argue for coarse recycled concrete aggregates as the type of recycled aggregate that has better properties for use as an aggregate for the production of concrete. Nonetheless, this type of recycled aggregate still has worse properties than natural aggregates and this has detrimental effects on the mechanical behavior of concrete. For the same compressive strength, the Young' modulus, tensile strength, shear strength and fracture energy of concrete tend to be smaller when recycled aggregates are used. This is mostly due to the fact that recycled aggregates are partly composed of old mortar, which is typically more deformable and weaker than stone.

Due to these reasons, the structural behavior of recycled aggregate concrete is worse than that of natural aggregate concrete. This is understood when analyzing experimental programs in which the structural behavior of analogue natural and recycled aggregate concrete specimens are compared: deflections and resistance tend to decrease when recycled aggregates are used, namely in the case of resistance mechanisms that do not rely on reinforcement significantly. However, the chapter also argues that analyses of single experimental programs is not a scientifically-sound way of assessing the influence of recycled aggregates on the structural behavior of concrete due to specificities in concrete mix design, quality of raw materials (including the recycled aggregates used), specimen detailing and test setup.

When several publications are gathered and jointly analyzed, the aforementioned specificities are averaged out and more reliable conclusions are attained. Therefore, an approach based on the assessment of model uncertainties developed from all publications that complied with stringent criteria was carried out in order to evaluate the influence of recycled aggregates on the structural behavior of reinforced concrete. The model uncertainties were determined for the following resistance mechanisms and models:

- The bond strength, using the model of *fib* Bulletin 72 (fib, 2014);
- The bending resistance, using the ultimate model resistance calculations that result from the provisions of EN1992, Part 1 (EN-1992-1-1, 2008);
- The shear resistance of elements without shear-specific reinforcement, using the respective resistance model of EN1992, Part 1 (EN-1992-1-1, 2008);
- The shear resistance of elements with shear-specific reinforcement, using the respective resistance model of EN1992, Part 1 (EN-1992-1-1, 2008).

The methodologies used for the development of the model uncertainties are fully conforming with Annex C of ISO2394 (ISO-2394, 2014). Two databases were built for each resistance model, one with natural aggregate concrete and the other with recycled aggregate concrete with full incorporation of coarse recycled aggregates produced from concrete waste.

From the analysis of the model uncertainties, the main findings of this chapter are that, for full incorporation of coarse recycled concrete aggregates in concrete:

- The bond strength decreases when recycled aggregates are used (on average, 8%). Furthermore, the resistance model of the bond strength is more scattered when recycled aggregates are incorporated;
- The bending resistance is unaffected by the incorporation of recycled aggregates and the design formulae of natural aggregate concrete design may be used without reservations;
- The shear strength of elements without shear reinforcement is detrimentally affected by the incorporation of recycled aggregates (on average, 8%), especially when the compressive strength of concrete is large;
- It seems that the shear strength of elements with shear reinforcement is also affected (average decreases of resistance of 6%) but additional research should confirm this hypothesis;
- The scatter of the resistance models related to bending resistance and shear resistance without shear reinforcement is unaffected by the incorporation of recycled aggregates.

These findings show that the structural design of recycled aggregate concrete may be unsafe when code provisions developed for natural aggregate concrete are used, namely when design concerns the bond length of anchorages or lap splices or shear resistance.

As future studies, the authors recommended additional research on the shear strength of elements with shear reinforcement; lap splice experiments with different types of coarse aggregates (and in which the aggregate crushing value of the aggregates is also tested, in order to calibrate a physically-based parameter that accounts for the influence of aggregate strength on bond strength); experiments on the uniaxial compression of recycled aggregate concrete columns (which was not studied due to the small number of tests presently available); and reliability analyses in order to propose changes to the design formulae of codes in order to provide design guidelines that ensure safe structural design of recycled aggregate concrete.

Acronyms and Symbols

CoV	Coefficient of variation
GF	Fracture energy
ITZ	Interfacial transition zone
NA	Natural aggregates
NAC	Natural aggregate concrete
RA	Coarse recycled aggregate produced from concrete waste
RAC	Concrete made with partial or full incorporation of coarse recycled aggregates produced from concrete waste
RAC100	Concrete made with full incorporation of coarse recycled aggregates produced from concrete waste
s	Longitudinal distance between webs of the shear reinforcement
a/d	Ratio of the shear span to the effective depth
b	Width of structural element
d	Effective depth of the reinforcement under tension
f_b	Bond stress
f_c	Compressive strength of concrete
f_{cm}	Mean cylindrical compressive strength
f_{stm}	Stress mobilized in the bar through bond
f_y	Yield strength of the reinforcement
f_{yw}	Yield stress of the shear reinforcement
l_b	Development length of the reinforcement
A_{sl}	Area of longitudinal reinforcement
A_{sw}	Area of shear reinforcement
D_{max}	Maximum aggregate diameter
M_R	Ultimate moment resistance
V_R	Shear resistance
$V_{R,max}$	Resistance of the compression strut
β	Ratio of the compression to tensile longitudinal reinforcement
θ_R	Model uncertainty
μ_{MUexp}	Experimental dimensionless ultimate moment
ρ_l	Geometric ratio of the longitudinal reinforcement under tension
$\rho_{w,min}$	Minimum geometric ratio of the shear reinforcement
Φ	Diameter of the reinforcement
Ω	Angle of the compression strut with the longitudinal axis of the element

References

Abbas, A., Fathifazl, G., Burkan Isgor, O., Razaqpur, A. G., Fournier, B., & Foo, S. (2008). Proposed method for determining the residual mortar content of recycled concrete aggregates. *Journal of ASTM International, 5*(1).

Abbas, A., Fathifazl, G., Fournier, B., Isgor, O. B., Zavadil, R., Razaqpur, A. G., & Foo, S. (2009). Quantification of the residual mortar content in recycled concrete aggregates by image analysis. *Materials Characterization, 60*(7), 716–728.

ACI-408. (2003). *Bond and development of straight reinforcing bars in tension*. Farmington Hills, MI: the American Concrete Institute.

Agrela, F., Sánchez de Juan, M., Ayuso, J., Geraldes, V. L., & Jiménez, J. R. (2011). Limiting properties in the characterisation of mixed recycled aggregates for use in the manufacture of concrete. *Construction and Building Materials, 25*(10), 3950–3955.

Ajdukiewicz, A., & Kliszczewicz, A. (2002). Influence of recycled aggregates on mechanical properties of HS/HPC. *Cement and Concrete Composites, 24*(2), 269–279.

Ajdukiewicz, A. B., & Kliszczewicz, A. T. (2007). Comparative tests of beams and columns made of recycled aggregate concrete and natural aggregate concrete. *Journal of Advanced Concrete Technology, 5*(2), 259–273.

Akbarnezhad, A., Ong, K. C. G., Zhang, M. H., Tam, C. T., & Foo, T. W. J. (2011). Microwave-assisted beneficiation of recycled concrete aggregates. *Construction and Building Materials, 25*(8), 3469–3479.

Albuquerque, A., Pacheco, J. N., & de Brito, J. (2021). Eurocode design of recycled aggregate concrete for chloride environments: Stochastic modeling of chloride migration and reliability-based calibration of cover. *Crystals, 11*(3), 284.

Alexander, M. G. (1996). Aggregates and the deformation properties of concrete. *Materials Journal, 93*(6), 569–577.

Alexander, M. G., & Mindess, S. (2005). *Aggregates in concrete. Modern concrete technology series*. New York: Taylor & Francis.

Allen, T. M., Nowak, A. S., & Bathurst, R. J. (2005). Calibration to determine load and resistance factors for geotechnical and structural design. In *TRB circular E-C079* (p. 82). Transportation Research Board: Washington DC.

fib, Bulletin 65: Model Code 2010 final draft. (2010). Switzerland: Lausanne. Vol 1.

Arezoumandi, M., Drury, J., Volz, J. S., & Kamal, H. K. (2015). Effect of recycled concrete aggregate replacement level on the shear strength of reinforced concrete beams. *Materials Journal, 112*(4), 559–568.

Arezoumandi, M., Smith, A., Volz, J. S., & Khayat, K. H. (2015). An experimental study on flexural strength of reinforced concrete beams with 100% recycled concrete aggregate. *Engineering Structures, 88*, 154–162.

Artoni, R., Cazacliu, B., Hamard, E., Cothenet, A., & Parhanos, R. S. (2016). Resistance to fragmentation of recycled concrete aggregates. *Materials and Structures, 50*(1), 11.

Bažant, Z. P., & Yavari, A. (2005). Is the cause of size effect on structural strength fractal or energetic–statistical? *Engineering Fracture Mechanics, 72*(1), 1–31.

Benjamin, J., & Cornell, C. A. (1970). *Probability, statistics and decision for civil engineers*. McGraw Hill.

Bordelon, A., Cervantes, V., & Roesler, J. R. (2009). Fracture properties of concrete containing recycled concrete aggregates. *Magazine of Concrete Research, 61*(9), 665–670.

Breccolotti, M., & Materazzi, A. L. (2013). Structural reliability of bonding between steel rebars and recycled aggregate concrete. *Construction and Building Materials, 47*(10), 927–934.

de Brito, J., Ferreira, J., Pacheco, J., Soares, D., & Guerreiro, M. (2016). Structural, material, mechanical and durability properties and behaviour of recycled aggregates concrete. *Journal of Building Engineering, 6*, 1–16.

Butler, L. (2012). *Evaluation of recycled concrete aggregate performance in structural concrete*. University of Waterloo: Ontario, Canada: Department of Civil Engineering.

Butler, L., West, J. S., & Tighe, S. L. (2011). The effect of recycled concrete aggregate properties on the bond strength between RCA concrete and steel reinforcement. *Cement and Concrete Research, 41*(10), 1037–1049.

Butler, L., West, J. S., & Tighe, S. L. (2013). Effect of recycled concrete coarse aggregate from multiple sources on the hardened properties of concrete with equivalent compressive strength. *Construction and Building Materials, 47*, 1292–1301.

Butler, L. J., West, J. S., & Tighe, S. L. (2015). Bond of reinforcement in concrete incorporating recycled concrete aggregates. *Journal of Structural Engineering, 141*(3), B4014001.

Cairns, J., & Plizzari, G. A. (2003). Towards a harmonised European bond test. *Materials and Structures, 36*(8), 498.

Casuccio, M., Torrijos, M. C., Giaccio, G., & Zerbino, R. (2008). Failure mechanism of recycled aggregate concrete. *Construction and Building Materials, 22*(7), 1500–1506.

CEN/TC-250/SC-2. (2020). *prEN 1992-1-1 D6 Working file (2020-10-05 Rev. 7).*

Choi, H. B., & Kang, K. I. (2008). Bond behaviour of deformed bars embedded in RAC. *Magazine of Concrete Research, 60*(6), 399–410.

Choi, H. B., Yi, C. K., Cho, H. H., & Kang, K. I. (2010). Experimental study on the shear strength of recycled aggregate concrete beams. *Magazine of Concrete Research, 62*(2), 103–114.

Choi, W.-C., & Yun, H.-D. (2017). Shear strength of reinforced recycled aggregate concrete beams without shear reinforcements. *Journal of Civil Engineering and Management, 23*(1), 76–84.

Choi, W.-C., Yun, H.-D., & Kim, S.-W. (2012). Flexural performance of reinforced recycled aggregate concrete beams. *Magazine of Concrete Research, 64*(9), 837–848.

Cladera, A., & Marí, A. R. (2004). Shear design procedure for reinforced normal and high-strength concrete beams using artificial neural networks. Part II: Beams with stirrups. *Engineering Structures, 26*(7), 927–936.

Collins, M. P., & Kuchma, D. (1999). How safe are our large, lightly reinforced concrete beams, slabs, and footings? *Structural Journal, 96*(4).

Corinaldesi, V., Letelier, V., & Moriconi, G. (2011). Behaviour of beam-column joints made of recycled-aggregate concrete under cyclic loading. *Construction and Building Materials, 25*(4), 1877–1882.

Darwin, D., Barham, S., Kozul, R., & Luan, S. (2001). Fracture energy of high-strength concrete. *Materials Journal, 98*(5).

EN-10002-1. (2001). *Metallic materials—tensile testing—Part 1: Method of test at ambient temperature.* Brussels, Belgium: CEN.

EN-12620+A1. (2008). *Aggregates for concrete, with 2008 amendment.* CEN.

EN-1992-1-1. (2008). *Eurocode 2—design of concrete structures: Part 1-1: General rules and rules for buildings.* Brussels, Belgium: Comité Européen de Normalisation (CEN).

Etxeberria, M. (2004). *Experimental study on microstructure and structural behaviour of recycled aggregate concrete.* Universitat Politècnica de Catalunya: Catalunya, Spain: Department of Civil Engineering.

Etxeberria, M., Vazquez, E., Mari, A., & Barra, M. (2007). Influence of amount of recycled coarse aggregates and production process on properties of recycled aggregate concrete. *Cement and Concrete Research, 37*(5), 735–742.

Evangelista, L., & de Brito, J. (2017). Flexural behaviour of reinforced concrete beams made with fine recycled concrete aggregates. *KSCE Journal of Civil Engineering, 21*(1), 353–363.

Fathifazl, G., Razaqpur, A. G., Burkan Isgor, O., Abbas, A., Fournier, B., & Simon, F. (2009). Flexural performance of steel-reinforced recycled concrete beams. *Structural Journal, 106*(06), 858–867.

Fathifazl, G., Razaqpur, A. G., Isgor, O. B., Abbas, A., Fournier, B., & Foo, S. (2009). Shear strength of reinforced recycled concrete beams without stirrups. *Magazine of Concrete Research, 61*(7), 477–490.

Fathifazl, G., Razaqpur, A. G., Isgor, O. B., Abbas, A., Fournier, B., & Foo, S. (2010). Shear strength of reinforced recycled concrete beams with stirrups. *Magazine of Concrete Research, 62*(10), 685–699.

Fathifazl, G., Razaqpur, A. G., Isgor, O. B., Abbas, A., Fournier, B., & Foo, S. (2012). Bond performance of deformed steel bars in concrete produced with coarse recycled concrete aggregate. *Canadian Journal of Civil Engineering, 39*(2), 128–139.

Ferreira, L., de Brito, J., & Barra, M. (2011). Influence of the pre-saturation of recycled coarse concrete aggregates on concrete properties. *Magazine of Concrete Research, 63*(8), 617–627.

fib. (2009). *Bulletin 51: Structural concrete. Textbook on behaviour, design and performance* (2nd ed., Vol 1). Switzerland: Lausanne.

fib. (2013). *Bulletin 70: Code-type models for concrete behaviour*. Switzerland: Lausanne.

fib. (2014). *Bulletin 72: Bond and anchorage of embedded reinforcement: Background to the fib model code for concrete structures 2010*. Switzerland: Lausanne.

fib. (2016). *Bulletin 80: Partial factor methods for existing concrete structures*. Switzerland: Lausanne.

García-González, J., Barroqueiro, T., Evangelista, L., de Brito, J., De Belie, N., Morán-del Pozo, J., & Juan-Valdés, A. (2016). Fracture energy of coarse recycled aggregate concrete using the wedge splitting test method: Influence of water-reducing admixtures. *Materials and Structures, 50*(2), 120.

Gaurav, G., & Singh, B. (2018). Experimental investigation of bond behavior with tension lap splice for deformed steel bars in recycled aggregate concrete. *Journal of Sustainable Cement-Based Materials, 7*(2), 99–121.

Ghorbel, E., & Wardeh, G. (2017). Influence of recycled coarse aggregates incorporation on the fracture properties of concrete. *Construction and Building Materials, 154*, 51–60.

Ghorbel, E., Wardeh, G., & Fares, H. (2019). Mechanical and fracture properties of recycled aggregate concrete in design codes and empirical models. *Structural Concrete, 20*, 2156–2170.

González-Fonteboa, B., & Martínez-Abella, F. (2007). Shear strength of recycled concrete beams. *Construction and Building Materials, 21*(4), 887–893.

González-Fonteboa, B., Martinez-Abella, F., Carro Lopez, D., & Seara-Paz, S. (2011). Stress-strain relationship in axial compression for concrete using recycled saturated coarse aggregate. *Construction and Building Materials, 25*(5), 2335–2342.

González-Fonteboa, B., Martínez-Abella, F., Eiras-Lopez, J., & Seara-Paz, S. (2011). Effect of recycled coarse aggregate on damage of recycled concrete. *Materials and Structures, 44*(10), 1759–1771.

Gonzalez, V. C., & Moriconi, G. (2014). The influence of recycled concrete aggregates on the behavior of beam–column joints under cyclic loading. *Engineering Structures, 60*, 148–154.

Guo, M., Alam, S. Y., Bendimerad, A. Z., Grondin, F., Rozière, E., & Loukili, A. (2017). Fracture process zone characteristics and identification of the micro-fracture phases in recycled concrete. *Engineering Fracture Mechanics, 181*, 101–115.

Hamad, B. S., Dawi, A. H., Daou, A., & Chehab, G. R. (2018). Studies of the effect of recycled aggregates on flexural, shear, and bond splitting beam structural behavior. *Case Studies in Construction Materials, 9*, e00186.

Henry, M., Hagiwara, K., Nishimura, T., & Kato, Y. (2011). Effect of recycled aggregate quality on variation and estimation of concrete strength. *Proceedings of the Japan Concrete Institute, 33*(1), 1535–1540.

Holický, M., Retief, J. V., & Sýkora, M. (2016). Assessment of model uncertainties for structural resistance. *Probabilistic Engineering Mechanics, 45*, 188–197.

Holicky, M., Sykora, M., Barnardo-Viljoen, C., Mensah, K. K., & Retief, J. V. (2013). Model uncertainties in reliability analysis of reinforced concrete structures. In *The 5th international conference on structural engineering* (pp. 741–748). Cape Town, South Africa: Mechanics and Computation.

Ignjatović, I. S., Marinković, S. B., Mišković, Z. M., & Savić, A. R. (2012). Flexural behavior of reinforced recycled aggregate concrete beams under short-term loading. *Materials and Structures, 46*(6), 1045–1059.

Ignjatović, I. S., Marinković, S. B., & Tošić, N. (2017). Shear behaviour of recycled aggregate concrete beams with and without shear reinforcement. *Engineering Structures, 141*, 386–401.

ISO-2394. (2014). *General principles on reliability for structures*. Geneve, Switzerland: ISO.

JCSS. (2001). Probabilistic model code. In *Part 3: Material properties*.

JIS-5021. (2011). *Recycled aggregate for concrete-class H (in Japanese)*. Tokyo, Japan: Japan Standards Association.

JIS-5022. (2012). *Recycled aggregate for concrete-class M (in Japanese)*. Tokyo, Japan: Japan Standards Association.

JIS-5023. (2012). *Recycled aggregate for concrete-class L (in Japanese)*. Tokyo, Japan: Japan Standards Association.

Kang, T. H.-K., Kim, W., Kwak, Y.-K., & Sung-Gul, H. (2014). Flexural testing of reinforced concrete beams with recycled concrete aggregates (with appendix). *ACI Structural Journal, 111*(3), 607–616.

Katkhuda, H., & Shatarat, N. (2016). Shear behavior of reinforced concrete beams using treated recycled concrete aggregate. *Construction and Building Materials, 125*, 63–71.

Kheder, G. F., & Al-Windaw, S. A. (2005). Variation in mechanical properties of natural and recycled aggregate concrete as related to the strength of their binding mortar. *Materials and Structures, 38*(281), 701–709.

Khoury, E., Ambrós, W., Cazacliu, B., Sampaio, C. H., & Remond, S. (2018). Heterogeneity of recycled concrete aggregates, an intrinsic variability. *Construction and Building Materials, 175*, 705–713.

Kim, S.-W., Jeong, C.-Y., Lee, J.-S., & Kim, K.-H. (2013). Size effect in shear failure of reinforced concrete beams with recycled aggregate. *Journal of Asian Architecture and Building Engineering, 12*(2), 323–330.

Kim, Y., Sim, J., & Park, C. (2012). Mechanical properties of recycled aggregate concrete with deformed steel re-bar. *Journal of Marine Science and Technology, 20*(3), 274–280.

Knaack, A., & Kurama, Y. (2015). Behavior of reinforced concrete beams with recycled concrete coarse aggregates. *Journal of Structural Engineering, 141*(3).

Kou, S. C., & Poon, C. S. (2012). Enhancing the durability properties of concrete prepared with coarse recycled aggregate. *Construction and Building Materials, 35*, 69–76.

Li, W., Xiao, J., Sun, Z., & Shah, S. P. (2012). Failure processes of modeled recycled aggregate concrete under uniaxial compression. *Cement and Concrete Composites, 34*(10), 1149–1158.

LNEC-E471. (2009). *Guide for the use of coarse recycled aggregates in concrete (in Portuguese)*. Portugal: National Laboratory of Civil Engineering - LNEC.

Lubell, A., Sherwood, E., Bentz, E., & Collins, M. P. (2004). Safe shear design of large wide beams. *Concrete International, 26*(1).

Malešev, M., Radonjanin, V., & Marinković, S. (2010). Recycled concrete as aggregate for structural concrete production. *Sustainability, 2*(5), 1204–1225.

Mechtcherine, V., & Müller, H. S. (1998). Effect of the test set-up on fracture mechanical parameters of concrete. In *FraMCoS-3: Fracture mechanics of concrete structures*. Gifu, Japan: Aedificatio Publishers.

Mehta, P. K., & Monteiro, P. J. (2006). *Concrete: Microstructure, properties and materials*. McGraw-Hill.

Muttoni, A., & Fernández Ruiz, M. (2010). Shear in slabs and beams: Should they be treated in the same way?. In *Fib Bulletin 57: Shear and punching shear in RC and FRC elements*. Switzerland: fib.

Neville, A. M., & Brooks, J. J. (2010). *Concrete technology*. Harlow, United Kingdom: Pearson.

Noguchi, T., Tomosawa, F., Nemati, K. M., Chiaia, B. M., & Fantilli, A. P. (2009). A practical equation for elastic modulus of concrete. *ACI Structural Journal, 106*(5), 690−696.

Ozbakkaloglu, T., Gholampour, A., & Xie, T. (2018). Mechanical and durability properties of recycled aggregate concrete: Effect of recycled aggregate properties and content. *Journal of Materials in Civil Engineering, 30*(2), 04017275.

Pacheco, J., & de Brito, J. (2021a). Recycled aggregate concrete: Properties and behaviour, applications and production challenges. In *Fib Symposium 2021: Concrete structures: New trends for eco-efficiency and performance* (Lisbon, Portugal).

Pacheco, J., & de Brito, J. (2021b). Recycled aggregates produced from construction and demolition waste for structural concrete: Constituents, properties and production. *Materials, 14*(19), 5748.

Pacheco, J., de Brito, J., Chastre, C., & Evangelista, L. (2019a). Scatter of constitutive models of the mechanical properties of concrete: Comparison of major international codes. *Journal of Advanced Concrete Technology, 17*(3), 102−125.

Pacheco, J., de Brito, J., Chastre, C., & Evangelista, L. (2019b). Experimental investigation on the variability of the main mechanical properties of recycled aggregate concrete. *Construction and Building Materials, 201*, 110−120.

Pacheco, J., de Brito, J., Chastre, C., & Evangelista, L. (2019c). Probabilistic conversion of the compressive strength of cubes to cylinders of natural and recycled aggregate concrete specimens. *Materials, 12*(2).

Pacheco, J., de Brito, J., Chastre, C., & Evangelista, L. (2019d). Uncertainty models of reinforced concrete beams in bending: Code comparison and recycled aggregate incorporation. *Journal of Structural Engineering, 145*(4), 04019013.

Pacheco, J., de Brito, J., Chastre, C., & Evangelista, L. (2020). Uncertainty of shear resistance models: Influence of recycled concrete aggregate on beams with and without shear reinforcement. *Engineering Structures, 204*(1), 109905.

Pacheco, J., de Brito, J., Chastre, C., & Evangelista, L. (2021). Bond of recycled coarse aggregate concrete: Model uncertainty and reliability-based calibration of design equations. *Engineering Structures, 239*, 112290.

Pacheco, J., de Brito, J., Ferreira, J., & Soares, D. (2015a). Destructive horizontal load tests of full-scale recycled aggregate concrete structures. *ACI Structural Journal, 112*(6), 815−826.

Pacheco, J., de Brito, J., Ferreira, J., & Soares, D. (2015b). Flexural load tests of full-scale recycled aggregates concrete structures. *Construction and Building Materials, 101*, 65−71.

Pedro, D., de Brito, J., & Evangelista, L. (2014). *Performance of concrete made with aggregates recycled from precasting industry waste: Influence of the crushing process*. Materials and Structures.

Pepe, M., Grabois, T. M., Silva, M. A., Tavares, L. M., & Toledo Filho, R. D. (2018). Mechanical behaviour of coarse lightweight, recycled and natural aggregates for concrete. In *Proceedings of the institution of civil engineers—construction materials* (pp. 1−9).

Pepe, M., Toledo Filho, R. D., Koenders, E. A. B., & Martinelli, E. (2016). A novel mix design methodology for recycled aggregate concrete. *Construction and Building Materials, 122*, 362–372.

Peterson, P. E. (1981). Crack growth and development of fracture zones in plain concrete and similar materials. In *Division of building materials*. Lund, Sweden: Lund University.

Prince, M. J. R., Gaurav, G., & Singh, B. (2017). Splice strength of deformed steel bars embedded in recycled aggregate concrete. *Structures, 10*, 130–138.

PT-SC2-T1, D3BG. (2018). *Background documents to the final PT1 draft prEN 1992-1-1:2018.*

Rahal, K. N., & Alrefaei, Y. T. (2017). Shear strength of longitudinally reinforced recycled aggregate concrete beams. *Engineering Structures, 145*, 273–282.

Rahal, K. N., & Alrefaei, Y. T. (2018). Shear strength of recycled aggregate concrete beams containing stirrups. *Construction and Building Materials, 191*, 866–876.

Rao, M. C., Bhattacharyya, S. K., & Barai, S. V. (2011). Influence of field recycled coarse aggregate on properties of concrete. *Materials and Structures, 44*(1), 205–220.

Sadati, S., Arezoumandi, M., Khayat, K. H., & Volz, J. S. (2016). Shear performance of reinforced concrete beams incorporating recycled concrete aggregate and high-volume fly ash. *Journal of Cleaner Production, 115*, 284–293.

Sanchez de Juan, M., & Alaejos Gutierrez, P. (2009). Study on the influence of attached mortar content on the properties of recycled concrete aggregate. *Construction and Building Materials, 23*(2), 872–877.

Sato, R., Maruyama, I., Sogabe, T., & Sogo, M. (2007). Flexural behavior of reinforced recycled concrete beams. *Journal of Advanced Concrete Technology, 5*(1), 43–61.

Seara-Paz, S., González-Fonteboa, B., Eiras-López, J., & Herrador, M. (2013). Bond behavior between steel reinforcement and recycled concrete. *Materials and Structures, 47*(1–2), 323–334.

Shima, H., Tateyashiki, H., Matsuhashi, R., & Yoshida, Y. (2005). An advanced concrete recycling technology and its applicability assessment by the input-output analysis. *Journal of Advanced Concrete Technology, 3*(1), 53–67.

Sigrist, V., Bentz, E., Ruiz, M. F., Foster, S., & Muttoni, A. (2013). Background to the fib model code 2010 shear provisions—part I: Beams and slabs. *Structural Concrete, 14*(3), 195–203.

Silva, R. V., de Brito, J., & Dhir, R. K. (2014). Properties and composition of recycled aggregates from construction and demolition waste suitable for concrete production. *Construction and Building Materials, 65*, 201–217.

Silva, R. V., de Brito, J., & Dhir, R. K. (2015a). Comparative analysis of existing prediction models on the creep behaviour of recycled aggregate concrete. *Engineering Structures, 100*, 31–42.

Silva, R. V., de Brito, J., & Dhir, R. K. (2015b). Prediction of the shrinkage behavior of recycled aggregate concrete: A review. *Construction and Building Materials, 77*, 327–339.

Soares, D., de Brito, J., Ferreira, J., & Pacheco, J. (2014). Use of coarse recycled aggregates from precast concrete rejects: Mechanical and durability performance. *Construction and Building Materials, 71*, 263–272.

Sýkora, M., & Holický, M. (2012). Assessment of the model uncertainty in shear resistance of reinforced concrete beams without shear reinforcement. In *Transactions of the VŠB – technical university of ostrava* (p. 195). Civil Engineering Series.

Sýkora, M., & Holicky, M. (2013). Assessment of uncertainties in mechanical models. *Applied Mechanics and Materials, 378*, 13–18.

Sykora, M., Holicky, M., Prieto, M., & Tanner, P. (2014). Uncertainties in resistance models for sound and corrosion-damaged RC structures according to EN 1992-1-1. *Materials and Structures, 48*(10), 3415–3430.

Sykora, M., Krejsa, J., Mlcoch, J., Prieto, M., & Tanner, P. (2018). Uncertainty in shear resistance models of reinforced concrete beams according to fib MC2010. *Structural Concrete, 19*(1), 284–295.

Tam, V., Tam, C. M., & Le, K. N. (2007). Removal of cement mortar remains from recycled aggregate using pre-soaking approaches. *Resources, Conservation and Recycling, 50*(1), 82–101.

Tošić, N., de la Fuente, A., & Marinković, S. (2018). Shrinkage of recycled aggregate concrete: Experimental database and application of fib model code 2010. *Materials and Structures, 51*(5), 126.

Tošić, N., de la Fuente, A., & Marinković, S. (2019). Creep of recycled aggregate concrete: Experimental database and creep prediction model according to the fib Model Code 2010. *Construction and Building Materials, 195*, 590–599.

Tošić, N., Marinković, S., & Kurama, Y. (2020). Improved serviceability and environmental performance of one-way slabs through the use of layered natural and recycled aggregate concrete. *Sustainability, 12*(24), 10278.

Tsujino, M., Noguchi, T., Tamura, M., Kanematsu, M., & Maruyama, I. (2007). Application of conventionally recycled coarse aggregate to concrete structure by surface modification treatment. *Journal of Advanced Concrete Technology, 5*(1), 13–25.

Vrouwenvelder, A., & Siemes, A. (1987). Probabilistic calibration procedure for the derivation of partial safety factors for The Netherlands building codes. *HERON, 32*(4), 9–29.

Walraven, J. (1980). Aggregate interlock: A theoretical and experimental analysis. In *Concrete structure's group. 1980*. Delft, The Netherlands: TU Delft.

Wang, H.-L. (2016). Steel–concrete bond behaviour of self-compacting concrete with recycled aggregates. *Magazine of Concrete Research, 68*(13), 678–691.

Wang, C., & Xiao, J. (2013). Study of the seismic response of a recycled aggregate concrete frame structure. *Earthquake Engineering and Engineering Vibration, 12*(4), 669–680.

Wardeh, G., Ghorbel, E., Gomart, H., Fiorio, B., & Pliya, P. (2018). Shear behavior of reinforced recycled aggregate concrete beams. In D. A. Hordijk, & M. Luković (Eds.), *High tech concrete: Where technology and engineering meet: Proceedings of the 2017 fib symposium, held in maastricht, The Netherlands, June 12–14, 2017* (pp. 508–515). Cham: Springer International Publishing.

Waseem, S. A., & Singh, B. (2016). Shear transfer strength of normal and high-strength recycled aggregate concrete – an experimental investigation. *Construction and Building Materials, 125*, 29–40.

Waseem, S. A., & Singh, B. (2018). An experimental study on shear capacity of interfaces in recycled aggregate concrete. *Structural Concrete, 19*(1), 230–245.

Xiao, J., & Falkner, H. (2007). Bond behaviour between recycled aggregate concrete and steel rebars. *Construction and Building Materials, 21*(2), 395–401.

Xiao, J., Li, W., Sun, Z., Lange, D. A., & Shah, S. P. (2013). Properties of interfacial transition zones in recycled aggregate concrete tested by nanoindentation. *Cement and Concrete Composites, 37*, 276–292.

Xiao, J., Li, W., Sun, Z., & Shah, S. (2012). Crack propagation in recycled aggregate concrete under uniaxial compressive loading. *Materials Journal, 109*(4).

Xiao, J. Z., Li, J., & Zhang, C. (2005). Mechanical properties of recycled aggregate concrete under uniaxial loading. *Cement and Concrete Research, 35*(6), 1187–1194.

Xiao, J., Pham, T. L., Wang, P. J., & Gao, G. (2014). Behaviors of semi-precast beam made of recycled aggregate concrete. *The Structural Design of Tall and Special Buildings, 23*, 692–712.

Xiao, J., Sun, Y., & Falkner, H. (2006). Seismic performance of frame structures with recycled aggregate concrete. *Engineering Structures, 28*(1), 1–8.

Xiao, J., Sun, C., & Jiang, X. (2015). Flexural behaviour of recycled aggregate concrete graded slabs. *Structural Concrete, 16*(2), 249–261.

Xiao, J., Sun, C., & Lange, D. A. (2016). Effect of joint interface conditions on shear transfer behavior of recycled aggregate concrete. *Construction and Building Materials, 105*, 343–355.

Xue, J., Zhai, L., Bao, Y., Ren, R., & Zhang, X. (2017). Cyclic loading tests and shear strength of steel reinforced recycled concrete beam–column inner joints. *The Structural Design of Tall and Special Buildings*, e1454.

Yagishita, F., Sano, M., & Yamada, M. (1993). Behaviour of reinforced concrete beams containing recycled coarse aggregate. In E. K. Lauritzen (Ed.), *Third international RILEM symposium on demolition and reuse of concrete and masonry* (pp. 384–399). Denmark: RILEM Publications SARL: Odense.

Yang, Y., Xue, Y., Yu, Y., Ma, N., & Shao, Y. (2017). Experimental study on flexural performance of partially precast steel reinforced concrete beams. *Journal of Constructional Steel Research, 133*, 192–201.

Yousefpour, H., & Bayrak, O. (2016a). *Shear design of beams without stirrups: Survey of participants in the fib workshop on beam shear.* Zurich, Switzerland: fib.

Recycled concrete for nonstructural applications

7

Wei Chen [1] *and Ruoyu Jin* [2]
[1]School of Civil Engineering and Architecture, Ningbo Technical University, Ningbo, Zhejiang Province, People's Republic of China; [2]School of Built Environment and Architecture, London South Bank University, London, United Kingdom

7.1 Introduction

Construction and demolition (C&D) wastes account for around 30%–40% of urban solid wastes (Rodríguez et al., 2015). Among the C&D wastes, concrete waste contribute to 50%–70% by weight (Tam, 2008). These different forms of C&D wastes, including concrete, bricks, and tiles, are causing pressures on urban landfilling spaces (Duan & Li, 2016). With the growing attention on the development of circular economy to recover used resources (Ghisellini et al., 2016), recycling and reusing C&D wastes is becoming the trend to meet the sustainability needs of the construction industry (Chen et al., 2019). Using recycled aggregate (RA) obtained from C&D wastes is one of the main studied topics in the domain of C&D waste management as identified by Jin et al. (2019). Fig. 7.1 illustrates the interconnections between RA, C&D waste management, and circular economy. RAs are produced by crushing C&D wastes (e.g., concrete). Using RAs including RCA can achieve multiple sustainable goals in the civil and construction industries, including reducing the consumption of natural aggregates, reusing C&D wastes by saving landfilling spaces, and other environmental benefits.

Fig. 7.1 also shows the common sources of C&D wastes, such as concrete. The practices of circular economy involve recovery, recycle, reuse, and other circular uses of wastes. In the context of C&D waste management, adopting RA is one of the key practices for achieving sustainability. Indeed, RA is one major utilization of C&D materials (Kianimehr et al., 2019). The end products or the applications of RAs can be divided into structural and nonstructural applications. RA may come from clean C&D wastes or dirty and polluted sources. Although the policies and practices could vary in different countries, RAs are typically used in the construction of pavements, fills and embankments (Rodríguez et al., 2020). According to Rodríguez et al. (2020), RAs only account for about 9% of all total aggregate produced in European Union countries for manufacturing structural or nonstructural concrete. However, shortages of natural or virgin aggregates are becoming a crisis or issue. For example, Construction Leadership Council in the United Kingdom has revealed that aggregates are now in the growing lists of items in short supply together with cements (Stein, 2021). PMR (2017, p. 235) revealed that the global demand for aggregates would

Recycled Concrete. https://doi.org/10.1016/B978-0-323-85210-4.00006-0

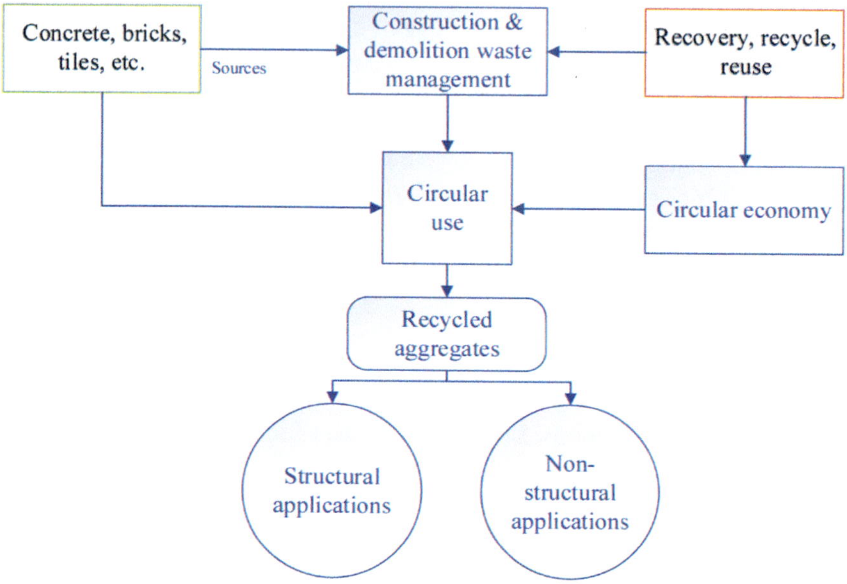

Figure 7.1 Interconnections between C&D waste management, circular economy, and recycled aggregates.

increase from 45 billion tonnes in 2017 to 66 billion tonnes by 2025. There is a growing need worldwide to adopt RAs to address the shortage of virgin aggregates.

This chapter focuses on nonstructural applications of RAs from C&D wastes, by introducing the current research and practices of adopting RAs, adopting RAs specifically for nonstructural uses, and relevant case studies. RAs can be used in the mix design of new concrete. Nonstructural concrete is the type of concrete with generally low strength (i.e., lower than 25 MPa) and is typically used when only small compression or temporary loading is involved (The Concrete Society, 2021). Examples of nonstructural applications of RAs, as referred by literature (e.g., Ahmed et al., 2020; Kovacs et al., 2019; Yap et al., 2018), include but are not limited to road sub-bases, new concrete production in nonload-bearing building elements, and sidewalk, etc.

7.2 Use of recycled concrete in nonstructural applications

7.2.1 *Technical studies of adopting recycled concrete aggregates*

There have been extensive studies on how adopting RAs would affect the properties of cementitious products. RAC or recycled aggregate concrete is one of the most widely studied cementitious products adopting RAs from different waste sources such as demolished concrete. Adapted from Chen et al. (2019), Table 7.1 shows several selected studies on how adoption of RAs affected the properties of cementitious products (e.g., concrete).

Table 7.1 Studies investigating the influences of RAs on cement composites' properties.

Referenced study	Type or source of RA	Substitution rates of RA to virgin aggregate in mix design	Properties tested for cementitious products	Applications of the studied products containing RAs	Summary of findings
Alexandridou et al. (2018)	RA from different Greek recycling plants	Three different substitution rates of RA to coarse virgin aggregates (VA) at 0%, 25%, and 75% respectively	Compressive strength, concrete absorption, sorptivity, and carbonation resistance of RAC	Concrete specimens for laboratory tests	The compressive strength of RAC ranged from significantly lower (37% reduction) than that of normal concrete. Higher water absorption of coarse RA was their most negative physical characteristic. Coarse RA reduced the durability of hardened concrete.

Continued

Table 7.1 Continued

Referenced study	Type or source of RA	Substitution rates of RA to virgin aggregate in mix design	Properties tested for cementitious products	Applications of the studied products containing RAs	Summary of findings
Dimitriou et al. (2018)	Coarse RAs from different sources of crushed concrete	Different substitution rates of RA to VA at 50% and 100% respectively	Compressive strength, flexural strength, splitting tensile strength, modulus of elasticity, porosity, sorptivity, and permeability of RAC	Concrete specimens for the laboratory tests	Increasing the substitution rate of RA to VA resulted in lower quality of concrete. Both mechanical and durability properties are negatively affected by the increase of the substitution rate. But a simple treatment method to reduce the adhered mortar at RA surface could diminish the negative effects of RAs and create a better quality of RAC, which could be competitive to normal concrete.

Ozbakkaloglu et al. (2018)	Coarse RAs in two different sizes (i.e., 7 and 12 mm)	RAs used to replace VA at different substitution rates, including 0, 25%, 50%, and 100%	Compressive strength, elastic modulus, flexural strength, splitting tensile strength, workability, drying shrinkage, and water absorption of RAC	Specimens for testing, including cylinder specimens and prism specimens	An increase in the coarse aggregate size led to an increase in the 28-day elastic modulus and a decrease in the 28-day flexural and splitting tensile strengths. Coarser RA caused higher drying shrinkage and water absorption in concrete mix. RACs with up to 25% RA content exhibited slightly inferior mechanical and durability-related properties compared to the conventional concrete with the same compressive strength. But replacement of 100% VA would cause significant reductions in concrete properties.

Continued

Table 7.1 Continued

Referenced study	Type or source of RA	Substitution rates of RA to virgin aggregate in mix design	Properties tested for cementitious products	Applications of the studied products containing RAs	Summary of findings
Thomas et al. (2013)	Fine and Coarse RAs from crushed test concrete specimens	20% of replacement of RA to the coarse VA, and 100% replacement to both fine and coarse VAs	Compressive and tensile strength, permeability, water penetration, chloride penetration	Mortar and concrete specimens for the laboratory tests	The sulfur within RA did not significantly affect the mechanical or physical performance of mortar or RAC. But using RA from crushed concrete, with or without sulfur, was viable for the manufacture of recycled structural concretes for applications without exposure to high temperatures. The use of the fine fraction in RAs caused a significant loss of properties.

Etxeberria et al. (2007)	Coarse RA from crushed concrete	Four different substitution rates of RA to VAs at 0%, 25%, 50% and 100% respectively	Compressive and tensile strength, modulus of elasticity of RAC	Concrete specimens for laboratory tests	Concrete crushed by an impact crusher achieves a high percentage of RAs without adhered mortar. Adhered mortar in RA caused the weak point in the RAC microstructure. RAC made with 100% of coarse RA had significantly lower compression strength than conventional concrete, or required more cement in mix design to achieve higher strength.

Continued

Table 7.1 Continued

Referenced study	Type or source of RA	Substitution rates of RA to virgin aggregate in mix design	Properties tested for cementitious products	Applications of the studied products containing RAs	Summary of findings
Evangelista and de Brito (2007)	Fine RA from crushed concrete	Five different replacement ratios of fine RA to fine VA, namely 10%, 20%, 30%, 50%, and 100%	Compressive strength, split tensile strength, modulus of elasticity and abrasion resistance of RAC	Structural concrete specimens for laboratory tests	It was viable to produce concrete made with fine RA for structural concrete. Up to 30% replacement of fine RA to fine VA did not seem affecting the compressive strength of RAC. Both tensile splitting and modulus of elasticity were reduced with the increase of the replacement ratio. The abrasion resistance seemed to increase with the replacement of fine VA with fine RA.

Tam et al. (2005)	RAs collected from local recycling plants, with sizes at 10 and 20 mm respectively	Substitution rates at 0%, 10%, 15%, 20%, 25% and 30% of RA to VA	Compressive strength of RAC specimens at different curing ages by using the normal mixing approach and the two-stage mixing approach	RAC specimens for laboratory tests	The two-stage mixing approach gives way for the cement slurry to gel up the RA, providing a stronger interfacial transition zone by filling up the cracks and pores within RA. This two-stage mixing approach can provide an effective method for enhancing the compressive strength and other mechanical performance of RAC.
Xiao et al. (2005)	Coarse RA from waste concrete brought from runway	Substitution rates of RA to VA at 0%, 30%, 50%, 70% and 100% respectively	Compressive strength, the elastic modulus, the peak and the ultimate strains of RAC	RAC specimens for laboratory tests	RAC specimens failed in a shear mode. The stress–strain curves of RAC indicated an increase in the peak strain and a significant decrease in the ductility. The compressive strength, elastic modulus of RAC generally decreased as the replacement ratio of RA increased. The peak strain of RAC also increased with the increase of RA contents.

Continued

Table 7.1 Continued

Referenced study	Type or source of RA	Substitution rates of RA to virgin aggregate in mix design	Properties tested for cementitious products	Applications of the studied products containing RAs	Summary of findings
Poon, Shui, and Lam (2004)	Coarse RAs from two different type of crushed concrete, namely normal-strength concrete (NC), and high-performance Concrete (HPC)	Full replacement of VA by RA from NC, and RA from HPC respectively	Microstructure and compressive strength of RAC	RAC specimens for laboratory tests	RAC prepared with the RA from HPC developed higher compressive strength than RAC prepared with RA from NC at all tested ages. In particular, the strength of RAC prepared with the RA from HPC was comparable to that of conventional concrete. The difference in strength development between the RAC with HPC and with NC aggregates was due to the differences in both the strength of the coarse aggregates and the microstructural properties of the interfacial transition zones.

Poon, Shui, Lam, Fok, and Kou (2004)	Coarse RA from crushed and graded unwashed concrete from a single source, sized at 10 and 20 mm	Various replacement ratios of RA to VA, were adopted, namely 0%, 20%, 50%, and 100%; The moisture states of RAs were controlled at air-dried (AD), oven-dried (OD) and saturated surface-dried (SSD) states prior to use.	Slump and compressive strength of RAC	RAC specimens for laboratory tests	The moisture states of the RAs affected the change of slump of the fresh RAC. RA with OD led to a higher initial slump and quicker slump loss, while RAs with SSD and AD had normal initial slumps and slump losses. RAC from RA with AD exhibited the highest compressive strength. Aggregates in the AD state containing not more than 50% RA should be optimum for normal strength RAC production.

Adapted from Chen, W., Jin, R., Xu, Y., Wanatowski, D., Li, B., Yan, L., Pan, Z., & Yang, Y. (2019). Adopting recycled aggregates as sustainable construction materials: A review of the scientific literature. *Construction and Building Materials*, *218*, 483–496.

Table 7.1 summarizes a few selected studies of adopting RAs to produce cementitious products such as RAC. It is seen that many prior studies focused on laboratory investigations on how RAC properties would be affected by the different substitution rates of RAs to virgin aggregates. RAs can be categorized into coarse and fine types, and the coarse RA (i.e., the aggregate size of over 5 mm) has been more widely studied. Intense utilization of RCA in the large-scale industrial production of new concrete has not been widely accepted by international standards due to quality issues of RAs such as high porosity, uncertainty of source concrete, and lack of territorial field studies (Ahmed et al., 2020). Extending from the findings of studies listed in Table 7.1, it is also recognized that new concrete containing RAs, such as recycled aggregate concrete (RAC), typically has inferior properties (e.g., lower strength) compared to normal aggregate concrete. Nevertheless, lower quality of RAC does not mean that there is little application of it, for example, RAC with low strength can be applied in nonstructural members (Zaben et al., 2021). Ahmed et al. (2020) found that RCA has significant potential to substitute virgin aggregates in producing RAC between 25 and 32 MPa. It was consistently found by various studies including Ahmed et al. (2020) and Jin et al. (2018) that RAC containing RAs, although suffered from some early age strength reduction as compared to the normal concrete with virgin aggregates, its longer-term compressive strength is generally comparable to normal concrete. Several studies listed in Table 7.1 also analyzed the features of RCAs from the microscopic view. Fig. 7.2 from Chen et al. (2019) demonstrates typical surface appearance of RCA under the observation of microscope.

As seen in Fig. 7.2, RAs collected from recycled concrete (i.e., RCA) often has cracks and mortar adhered to the surface. The surface cracks could cause potential inferior mechanical properties when used as raw concrete materials in the mix design. The mortar from old concrete makes RAs with higher water absorption, and could cause inferior quality of new concrete containing these RAs. It is hence widely believed that RAs cause negative effects on the end products' engineering properties. However, researchers have been keeping exploring how to minimize the inferior features of RAs in order to increase the reuse rate. Several studies, such as Jin et al. (2018) and Xu et al. (2018), suggested that RAs, based on its higher water absorption nature, could be prewetted prior to concrete mix to improve the fresh and hardened concrete properties. Xu and Sun (2011) suggested that RAs had its both positive and negative effects on the properties of RAC. Somehow similar to lightweight aggregates, which have internal voids that can store moisture, the prewetting method suggested by Jin et al. (2018) and Xu et al. (2018) also allows RAs to store moisture prior to concrete mix, and then slowly releasing the prestored moisture in the early age of concrete curing. This slow release of stored moisture, in the bespoke "internal curing", can enhance concrete properties including strength development and longer-term durability. Xu et al. (2018) further tested the proper substitution rates of RAs to virgin aggregate to maintain or even enhance RCA strength. Substitution rates between 30% and 50% of coarse RA to virgin aggregate by weight could maintain the concrete mechanical properties as found by Xu et al. (2018). During this range of substitution rate, it was believed that the "internal curing" feature of RAs by prestoring moisture to be

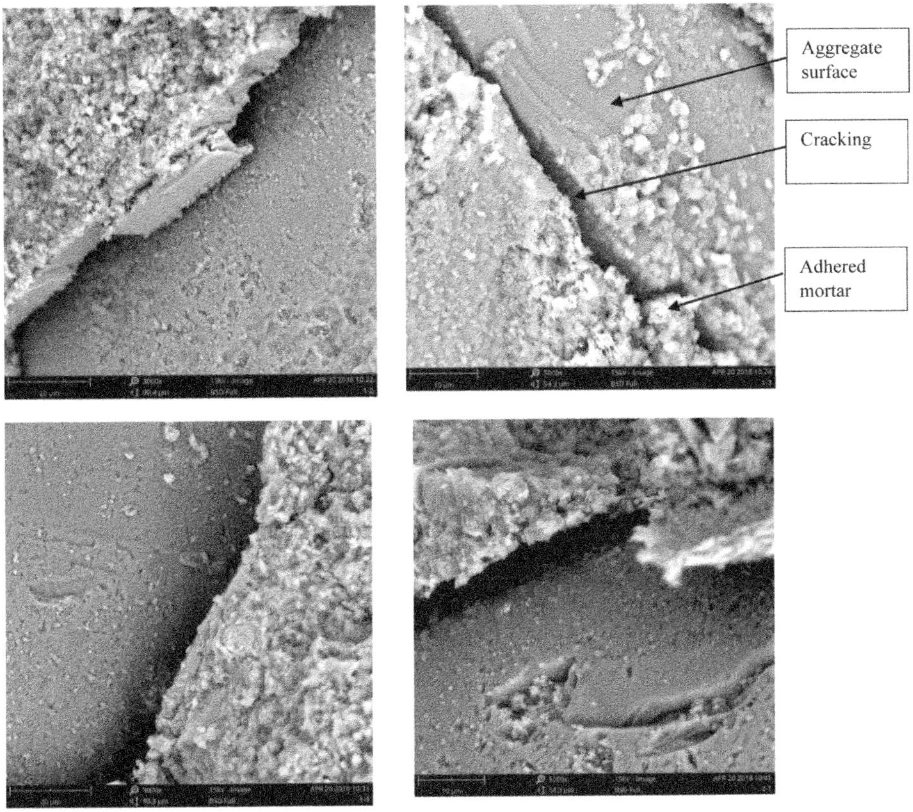

Figure 7.2 Microstructure of RAs from crushed concrete.
From Chen et al. (2019).

released to assist secondary cement hydration in early curing stage of concrete could compensate the inferior properties of RAs as shown in Fig. 7.2.

Nevertheless, it is worth considering several issues of extending laboratory tests to real world applications. The first issue is the source of wastes, as C&D wastes are often mixed with different sources, for example, old concrete, red bricks, and tiles, etc. Even the concrete wastes can be with different sources of concrete, such as structural concrete from columns, and nonstructural concrete, which may be with lower strength. RA properties can differ according to their waste sources, which can be called the "parent" concrete. It has been found from different studies (Chakradhara Rao, 2018; Kou & Poon, 2015) that RAs from higher strength of "parent" concrete could result in comparable properties with normal concrete, and even with high performance concrete. A second issue is the prewetting method for enhancing RA properties. It was performed at laboratory conditions in the published work (e.g., Jin et al., 2018). However, it remains a concern of how that can be scaled up for large production of prewetted RAs to produce RAC members in real-world applications. There is a need to review the existing applications of RAs in construction and civil engineering, including nonstructural members.

7.3 Practices and issues of adopting recycled aggregates in new applications

It has been commonplace to adopt RAs from crushed C&D wastes for downcycling purposes, for example, localized fill, granular fill, backfill, and pavement base as identified in different countries such as USA (Jin & Chen, 2015) and Spain (Rodríguez et al., 2020). Besides the technical barriers of adopting RAs summarized in Section 2.1, several other barriers or challenges have been studied in the literature. Table 7.2 lists several widely encountered barriers of adopting RAs in new applications.

RAs from recycled concrete (i.e., RCA) and other waste sources, have been regulated in standards worldwide, especially for being applied in new concrete production. Tam et al. (2018) summarized a total of 18 different countries' or regions' standards regulating the use of RAs for new concrete productions. The majority of the standards evaluated by Tam et al. (2018) specified the applications of RAs, such as structural reinforced concrete, prestressed concrete, and nonstructural concrete. The restrictions on RAs, for example, maximum allowed replacement rates of RAs to virgin aggregates

Table 7.2 Barriers or challenges in adopting RAs for new applications.

Category	Barrier or recommendation in promoting RA usage	References
Legislation, standard, regulation	Outdated or insufficient legislation	Rodríguez et al. (2020)
	Lack of technical knowledge, standards and procedure	Zhang et al. (2019)
	Government support	Zhang et al. (2019)
	Control and certifications of end products containing RAs	Rodríguez et al. (2020)
Market	Virgin aggregates widely used in the market, underdeveloped reuse market	Jin and Chen (2019), Rodríguez et al. (2020)
	Availability and sufficiency of RAs and useable materials	Kasai (1998), Tam et al. (2018)
Social	Social and cultural acceptance of adopting RAs	Jin et al. (2017), Jin and Chen (2019), Zhang et al. (2019)
Technical	Uncertainty of the C&D waste sources affecting RA properties	Meddah (2017), Chen et al. (2019)
	Labor and energy inputs in improving the physical properties of RAs (e.g., "cleaning" the surface impurities or other treatments)	Pepe et al. (2014), Tang et al. (2019)
	Lack of well-developed collection and processing facilities	Tam et al. (2018)
Financial	Cost effectiveness and financial return	Zhang et al. (2019)
Education/ training	Lack of professional training	Jin and Chen (2015), Zhang et al. (2019)

Table 7.3 Selected regulations/standards regarding adopting RAs in nonstructural applications.

Country or region	RA type	Restrictions for nonstructural application	Percentage of RA allowed to replace virgin aggregates
Hong Kong	Recycled concrete aggregates (RCA)	28 day's compressive strength to reach 20 MPa	100% for nonstructural concrete
Japan	Coarse recycled mixed aggregate (RMA)	Backfill concrete, blinding concrete & concrete filled in steel tubes	N/A
Italy	Coarse RMA	28 day's compressive strength to reach 10 MPa	100%
Brazil	Coarse and fine RCAs	Both coarse and fine RCAs are allowed in nonstructural concrete	20%

Adapted from Tam, V. W., Soomro, M., & Evangelista, A. C. J. (2018). A review of recycled aggregate in concrete applications (2000–2017). *Construction and Building Materials, 172*, 272–292.

are specified. The type of RAs related to the source of wastes are also defined, such as RCA and mixed C&D wastes. Table 7.3 filtered the summary of Tam et al. (2018) related to nonstructural applications only.

Only four out of the 18 countries listed by Tam et al. (2018) specifies the usage of RAs in nonstructural concrete. The requirements of other RA properties can be found in these standards, including dry density, water absorption, CI content, and limitations of applications (e.g., not allowed in prestressed concrete or reinforced concrete in some European countries). Compared to RAs for structural uses, fewer standards found related to nonstructural applications of RAs do not indicate that RAs have been more commonly adopted for structural purposes. In contrast, RCA has been predominantly used in nonstructural members such as filler, road sub-base, drainage material, and low quality concrete, but with limitations on structural applications (Bank et al., 2018).

Adoption of RAs often cause the changes of mix design in new concrete production, for example, increased cement consumption and usage of other supplementary cementitious materials (SCMs). Usage of SCMs to partly replace Portland cement would also enhance the sustainability of concrete production with reduced carbon emission. SCMs and RAs can be applied in either ready-mix concrete or prefabricated concrete. The survey of Jin et al. (2015) in the USA found that a higher percentage of ready-mix concrete suppliers adopted SCMs compared to prefabricators for nonstructural applications. But between the ready-mix and prefabricated concrete suppliers, there were no significantly different perceptions found regarding the benefits and barriers of using alternative aggregates (e.g., RAs). These barriers are generally consistent with these listed in Table 7.2, including cost barrier, local availability, technical barriers, lower qualities, and lack of data on concrete properties, etc.

7.4 Types of nonstructural members

Besides these practical issues and existing standards on applying RAs, there have also been numerous scholarly studies investigating how RAs affect the properties of building and civil engineering products. These scholarly works extend the practical guides aiming to explore the variety of RA sources or the extended applications of using RAs. For example, Kianimehr et al. (2019) collected RAs from concrete debris obtained by demolishing building blocks, and applied RAs for light-stabilization of clay soils. It was found that RAs added to clay soils could enhance several mechanical properties (e.g., dry density, shear strength) for being applied in sub-base/subgrade of road pavements. Other than the pavement, application in infrastructure and building construction as load-bearing members, studies could also be found on applying RAs in concrete production for nonstructural use. Generally, structural applications of RA in concrete members are with more stringent requirements on waste sources and mix design. To meet durability requirements, laboratory studies revealed that 35% substitution rate of recycled coarse aggregates together with 6% rate of recycled fine aggregate could be utilized for above-ground structures (Zhang et al., 2019). In comparison, 70% rate of recycled coarse aggregates together with 30% rate of recycled fine aggregate can be allowed for nonstructural concrete (Zhang et al., 2019).

Since nonstructural applications are typically with lower property requirements compared to structural applications, RAs can be with wider usage in constructing nonstructural members or those with reduced specifications as suggested by Agrela et al. (2018). As indicated by Agrela et al. (2018), nonstructural usage by adopting RAs could allow a wider source of wastes, for example, concrete wastes and other mixed C&D wastes. RAs for nonstructural applications could also increase the substitution rates of RAs to virgin aggregates, even as high as 100% replacement rate as implied by Silgado et al. (2018). Similarly, the 5-year research of The University of British Columbia suggested that recycled concrete can be produced with 100% substitution of RA for nonstructural applications such as sidewalk (AC Business Media, 2021). For example, it was reported that compared to conventional concrete made with virgin aggregates, RAC could have a higher rate of compressive strength after 28 days of curing while maintaining a greater or equal strength during the period of the research (AC Business Media, 2021). Table 7.4 presents several selected examples of RAs applied in the nonstructural members in construction.

It is worth noticing that using RAs for nonstructural applications does not mean the strength or other mechanical properties are unimportant. Instead, as indicated in Table 7.4, the typical mechanical and durability properties of products containing RAs are also tested for nonstructural members. It is common practice to limit the substitution rates of RAs to virgin aggregates in order to control the quality of cementitious products containing RAs. Different substitution rates have been tried in various studies summarized in Table 7.4. Depending on the source of RAs (e.g., C&D wastes) and the end application of RAs (e.g., sidewalk), the recommended substitution rates of RA to virgin aggregates could vary. For example, Saeed et al. (2017) recommended 10% replacement rate of virgin aggregates by RAs as the optimum.

Table 7.4 Summary of existing studies on applying RAs for nonstructural applications.

Referred study	Source of RA	Nonstructural application	Key properties tested	Major findings
Ahmed et al. (2020)	RAs were collected from the landfill to represent the real-world situation of commercial construction. RAs came from mixed strength of concrete from different demolition projects. To produce RAs, firstly demolished concrete was visually inspected and any impurities like steel, wood, gypsum, or organic compounds were manually removed. Then jaw crusher was used to crush demolished concrete. RAs at the range of 5–20 mm were screened through sieve analysis.	Municipal sidewalk with target strength of 32 MPa at Day 28, and different substitution rates of RAs to virgin aggregates at 0%, 30%, 50%, 70%, and 100%	Compressive strength at testing age from Day 7 to Year 5.	It was established that long term strength development among all substitution rates used in sidewalk mixtures were comparable to the conventional concrete with 0% RAs. The long term in-place performance of RAC was adequately comparable to conventional in terms of compressive strength as well as its serviceability. No issues had been reported since its placement at site i.e., five years. In addition to this, commercially produced RAC in a ready-mix plant has no difficulties in its production and placement.

Continued

Table 7.4 Continued

Referred study	Source of RA	Nonstructural application	Key properties tested	Major findings
Xu et al. (2020)	Coarse RAs recycled from local concrete demolition, with sizes between 5 and 10 mm	Pervious concrete with substitution rates of RAs to virgin aggregates at 30%, 50%, and 70% respectively. Coarse RAs were soaked with nano particles of TiO_2 for photocatalytic applications of pervious concrete	NO degradation of pervious concrete adopting RAs coated with nano particles of TiO_2; compressive strength	RAs' internal voids can be positively utilized for storages of photocatalytic nanoparticles for enhancing the air purification functions of pervious concrete. Utilizing coarse RAs to partly replace virgin aggregates in pervious concrete could achieve two sustainability purposes simultaneously, namely reusing wastes and degrading air pollutants such as NO_x. RAs with substitution rate below 50% to virgin aggregates, were found with positive effects on compressive strength.

Yap et al. (2018)	Recycled coarse aggregates from demolition site, with large-sized C&D wastes crushed and sieved to obtain the aggregate sized from 4.5 to 9.5 mm.	Pervious concrete with substitution rates of RAs to virgin aggregates at 20%, 40%, 60%, 80% and 100% respectively	Mechanical strength, permeability, skid and abrasion resistance	RCA could be used to replace up to 100% of virgin aggregates in pervious concrete mixture, by achieving maximum waste recycle and 24% reduction of CO_2 reduction; RCA reduced the compressive, splitting tensile and flexural strengths of pervious concrete due to adhered mortar to RAs and the poor aggregate-paste bonding; RCAs increased the permeability in pervious concrete.
Marie (2017)	Coarse RAs obtained from crushing test concrete samples; Crumb rubber from a local industrial plant	Hybrid recycled aggregate-rubberized concrete (RARC) for nonstructural thermal insulating material	Compressive strength and the thermal conductivity	The experimental study noted that a 20% replacement of virgin coarse aggregates with RA by weight, along with a 10% replacement of sand with crumb rubber by volume to produce RARC reduces the thermal conductivity of concrete relative to that of conventional concrete

Continued

Table 7.4 Continued

Referred study	Source of RA	Nonstructural application	Key properties tested	Major findings
López-Uceda, Ayuso, Jiménez, et al. (2016)	Mixed RAs sourced directly from a C&D waste recycling plant, consisting of asphalt, ceramics, mortar and concrete, gypsum, and others	Nonstructural concrete cores, with substitution rates of RAs to virgin aggregates at 20%, 40%, and 100% respectively, with low cement consumption	Compressive strength, splitting tensile strength, water penetration under pressure, porosity and water sorptivity	The use of mixed RA in concrete has a significant influence on the physical properties related to durability. Nevertheless, this material can be used without steel reinforcement in nonstructural applications.
López-Uceda, Ayuso, López, et al. (2016)	Two mixed recycled aggregates with different percentages of ceramic particles collected on different days from a C&D waste treatment plant	A nonstructural, low-strength concrete (15 MPa) with a mixed RA substitution rate of up to 100% for 200 kg/m^3 of cement	Compressive and splitting tensile strength, modulus of elasticity, density, porosity, water penetration, and sorptivity	The type of aggregate for this study did not result in significantly different effects on concrete properties. This type of concrete could be potentially applied in the construction of ditches, sidewalks, and other similar civil works.
Etxeberria et al. (2013)	Two types of recycled coarse aggregates from two different recycling plants were used in the substitution of raw coarse aggregates for concrete production. One was with a nominal size of 20 mm, and the other one was with a nominal size of 16 mm.	Two pilot tests of sidewalks as nonstructural applications of RAs	Physical, mechanical and durability properties, including workability, compressive strength, resistance to sulfate attack, etc.	The laboratory tests showed that the concretes produced with 50% of RAs obtained adequate physical, mechanical and durability properties for their use in nonstructural concrete production.

Mas et al. (2012)	Mixed RAs from C & D wastes treated, included concrete waste, unbound aggregates and ceramic products as predominant components	Low-strength blinding concrete (15 MPa), medium-strength blinding/backfill concrete (25 MPa) and nonstructural concrete for high-resistance precast elements (65 MPa)	Compressive, tensile and flexural strength	It was feasible to substitute about 40% of coarse aggregate to still comply with all the regulatory requirements. Thus, it could be viable to manufacture nonstructural concrete with mixed RAs.

Xu et al. (2018) suggested that substitution rate can be between 30% and 50% without significantly affecting concrete strength. Jayakody et al. (2018) suggested that it was feasible to substitute up to 40% of coarse virgin aggregates with RAs in conventional low-intermediate strength concrete applications. Yap et al. (2018) found that up to 100% of substitution rate could be applicable.

It may be widely viewed that nonstructural applications of RAs, such as in new concrete production or road subbase, are limited to less important members in building or civil engineering. Nevertheless, some experimental investigations listed in Table 7.4 show the potential of RAs for innovative or new applications in nonstructural members. For instance, through invesigation of the thermal properties of concrete products adopting RAs, Marie (2017) evaluated the feasibility of applying RAs for thermal insulation purposes. RAs can be mixed with different sources and different sizes (i.e., coarse and fine RAs) in order to maximize the reuse of wastes. It is not uncommon to have mixed RAs for reuse, because it may not be cost effective or labor efficient to sort wastes. In the study of López-Uceda, Ayuso, Jiménez, et al. (2016), mixed RAs were used in concrete cores by testing the long term key mechanical properties and durability. López-Uceda, Ayuso, Jiménez, et al. (2016) demonstrated the feasibility of using concrete with full coarse mixed RA incorporation and low cement content in nonstructural applications, e.g., bike lanes, gutters, ground slabs, leveling surfaces, subgrades for foundations and similar civil works. C&D wastes are generally the most available sources for mixed RAs, although different sources could be identified other than C&D activities. For example, Bank and Yazdanbakhsh (2016) investigated concrete containing RAs recycled from Glass Fiber Reinforced Polymer (GFRP), and concluded that RAs made from GFRP could be a viable option for nonstructural concrete production. The mechanical properties of concrete produced with GFRP were found comparable to concrete containing recycled concrete aggregates (Bank & Yazdanbakhsh, 2016).

Some other studies (e.g., Xu et al., 2020) applied coarse RAs in pervious concrete for environmental purification purposes. It is more traditionally believed that RAs, due to its inferior surface mortar and internal micro-structure features seen in Table 7.1 and Fig. 7.2, would typically negatively affect concrete properties. Scholars such as Xu and Sun (2011), and Xu et al. (2018) believed that RAs' internal voids could be utilized to preabsorb moisture, which could be released to assist cement hydration in concrete. Xu et al. (2020) further extended the prior studies by exploring how RAs could replace virgin aggregates and absorb photocatalytic nanoparticles such as TiO_2. Compared to conventional concrete with limited capability to absorb nanoparticles for the same purpose of air purification, RAs, as seen in Fig. 7.3, are with higher capability of being coated with nanoparticles. This feature of RAs (i.e., internal voids to absorb photocatalytic particles), enhances the durability of pervious concrete in degrading air pollutants such as NO. The experimental study of Xu et al. (2020) suggested that even after 10-min heavy rainwater wash, the pervious concrete containing RAs coated with TiO_2 particles still displayed the highest photocatalytic performance. This high durability under rain-wash of pervious concrete made with RAs indicated potential applications of RAs for pervious concrete surfaces (e.g., parking lot).

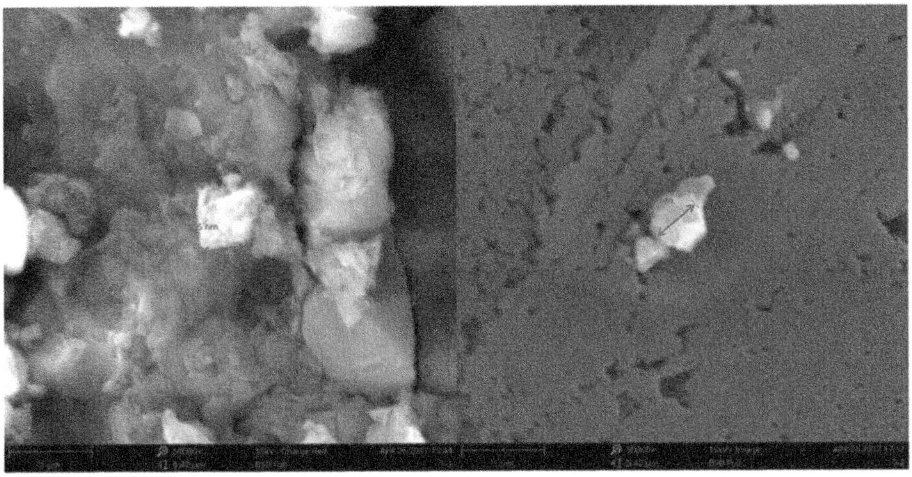

Figure 7.3 Microstructure of observing TiO_2 particles absorbed into the internal voids of RCAs (Scale at 50,000:1) (Xu et al., 2020).

7.5 Case studies

The experimental studies presented in Tables 7.1 and 7.4 are mostly from the laboratory research with limited real-world tests or applications. As suggested by Silva et al. (2019), the experimental studies may have not undergone the test under real life conditions and stakeholders would be more concerned in the recycled products' real-world circumstances before deploying them into the market. The real world applications of utilizing RAs from C&D wastes for infrastructure services including both unbound and hydraulically bound applications can be found in the review work of Silva et al. (2019), for example, applying RAs for base course of road or pavements. Reusing RAs for new concrete production in real world applications as described by Silva et al. (2019) were mainly in structural concrete, such as floor slabs, load-bearing walls, and beams, etc. Only one out of the 11 projects introduced by Silva et al. (2019) for reusing RAs in new concrete was for nonstructural applications, i.e., decorative concrete elements in Federal Gardening Exhibition (Bundesgartenschau) in Magdeburg, Germany for outdoor exposure. RAs in this project were prepared from crushed clay bricks and concrete. Special colors of cement (e.g., white and brown) was added to mix the coarse RAs to meet the decorative design purposes. It can be seen from existing research that RAs for new concrete have been more widely studied of their mechanical properties and durability-related properties for structural applications. Below more details of real-life case studies are described.

7.5.1 Case study: Chicago O'Hare Modernization Project

The Portland Cement Association (PCA) provided the case study of the field test of RCA in two lanes at one of the gates of O'Hare Modernization Project conducted in

2009. According to PCA (2019), laboratory tests based on a two-stage mixing method using coarse RCA from O'Hare International Project revealed that RCA reduced bleeding and segregation, and produced similar workability, compressive strength, and shrinkage as virgin aggregates. In the field tests of lanes using on-site recycled concrete at the airport, ready-mix concrete supplier treated RCA similar to lightweight aggregate, i.e., prewetting. Similar workability was achieved and the placement was in full service within 4 days (PCA, 2019).

PCA (2019) also reported the application of sensory technology at the time of placement to measure the internal relative humidity, temperature, and the lift-off of concrete member from the cement-treated permeable base, together with the monitoring of other properties, such as surface appearance and joint width. After 5 months of monitoring, the data between the two concrete lanes were statistically consistent, indicating no difference in behavior between the RCA concrete and virgin aggregate concrete (PCA, 2019).

The field study of concrete lanes of O'Hare Modernization Project highlighted the importance of pretreatment of RAs. The internal voids and high water absorption of RAs have been widely found lowering concrete mechanical properties and affecting workability. Both this case study and multiple prior laboratory research (e.g., Jin et al., 2018) recommended treating RAs following the similar treatment method to lightweight aggregate in order to achieve similar concrete properties as in conventional concrete. For example, PCA (2019) also recommended prewetting RCA and close to a saturated surface dry condition, like lightweight aggregates.

7.5.2 Case study: Pelican Self Storage

Pelican Self Storage, a new warehouse, is considered one of the world's first buildings made from recycled concrete and sets new standards for sustainable construction (Fibo Intercon, 2019a). Supported by Ministry of the Environment in Denmark, it demonstrates the circularity approach with reduced carbon emissions and with new concrete members containing RAs from crushed concrete & bricks (Fibo Intercon, 2019a). On-site reuse of demolished materials was the key of this project. According to Fibo Intercon (2019a), by reusing RAs from on-site demolished building materials, the project achieved these key benefits including reduced cost of purchasing virgin aggregates, lower transport costs, and promoting the concept of circular economy.

The demolished building material was sorted on site for further processing involving plant for concrete crushing, and sieve analysis for concrete mix design. The design of concrete mix was tested as most laboratory research of recycled aggregate concrete would perform, including: using test specimens crushed at different curing ages up to 28 days, as well as testing the moisture content of RAs. As RAs typically has higher moisture content and internal voids that affect new concrete fresh and hardened properties, it is also part of the processing to calculate additional water needed in the batching.

The project of Pelican Self Storage demonstrated the business model of on-site concrete production using site demolished materials. In this project, the main contractor selected Fibo Intercon (2019a) as the concrete supplier, who could set up the site concrete batching plant in a few hours for easy move and relocation. The cost and carbon saving of this project by reusing RAs were achieved by two critical factors in this business case, namely easy site setup and high flexibility without requiring a concrete base, as well as the reuse of RAs from site demolished concrete. Fibo Intercon (2019b) also provided business ideas of applying RAs for multiple new concrete members or other cementitious products including those for nonstructural purposes, for example, bed and surrounds for utility drains and service pipes, concrete lego blocks, lean mix oversite concrete, and general site concrete, etc. Practically, the sources of RAs could come from road sweepings, crushed concrete, crushed brick, crushed road surfacing, and crushed glass. The waste sources will determine the application of concrete (Fibo Intercon, 2019b).

It is often the business model that clients or owner in the construction industry have to pay for demolisher or recycler to remove C&D debris from site. Often the demolished C&D wastes end up landfilled. In the business case named *Concrete2Collect* as proposed by Fibo Intercon (2019b), the owner of the C&D wastes such as old concrete can be sold to the buyers or customers. Fibo Intercon developed its own software named Fibo Link that allows the seller to issue a receipt to the buyer for the quantity and type of old concrete. The receipt involves a code that enables logging into the batching plant, which then discharges the correct mix and quantity into an awaiting truck (Fibo Intercon, 2019b). Besides developing the software system, which enables online business of selling and purchase of recycled concrete, Fibo Intercon also sells the aforementioned site concrete batching plant, or the bespoke Mobile Batching Plant, for flexible onsite reuse of RAs from crushed and sieved C&D wastes. According to Fibo Intercon (2019b), the plant can store up to 50 concrete design mixes and allow the users to change the mix design with button control.

7.5.3 Case study of "green" masonry bricks in China

The recycled market to reuse C&D wastes remains under development in countries like China. The huge amount of C&D wastes generated has been largely landfilled. Traditionally, construction project clients have to pay for the removal of C&D wastes including the long-distance haul of urban wastes from demolition site into rural landfills. Fig. 7.4 shows a brick factory in the suburban area of south-eastern China. The business owner of the factory received the dumped C&D wastes and charged the dumpers or the waste generators (e.g., project client). On the other hand, the C&D wastes are crushed into fine particles with site crushers. Further the processed fine RAs are reused as raw materials for producing "green" masonry products. The workflow can be illustrated in four steps demonstrated in Fig. 7.4.

The business model in this case study can be described that the factory owner charges both the C&D waste dumper/generator and the end-user of the masonry products containing RAs from crushing C&D wastes. The end products were not limited to the hollow bricks as seen in Fig. 7.4, but also other types (e.g., solid masonry bricks) depending on the applications. These "green" products were mainly used in nonstructural partition walls in buildings. Few similar brick factories could be found in the south-eastern region of China, utilizing recycled and crushed C&D wastes in the form of RAs as raw materials of masonry bricks or blocks. These "green" bricks or blocks containing RAs were limited to use in nonstructural purposes, for instance, partition walls. The mixed C&D wastes generally consisted of demolished concrete and red bricks. Business owners or

Figure 7.4 A case study of reusing crushed C&D wastes to produce masonry bricks. Photos in Fig. 7.4 belong to the authors' own work.

manufacturers would not sort the wastes, but instead crushed the dumped wastes into mixed RAs. Factories may have some compressive test equipment to have the data on strength of "green" bricks or blocks, but with limited research & development on other property tests, such as longer-term durability.

These reuses of RAs from unsorted C&D wastes had been a shift toward circularity from the traditional landfilling approach in the developing construction markets (e.g., China). Further research and development could be helpful to further extend the applications of the bricks and blocks for various nonstructural uses and even structural load-bearing purposes.

7.6 Summary

This chapter focused on the nonstructural applications of using recycled aggregates (RAs) from C&D wastes. Concrete is widely recognized as the major source of C&D wastes. Existing research on technical issues of adopting recycled concrete aggregates (RCAs) are reviewed for being reused, such as in new concrete production. For example, high water absorption and surface cracking are traditionally considered inferior qualities of RCAs that limit their reuse in building and civil engineering. Pretreatment of RCAs, such as treating RCAs as lightweight aggregate, is recommended in both scholarly research and practices.

It is worth a more comprehensive investigation from technical, economic, environmental, and other dimensions to evaluate the feasibility of adopting RAs for various nonstructural usages. Various factors for adopting RAs into new applications are listed in this chapter, including legislation, market, social, technical, financial, and education. It is hence important to evaluate using RAs by looking at both the technical research and practical implementation.

An overview of published technical literature worldwide indicated that more scholarly research has been focusing on structural applications of RAs, especially related to how RAs affected new concrete properties, such as mechanical strength and durability. Nonstructural applications could have lower specifications on RA properties or sources. It is widely recognized that RAs have been more widely applied in nonstructural uses in buildings or infrastructure, for example, road base, backfill, and sidewalk, etc. Nevertheless, many of existing practical uses of RAs for nonstructural applications have been for downcycling uses, there is a need to increase the usage of RAs in the upcycling approach, or alternative new applications for nonstructural uses. This would enhance the circularity value and reuse rate of RAs following the principles of Circular Economy, which deploys different levels of circular strategies, including recover, recycle, repurpose, remanufacture, refurbish, repair, reuse, reduce, rethink, and refuse (Kirchherr et al., 2017).

Besides the three case studies from real-world practices, some other practical cases can be found on the upcycling practices of using C&D wastes, although they may not involve adopting RAs. For example, the radical recycling approach introduced by Wilson (2019) to reuse recycled bricks to form the façade of Copenhagen housing project. The project, consisted of 92 homes, was driven by the architectural idea of upcycling old buildings into new with wall panels made of recycled brickwork. Another housing project was delivered by LafargeHolcim France in partnership with aggneo brand (Holcim, 2018). In this Bordeaux social housing project, 4000 m^3 of concrete made using RAs were produced. There can be more emerging practices of adopting RAs for various nonstructural uses by following the different circular strategies and by extending the traditional downcycling approach.

References

AC Business Media. (2021). *University research suggests recycled concrete as good, or better than traditional.* Available at https://www.forconstructionpros.com/concrete/news/21205415/university-research-suggests-recycled-concrete-as-good-or-better-than-traditional. (Accessed 29 June 2021).

Agrela, F., Beltran, M. G., Cabrera, M., López, M., Rosales, J., & Ayuso, J. (2018). Properties of recycled concrete manufacturing with all-in recycled aggregates and processed biomass bottom ash. *Waste and Biomass Valorization, 9*(7), 1247–1259.

Ahmed, H., Tiznobaik, M., Huda, S. B., Islam, M. S., & Alam, M. S. (2020). Recycled aggregate concrete from large-scale production to sustainable field application. *Construction and Building Materials, 262*, 119979.

Alexandridou, C., Angelopoulos, G. N., & Coutelieris, F. A. (2018). Mechanical and durability performance of concrete produced with recycled aggregates from Greek construction and demolition waste plants. *Journal of Cleaner Production, 176*, 745–757.

Bank, L. C., Yazdanbakhsh, A., Hamidi, A., Reichman, S., Weng, M., & Chen, C. (2018). *FRP strengthening of beams made with recycled concrete aggregate, 322* pp. 1–9). Special Publication.

Bank, L. C., & Yazdanbakhsh, A. (2016). *Concrete containing coarse aggregate recycled from scrap FRP rebars. Insights and innovations in structural engineering, mechanics and computation - Proceedings of the 6th international conference on structural engineering, mechanics and computation*. SEMC, 2016.

Chakradhara Rao, M. (2018). Properties of recycled aggregate and recycled aggregate concrete: Effect of parent concrete. *Asian Journal of Civil Engineering, 19*(1), 103–110.

Chen, W., Jin, R., Xu, Y., Wanatowski, D., Li, B., Yan, L., Pan, Z., & Yang, Y. (2019). Adopting recycled aggregates as sustainable construction materials: A review of the scientific literature. *Construction and Building Materials, 218*, 483–496.

Dimitriou, G., Savva, P., & Petrou, M. F. (2018). Enhancing mechanical and durability properties of recycled aggregate concrete. *Construction and Building Materials, 158*, 228–235.

Duan, H., & Li, J. (2016). Construction and demolition waste management: China's lessons. *Waste Management & Research, 34*(5), 397–398.

Etxeberria, M., Corominas, A., & Valero, I. (2013). Application of low-grade recycled aggregates for non-structural concrete production in Barcelona city. In *Third international conference on sustainable construction materials & technologies at Kyoto, Japan*.

Etxeberria, M., Vázquez, E., Marí, A., & Barra, M. (2007). Influence of amount of recycled coarse aggregates and production process on properties of recycled aggregate concrete. *Cement and Concrete Research, 37*(5), 735–742.

Evangelista, L., & de Brito, J. (2007). Mechanical behaviour of concrete made with fine recycled concrete aggregates. *Cement and Concrete Research, 29*(5), 397–401.

Fibo Intercon. (2019a). *Fibo Intercon helped build one of the world's first recycled concrete buildings*. https://fibointercon.com/business-cases/recycled-concrete-buildings/. (Accessed 30 June 2021).

Fibo Intercon. (2019b). *Recycled concrete production business ideas*. https://fibointercon.com/solutions/recycled-concrete-production/. (Accessed 30 June 2021).

Ghisellini, P., Cialani, C., & Ulgiati, S. (2016). A review on circular economy: The expected transition to a balanced interplay of environmental and economic systems. *Journal of Cleaner Production, 114*, 11–32.

Holcim. (2018). *aggneo®: building a circular economy in the heart of historic French cities*. Available at https://www.lafargeholcim.com/aggneo-building-circular-economy-avignon-bordeaux. (Accessed 29 June 2021).

Jayakody, S., Zimar, A. M. Z., & Ranaweera, R. A. L. M. (2018). Potential use of recycled construction and demolition waste aggregates for non- structural concrete applications. *Journal of the National Science Foundation of Sri Lanka, 46*(2), 205–216.

Jin, R., & Chen, Q. (2015). Investigation of concrete recycling in the US construction industry. *Procedia Engineering, 118*, 894–901.

Jin, R., & Chen, Q. (2019). Overview of concrete recycling legislation and practice in the United States. *Journal of Construction Engineering and Management, 145*(4), Article 05019004.

Jin, R., Chen, Q., & Soboyejo, A. (2015). Survey of the current status of sustainable concrete production in the US. *Resources, Conservation and Recycling, 105*, 148–159.

Jin, R., Li, B., Elamin, A., Wang, S., Tsioulou, O., & Wanatowski, D. (2018). Experimental investigation of properties of concrete containing recycled construction wastes. *International Journal of Civil Engineering, 16*(11), 1621–1633.

Jin, R., Li, B., Zhou, T., Wanatowski, D., & Piroozfar, P. (2017). An empirical study of perceptions towards construction and demolition waste recycling and reuse in China. *Resources, Conservation & Recycling, 126*, 86–98.

Jin, R., Yuan, H., & Chen, Q. (2019). Science mapping approach to assisting the review of construction and demolition waste management research published between 2009 and 2018. *Resources, Conservation & Recycling, 140*, 175–188.

Kasai, Y. (1998). By-products and the use of recycled aggregate in concrete in Japan. In *Sustainable Construction: Use of Recycled Concrete Aggregate*.

Kianimehr, M., Shourijeh, P. T., Binesh, S. M., Mohammadinia, A., & Arulrajah, A. (2019). Utilization of recycled concrete aggregates for light-stabilization of clay soils. *Construction and Building Materials, 227*, 116792.

Kirchherr, J., Reike, D., & Hekkert, M. (2017). Conceptualizing the circular economy: An analysis of 114 definitions. *Resources, Conservation & Recycling, 127*, 221–232.

Kou, S. C., & Poon, C. S. (2015). Effect of the quality of parent concrete on the properties of high performance recycled aggregate concrete. *Construction and Building Materials, 77*, 501–508.

Kovacs, R., Shamass, R., Limbachiya, V., & Datoo, M. (2019). Experimental investigation on the behaviour of recycled aggregate concrete. In *5th International conference on sustainable construction materials & technologies*.

López-Uceda, A., Ayuso, J., Jiménez, J. R., Agrela, F., Barbudo, A., & De Brito, J. (2016a). Upscaling the use of mixed recycled aggregates in non-structural low cement concrete. *Materials, 9*(2), 91.

López-Uceda, A., Ayuso, J., López, M., Jimenez, J. R., Agrela, F., & Sierra, M. J. (2016b). Properties of non-structural concrete made with mixed recycled aggregates and low cement content. *Materials, 9*(2), 74.

Marie, I. (2017). Thermal conductivity of hybrid recycled aggregate–Rubberized concrete. *Construction and Building Materials, 133*, 516–524.

Mas, B., Cladera, A., Del Olmo, T., & Pitarch, F. (2012). Influence of the amount of mixed recycled aggregates on the properties of concrete for non-structural use. *Construction and Building Materials, 27*(1), 612–622.

Meddah, M. S. (2017). Recycled aggregates in concrete production: Engineering properties and environmental impact. In *MATEC web of conferences, 101* p. 05021). EDP Sciences.

Ozbakkaloglu, T., Gholampour, A., & Xie, T. (2018). Mechanical and durability properties of recycled aggregate concrete: Effect of recycled aggregate properties and content. *Journal of Materials in Civil Engineering, 30*(2), 04017275.

PCA. (2019). *Recycled aggregates*. Available at https://www.cement.org/learn/concrete-technology/concrete-design-production/recycled-aggregates. (Accessed 29 June 2021).

Pepe, M., Toledo Filho, R. D., Koenders, E. A. B., & Martinelli, E. (2014). Alternative processing procedures for recycled aggregates in structural concrete. *Construction and Building Materials, 69*, 124–132.

PMR. (2017). *Global market study on construction aggregates: Crushed stone product type segment projected to register high value and Volume CAGR during 2017 - 2025*. New York, USA: Persistence Market Research.

Poon, C. S., Shui, Z. H., & Lam, L. (2004a). Effect of microstructure of ITZ on compressive strength of concrete prepared with recycled aggregates. *Construction and Building Materials, 18*(6), 461–468.

Poon, C. S., Shui, Z. H., Lam, L., Fok, H., & Kou, S. C. (2004b). Influence of moisture states of natural and recycled aggregates on the slump and compressive strength of concrete. *Cement and Concrete Research, 34*(1), 31–36.

Rodríguez, G., Medina, C., Alegre, F. J., Asensio, E., & Rojas, M. I. S. (2015). Assessment of construction and demolition waste plant management in Spain: In pursuit of sustainability and eco-efficiency. *Journal of Cleaner Production, 90*, 16–24.

Rodríguez, G., Sáez del Bosque, I. F., Asensio, E., Sánchez de Rojas, M. I., & Medina, C. (2020). Construction and demolition waste applications and maximum daily output in Spanish recycling plants. *Waste Management & Research, 38*(4), 423–432.

Saeed, D., Jahanzaib, M., & Khan, M. (2017). Production of recycled aggregate concrete for non-structural applications. In *Conference: The fifth International academic conference for graduates, NUAA, Nanjing, China*.

Silgado, S. S., Valdiviezo, L. C., Domingo, S. G., & Roca, X. (2018). Multi-criteria decision analysis to assess the environmental and economic performance of using recycled gypsum cement and recycled aggregate to produce concrete: The case of Catalonia (Spain). *Resources, Conservation and Recycling, 133*, 120–131.

Silva, R. V., De Brito, J., & Dhir, R. K. (2019). Use of recycled aggregates arising from construction and demolition waste in new construction applications. *Journal of Cleaner Production, 236*, 117629.

Stein, J. (2021). *UK material shortages: Cement and aggregates added to the list*. Available at https://www.constructionnews.co.uk/news/uk-material-shortages-cement-and-aggregates-added-to-the-list-08-04-2021/. (Accessed 20 June 2021).

Tam, V. W. Y. (2008). Economic comparison of concrete recycling: A case study approach. *Resources, Conservation & Recycling, 52*(5), 821–828.

Tam, V. W. Y., Gao, X. F., & Tam, C. M. (2005). Microstructural analysis of recycled aggregate concrete produced from two-stage mixing approach. *Cement and Concrete Research, 35*(6), 1195–1203.

Tam, V. W., Soomro, M., & Evangelista, A. C. J. (2018). A review of recycled aggregate in concrete applications (2000–2017). *Construction and Building Materials, 172*, 272–292.

Tang, W., Khavarian, M., Yousefi, A., Chan, R. W., & Cui, H. (2019). Influence of surface treatment of recycled aggregates on mechanical properties and bond strength of self-compacting concrete. *Sustainability, 11*(15), 4182.

The Concrete Society. (2021). *Non-structural concrete*. Available at http://www.concrete.org.uk/fingertips-nuggets.asp?cmd=display&id=5. (Accessed 30 June 2021).

Thomas, C., Cimentada, A., Polanco, J. A., Setién, J., Méndez, D., & Rico, J. (2013). Influence of recycled aggregates containing sulphur on properties of recycled aggregate mortar and concrete. *Composites Part B: Engineering, 45*(1), 474–485.

Wilson, R. (2019). *Old into new: Recycled bricks form facade of Copenhagen housing project*. Available at https://www.architectsjournal.co.uk/buildings/old-into-new-recycled-bricks-form-facade-of-copenhagen-housing-project. (Accessed 28 June 2021).

Xiao, J., Li, J., & Zhang, C. (2005). Mechanical properties of recycled aggregate concrete under uniaxial loading. *Cement and Concrete Research, 35*(6), 1187–1194.

Xu, Y., Chen, W., Jin, R., Shen, J., Smallbone, K., Yan, C., & Hu, L. (2018). Experimental investigation of photocatalytic effects of concrete in air purification adopting entire concrete waste reuse model. *Journal of Hazardous Materials, 353*, 421–430.

Xu, Y., Jin, R., Hu, L., Li, B., Chen, W., Shen, J., Wu, P., & Fang, J. (2020). Studying the mix design and investigating the photocatalytic performance of pervious concrete containing TiO_2-Soaked recycled aggregates. *Journal of Cleaner Production, 248*, 119281.

Xu, Y., & Sun, J. (2011). Influence of recycled aggregate on physical and mechanical properties of high performance recycled aggregate concrete. *Key Engineering Materials, 460–461*, 764–767.

Yap, S. P., Chen, P. Z. C., Goh, Y., Ibrahim, H. A., Mo, K. H., & Yuen, C. W. (2018). Characterization of pervious concrete with blended natural aggregate and recycled concrete aggregates. *Journal of Cleaner Production, 181*, 155–165.

Zaben, A., Maslehuddin, M., Al-Amoudi, O. S. B., & Al-Dulaijan, S. U. (2021). Influence of mix composition on the properties of recycled aggregate concrete. *Structural Concrete*. https://doi.org/10.1002/suco.202000062

Zhang, L. W., Sojobi, A. O., Kodur, V. K. R., & Liew, K. M. (2019). Effective utilization and recycling of mixed recycled aggregates for a greener environment. *Journal of Cleaner Production, 236*, 117600.

Durability of recycled concrete

8

Nariman J. Khalil
Department of Civil & Environmental Engineering, University of Balamand, Koura, Lebanon

8.1 Introduction

Recycled aggregates (RA) offer a solution to a worldwide problem related to construction sustainability and natural resources preservation. Utilizing RA in concrete production could save up to 60% of limestone resources and reduce CO_2 emissions by about 15%–20% (Guo et al., 2018). Extensive efforts have been devoted over the past two decades to study the behavior of recycled aggregate concrete (RAC). Researchers suggest that RAC is adequate as a structural building material (Bao et al., 2020; Hoffmann et al., 2012; Makul et al., 2021; Pedro et al., 2014; Thomas et al., 2013). However, the structure of RAC is more complex than the conventional concrete due to the presence of the adhered cement mortar on the recycled concrete aggregates (RCA) leading to the formation of multiple interfacial transition zones (ITZ), the old zone is between the adhered paste and the virgin aggregates, and the newly formed ITZ are between the old and the new cement matrices. As a result, RAC has different properties and exhibits long-term durability problems.

Durability is defined as the ability of concrete structures to maintain their appearance and function under service load. Deterioration of concrete can be caused by external factors resulting from the environmental conditions to which the concrete is exposed to or could be caused by internal factors within the concrete structure itself (Neville & Brooks, 1987). The external factors can generally be categorized as chemical due to liquid and gas attacks, physical such as the freeze-thaw cycles, or mechanical such as abrasion. The effect of these factors depends on the transport property of the elements migrating through concrete and on the physical properties of the concrete itself. Therefore, it is very essential to understand the mechanism by which such factors influence the service life of concrete structures and how RAC differs from concrete made with natural aggregates. It has been agreed upon that a replacement ratio of 30% of coarse aggregate produces concrete with comparable durability to that produced with virgin aggregate (Andal et al., 2016; Khalil et al., 2016; Khalil & Aouad, 2019; Limbachiya et al., 2000; Limbachiya et al., 2012b; Thomas et al., 2018). Generally, if the mix design for RAC is properly proportioned and good quality recycled aggregates are used, the resulting concrete can be highly durable (Environmental Council of Concrete Organizations, 1999; Safiuddin et al., 2013).

In the following sections, a detailed discussion is presented about RAC durability properties and its behavior under various environmental conditions.

Recycled Concrete. https://doi.org/10.1016/B978-0-323-85210-4.00005-9

8.2 Permeability

Permeability of concrete is defined as the ease by which liquids and gases can penetrate through concrete (Neville, 1981). While "Diffusivity" refers to "the ease with which dissolved ions move through concrete" (Kosmatka & Wilson, 2016). Permeability and diffusivity are dependent on the volume of voids (as percent of the total volume) inside the concrete matrix, which is known as "porosity". However, this relationship is not simple; the size and connectivity of the voids influence permeability. If the pores are discrete, then they have no impact on permeability; while if they are interconnected then permeability increases, as illustrated in Fig. 8.1.

Permeability and diffusivity are also affected by the properties of the substance that is penetrating through concrete such as the viscosity of the fluid, size of the molecules, and valence of the ions (Kosmatka & Wilson, 2016). Water-cement ratio (W/C) and curing conditions are among the factors that influence permeability.

Decrease of permeability improves concrete resistance to chemical attacks (such as by sulfates, chlorides, acids or by sea water) and the resistance to freeze and thaw cycles.

Recycled concrete aggregates are composed of natural aggregates enveloped by the hardened cement paste. This cement paste has a high absorptivity rate leading to more porous RCA with porosity 10 to 30 times higher than the porosity of the natural aggregates (Safiuddin et al., 2013; Théréné et al., 2020). The more the cement mortar attached to the aggregates, the lower their density and the higher their absorptivity and porosity (Hoffmann et al., 2012; Li et al., 2021; Pepe et al., 2014; Tam et al., 2015). Generally, the water absorption for RCA ranges between 3% and 10% and from less than 1%–5% for natural aggregates (NA) (Behera et al., 2014; Khalil et al., 2016; Pedro et al., 2014; Tam et al., 2008).

The permeability of RAC increases with the increase of the replacement ratio of natural aggregates and with the increase of the water to cement ratio. It can be 200%–500% higher than concrete made with natural coarse aggregates (Anderson

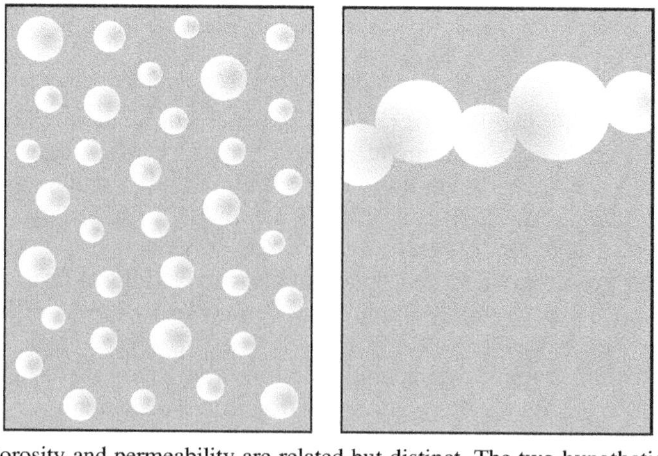

Figure 8.1 Porosity and permeability are related but distinct. The two hypothetical materials above have about same porosity but different permeabilities (Kosmatka and Wilson, 2016).

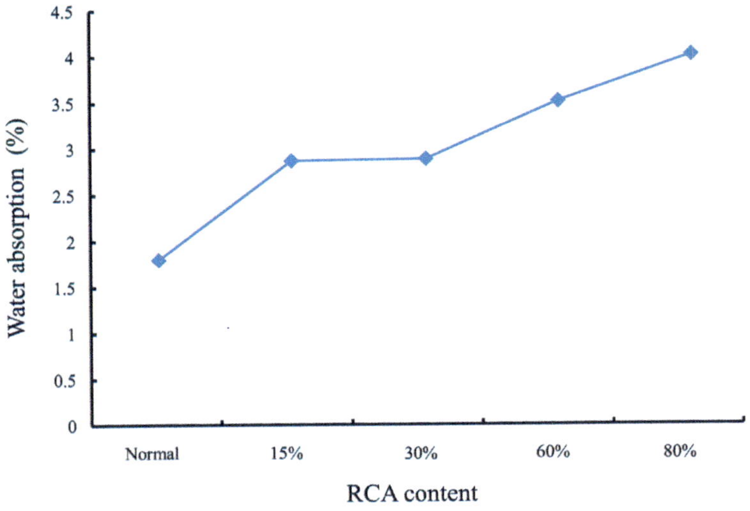

Figure 8.2 Water absorption of specimens at various replacement percentages of RCA (Kwan et al., 2012).

et al., 2009). This is explained by the higher rate of water absorption (Kwan et al., 2012; Thomas et al., 2018). Fig. 8.2 illustrates the relationship between water absorption and RCA content.

Using good quality recycled concrete aggregates improves the permeability (Gómez-Soberón, 2002; Bao et al., 2020; Thomas et al., 2013). A reduction in the water absorption is observed when the RCA were pretreated. Several enhancement methods are reported in the literature to improve RCA performance (García-González et al., 2014; Pepe et al., 2014; Santos et al., 2017; Shi et al., 2016; Xuan et al., 2017).

The effect of pores volume and size is more prominent at lower ages and diminishes as concrete ages (Gómez-Soberón, 2002). With continuous water curing the permeability decreases and the difference between the RAC with the highest replacement ratio and concrete made with natural aggregates would be significantly reduced (Kwan et al., 2012), as shown in Fig. 8.3. W/C ratio plays an important role in obtaining dense concrete with low permeability.

Different methods are used to measure the permeability of concrete, such as water permeability, capillary water absorption, air permeability, and oxygen permeability (Behera et al., 2014).

There is a concern about the relevance of implementing the existing standard test methods available for conventional concrete to determine the properties of RAC. For example, there are some drawbacks when applying the standard water absorption test of virgin aggregates to RCA; it is believed that it will lead to an overestimation of water absorption. When accounted for the absorbed water in the concrete mix, high workability can be obtained at start, depending on the saturation rate of aggregates. The concrete will be more porous in the hardened state (Tam et al., 2008). Alternative methods were proposed such as real-time assessment of water absorption (RAWA).

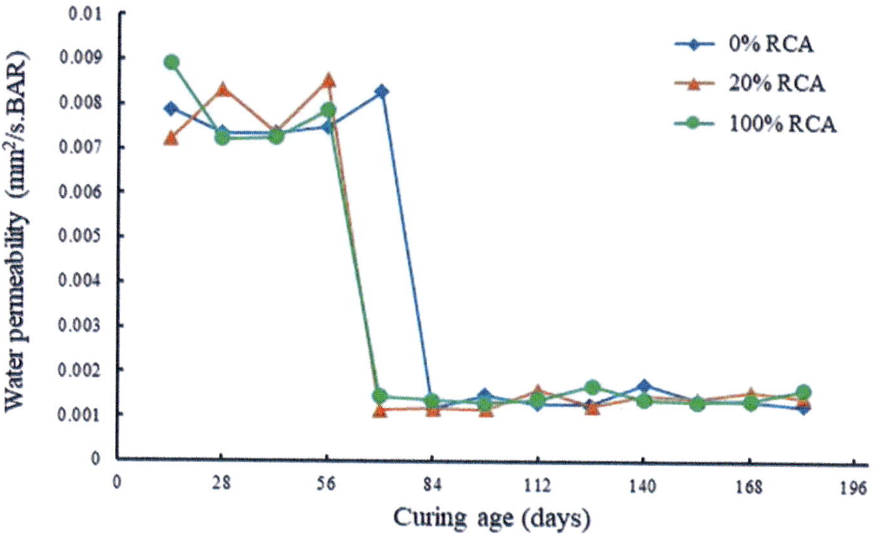

Figure 8.3 Effect of RCA on the water permeability of concrete (Tam & Tam, 2007).

8.3 Chemical attacks on concrete

Concrete is susceptible to chemical attacks by a variety of chemical substances. Compounds that are aggressive to concrete are mainly: sulfates, chlorides, and acids. The severity of the attack depends primarily on the chemical composition of the aggressive agent and its concentration, and on the other hand, depends on the porosity and permeability of concrete and the type of cement used. In the case of recycled aggregate concrete, the permeability and density effect on durability has a dominant influence over the type of cement used.

8.3.1 *Sulfate attack*

Concrete is subjected to salt attack when the salt is present in a solution and not in a solid format (Neville & Brooks, 1987). An ideal situation for such an attack, for example, is the groundwater of some clays that might contain sodium, calcium, or magnesium. That's when salts attack the cement paste. The result of the sulfate attack is cracking and spalling of the hardened concrete characterized by a whitish appearance. The reason behind this phenomenon is the formation of calcium sulfate (gypsum) and calcium sulfoaluminate (ettringite). Both products cause an expansion in the hardened concrete because of the large volume they occupy, leading to disruption of concrete. In some situations, magnesium attack could be very harmful as it causes cement binder decomposition.

The magnitude of the sulfate attack depends on the solution concentration and the permeability of hardened concrete. Therefore, concrete made with recycled aggregates is expected to exhibit low resistance to sulfate attack. Many researchers have investigated the performance of RAC when subjected to sulfate exposed environment (Khalil & Aouad, 2019; Kwan et al., 2012; Limbachiya et al., 2012a; Limbachiya et al., 2012b; Qi et al., 2017; Schmidt et al., 2009). Some reports claimed that the resistance of RAC to sulfates is comparable or a slightly inferior to that of conventional concrete (Environmental Council of Concrete Organizations, 1999; Zega et al., 2016). However, further investigations proved that the loss of the concrete mass in sulfate resistance test increases with the increase of RA replacement ratio due to the higher expansion rate (Andal et al., 2016; Behera et al., 2014; Hua & Song, 2007; Li, 2008; Limbachiya et al., 2000; Qi et al., 2017). In addition, the presence of multi-interfacial zones ITZ in RAC accelerates the sulfate ions accumulation (Zhang et al., 2020).

Proper mix proportioning using a low W/C ratio can produce denser RAC and enhance the resistance to sulfate attack. Researchers have attempted several methods to compensate for the RAC low quality such as using mineral additions or substitution of fly ash, ground granulated blast-furnace slag (Qi et al., 2017), or silica fume (Environmental Council of Concrete Organizations, 1999), partial replacement of the cement by pozzolanas, and incorporating fine-size crumb rubber particles.

8.3.2 Chloride penetration resistance

The chloride penetration resistance of concrete is of particular importance in situations where concrete is exposed to deicing salts or seawater, increasing the risk of reinforcement corrosion. Solution of some chlorides, such as ammonium, aluminum, and magnesium attack concrete. The severity of the attack depends on cement content, aggregate type, and concrete permeability (Medina et al., 2012).

Commonly, the chloride penetration resistance of RAC is poorer than that of conventional concrete, as the chloride conductivity tends to increase with the increasing content of porous aggregates (Bao et al., 2020; Guo et al., 2018; Hoffmann et al., 2012). Therefore, it is expected that, as the recycled aggregate replacement ratio increases, the chloride penetration increases (Andal et al., 2016; Bao et al., 2020; Safiuddin et al., 2013; Surya et al., 2015; Thomas et al., 2018; Xiao et al., 2013).

Fig. 8.4 illustrates the increase in the chloride diffusion coefficient measured in m^2/s as the RCA replacement ratio increases (Liang et al., 2021).

The loss of chloride penetration resistance is attributed to the permeable nature of RAC (Medina et al., 2016). Fig. 8.5 shows the relationship between the chloride diffusion coefficient and the water absorption of RCA. Another factor attributes to the rise of chloride diffusion, is the presence of cracks in the residual mortar; as the cracks width increases the chloride diffusion increases (Xiao et al., 2012).

The chloride permeability of RAC with 100% RA is reported to be 58.1% and 20.7% greater than that of conventional concrete at 28 and 90 days, respectively (Kou & Poon, 2010). This increase in chloride ingress results in steel corrosion risk in reinforced RAC. Again, a replacement ratio of recycled aggregates up to 30% has no significant impact on the chloride resistance (Limbachiya et al., 2012a).

Lowering W/C ratio in the mix design improves the performance of RAC in chloride environments; and when using average to high quality recycled aggregates, it is

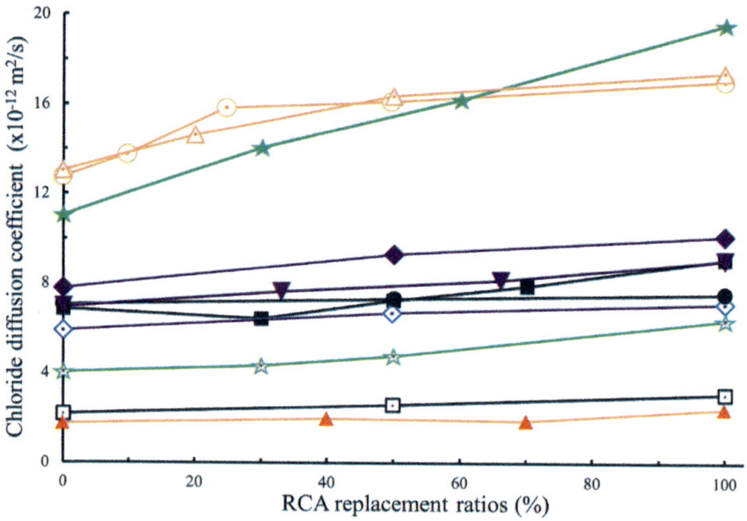

Figure 8.4 Chloride diffusivity of RAC as a function of RCA content (Liang et al., 2021).

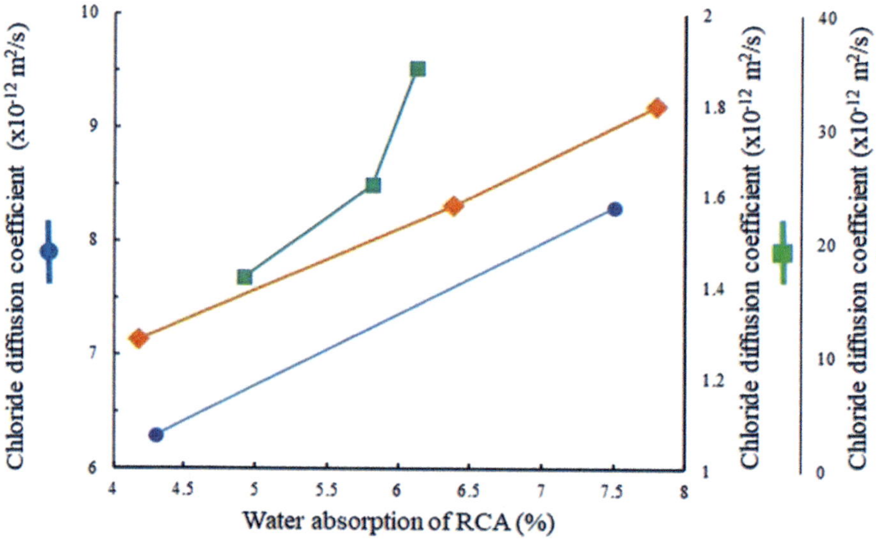

Figure 8.5 Chloride diffusivity of RAC as function of RCA water absorption (100% substitution rates) (Liang et al., 2021).

possible that the resulting concrete may exhibit a comparable performance to that of conventional concrete (Etxeberria et al., 2007; Guo et al., 2018; Limbachiya et al., 2000; Medina et al., 2016; Soares et al., 2014).

Incorporation of mineral admixtures is known to be effective in refining the pore structure of RAC through the formation of secondary calcium-silicate-hydrate (C–S–H) products, which fill the open pores and block the ingress path of chloride ions (Andal et al., 2016; Behera et al., 2014; Evangelista & de Brito, 2010; Kou & Poon, 2012; Safiuddin et al., 2013; Surya et al., 2015; Xiao et al., 2013; Xuan et al., 2017).

Two common test methods exist for measuring the chloride permeability of concrete, the long-term immersion test and the rapid chloride diffusion test.

8.3.3 Acid attack

Acids attack concrete by dissolving the hydrated and unhydrated compounds of cement paste and removing part of it, leaving soft and weak mass. The intensity of the attack depends on the acidity level. Usually, pH less than 4.5 leads to a severe damage. Concrete permeability, cementitious material content and water to cement ratio are the major factors that influence the resistance to acid attack (Kosmatka & Wilson, 2016).

Limited experimental work has been conducted to study the performance degradation of RAC under the condition of acid attack (Lu et al., 2020; Safiuddin et al., 2013), due to the presence of a wide variety of acids and numerous exposure conditions. The acid penetration into RAC samples is slightly higher than that of conventional concrete (Makul et al., 2021; Poon & Chan, 2007) and it increases as the RCA replacement ratio increases (Thomas et al., 2018). This is attributed to the high porosity and absorption of recycled concrete aggregates.

Generally, RAC is not recommended to be used in aggressive environments. Agricultural industrial structures are examples of weak organic acids. Some researchers (Joseph et al., 2016) found that recycled aggregates have a positive impact on concrete resistance to some weak acids. However, this cannot be generalized before conducting further studies.

Erosion initiated by acid rain attack caused an initial increase in the mechanical properties of RAC, at the onset of the exposure, but then a gradual decrease was observed as the acid rain attack continues on concrete (Lu et al., 2020). This decrease is proportional to the acidity of the solution. The higher replacement level of RCA can cause permanent damage to the acid rain resistance, particularly after long-term erosion.

Researchers have shown an improvement in the RAC performance by adding supplementary cementitious materials SCM, such as fly ash (Lu et al., 2020; Safiuddin et al., 2013), or by increasing the cement content, which improves the quality of the hydrated cement paste (Thomas et al., 2018).

More research is needed to better understand the structural performance and the durability aspects of RAC under various acid environments.

8.4 Freeze and thaw cycles

One of the main problems in concrete durability is the freeze and thaw cycles. When the temperature of the saturated hardened concrete drops, the water held in the capillary pores within the cement paste freezes causing expansion of the concrete. Therefore, the frost resistance is influenced by porosity, W/C ratio, aggregate type, and environmental conditions. Repeated cycles of freezing and thawing have a cumulative effect, in a similar way to the fatigue phenomenon (Neville, 1981). The freeze-thaw effect can be evaluated by measuring the rate of weight loss, strength loss, and the reduction in the dynamic elastic modulus (Guo et al., 2018; Luan et al., 2021).

Generally, RAC exhibits lower frost resistance as compared to conventional concrete, due again to its higher water absorption (Tam et al., 2008; Xiao et al., 2013). When the number of freeze-thaw cycles increases, there is a reduction in the mechanical properties and an increase in the mass loss.

The W/C ratio influences the frost resistance of RAC more than the type of aggregates used (Bogas et al., 2016; Guo et al., 2018), hence lowering the amount of mixing water can improve the frost resistance.

The freezing and thawing resistance decreases with the increase of the RCA replacement percentage. A noticeable difference was seen for coarse aggregate replacement ratio above 40%. Fig. 8.6 shows the mass loss rate for three replacement ratios: 20%, 40%, and 60% in comparison to a conventional concrete mix referred to as R0. The effect of fine recycled aggregates could be more remarkable.

Excellent frost resistance can be obtained by using presaturated RCA or aggregates originated from a high-strength concrete or from an air-entraining concrete. The incorporation of mineral admixtures such as fly ash, has a positive impact on RAC frost resistance (Guo et al., 2018; Sadati & Khayat, 2018; Xiao et al., 2013).

Table 8.1 below summarizes the influence of different parameters on the frost resistance of RAC.

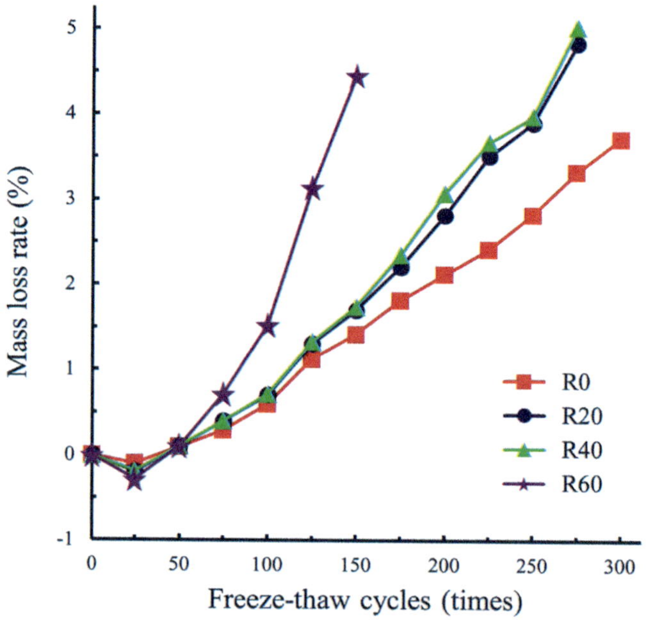

Figure 8.6 Mass loss rate (Xiao et al., 2020).

Table 8.1 Influencing factors on the frost resistance of RAC (Guo et al., 2018).

Parameter	Change	Effect
RA content	↑	Strong negative
W/C ratio	↑	Strong negative
Adhered mortar	↑	Strong negative
Strength of original concrete	↑	Mild positive
Air entraining agent	↓	Mild positive
Pozzolanic materials	↓	Mild positive
Presoaking RA	↓	Mild positive

Remark: ↑-increase, ↓-decrease.

There is some evidence in the literature that RAC has sufficient resistance to freeze-thaw cycles (Makul et al., 2021; Safiuddin et al., 2013), particularly if RA are of good quality (Behera et al., 2014). This demands further investigations to draw a valid conclusion on RAC resistance to freeze-thaw cycles. Additionally, the microstructure development of RAC under the combined action of both chloride attack and freeze-thaw cycles received little attention and requires more exploration (Liang et al., 2021).

8.5 Resistance to carbonation

Carbon dioxide gas is present in the environment with varying percentages. In the presence of moisture, this gas can form carbonic acid that reacts with the hydrated cement in concrete causing decomposition of cement compounds; a process known as "carbonation". Carbonation may lead eventually to steel corrosion. The ingress of carbon dioxide into concrete depends on many factors, the main of which are the porosity, permeability, and moisture content of concrete.

Recycled aggregate concrete has generally weaker carbonation resistance than conventional concrete as a result of its intrinsic porosity (Shicong & Poon, 2006; Xiao et al., 2013). In a similar trend to other properties, the carbonation depth increases as the replacement ratio of RA increases up to 70%; above this limit, the carbonation depth decreases (Evangelista & de Brito, 2010; Guo et al., 2018; Kou & Poon, 2012; Limbachiya et al., 2012b; Pedro et al., 2014; Xiao et al., 2013). For full replacement of coarse aggregate, the carbonation depth is as high as twice and a half that of conventional concrete and if recycled fine aggregate is incorporated, the carbonation depth could reach nine times higher (Guo et al., 2018). The exposure time plays a role in the progress of the carbonation depth. Researchers reported that after one year, the RA replacement ratio has an insignificant effect on the depth of carbonation (Kou & Poon, 2013).

W/C ratio affects the carbonation depth. As the ratio increases, the depth also increases (Kou & Poon, 2012; Thomas et al., 2013). Lowering W/C ratio and increasing the RA replacement level, result in concrete of comparable strength and carbonation depth to that of conventional concrete (Silva et al., 2015).

The rate of carbonation in RAC is also higher than that in conventional concrete, due to the higher porosity in the concrete matrix (Nobuaki et al., 2003). The rate could be four times greater than that in conventional concrete (Hansen, 2011).

Reducing W/C ratio in the mix, combined with proper curing, is shown to be an effective measure in reducing the risk of steel corrosion. The use of suitable supplementary cementitious materials SCM, such as fly ash, slag or metakaolin, could improve the RAC microstructure and thereof the resistance to CO_2 ingression (Kou & Poon, 2012). Modification of RA using pretreatment techniques, such as the removal of bonded mortar by hydrochloric acid (HCl) and sodium sulfate (Na_2SO_4) have yielded better carbonation resistance (Kim et al., 2018).

8.6 Resistance to corrosion

Reinforced concrete structures are susceptible to steel corrosion when favorable conditions exist. If the concrete is permeable enough for the carbonation depth to reach layers near the steel, or chloride solutions can penetrate straight to the reinforcement and in the presence of water and oxygen, corrosion will take place and rust is formed. The corrosion compounds are two to six times larger in volume than the original steel (Liang et al., 2021; Zhao et al., 2014) causing expansive pressure, which eventually leads to cracking and spalling of concrete. The durability properties of recycled aggregate concrete in terms of high porosity and lower carbonation resistance increase the risk of steel corrosion. Researchers have found that the steel corrosion in RAC starts at an earlier age and at a faster rate than the conventional concrete (Liang et al., 2021; Zhao et al., 2014). At lower levels of chlorides, the corrosion rate does not seem to depend on the type of aggregates used, but for higher chloride levels, the change is significant (Safiuddin et al., 2013). It is observed that the number of cracks and their width increase with RA incorporation when compared to the conventional concrete under the same level of chloride attack. The presence of multiple interfacial transition zones (ITZ) in RAC between the natural aggregates and the old mortar, between the natural aggregates and the new mortar, and between the old mortar and the new mortar, has led to a cracking path that is different from the cracking path in the conventional concrete. This is because the microstructure of these ITZ is irregular and loose forming a weak point where many cracks would develop and propagate.

Recycled aggregate concrete has another different aspect from the conventional concrete related to the chloride-induced corrosion. If RA is derived from an aged concrete that has been subjected to an aggressive environment, they may inevitably carry an amount of corrosion media, such as chloride (Cl^-), which internally affects their performance (Tian et al., 2021). A small amount of that Cl^- is diffused into cement mortar, while its vast majority is distributed in the ITZ. Although this quantity is small, it accelerates the cement hydration, resulting in an initial increase in the RAC strength. The increase in the corrosion degree with time will eventually cause a decrease in the concrete strength.

Several studies have been conducted to enhance the performance of RAC against corrosion. The addition of SCM was found to be helpful (Al-Yaqout et al., 2020; Liang et al., 2021; Safiuddin et al., 2013; Zeng et al., 2020). Researchers also investigated methods to remove the adhered mortar by physical or chemical techniques (Liang et al., 2021). Both approaches are proved to be efficient in modifying RA properties and in decreasing the chloride ingress into RAC. Mechanical enhancement, incorporating polypropylene or steel fibers, and particle shaping treatment, is used in a large-scale treatment of RA. The chemical methods comprise polymer treatment, pozzolanic material treatment, and presoaking in acid solution. New enhancement approaches are now available using CO_2-curing and nanoparticle modification. The CO_2-cured RA has refined microstructure and ITZ, which led to a lower porosity. Presoaking RA with nanoparticle solution improved the RAC mechanical properties and yielded stronger protection of steel against corrosion. It is believed that the nanoparticles have played a double role in accelerating cement hydration reaction and filling the pores in the concrete (Zeng et al., 2020).

8.7 Alkali-silica reaction

Alkali-silica reaction (ASR) is one of the most harmful distress that leads to early concrete deterioration. It happens when the active silica constituents of the aggregates react with the alkalis in the cement. The product of this reaction is the alkali-silicate gel. This gel attracts water from its surroundings by absorption or by osmosis and thus tends to increase in volume causing tensile stresses that lead eventually to cracking (Neville & Brooks, 1987). If the reaction continues, an important network of cracks forms and continues to propagate till concrete loses its physical integrity. If the original concrete was demolished before the ASR reaction had come to a completion, then RA may contain reactive aggregates and the RAC produced with this RA may exhibit further expansion and deterioration (secondary expansion). The amount of this secondary damage in RAC is related to the extent of the previous reaction (Trottier et al., 2021).

The severity of ASR is represented by the number and width of cracks produced. These cracks often develop within the aggregates then extend to the cement paste, negatively influencing the concrete shear and compressive strength.

The demolition method is found to have a significant influence on RA expansion. The smaller the coarse RA size, the greater the expansion (Johnson & Shehata, 2016). The sand (fine particles) in the residual mortar can sometimes be very reactive (Etxeberria & Vázquez, 2010). The crack propagation path in RAC differs with respect to the origin of the reaction. Reactive sand causes cracks to propagate from the virgin aggregates toward the residual cement paste; while reactive coarse aggregate causes cracks to propagate into the residual and new cement pastes (Trottier et al., 2021).

When RAC is composed of nonreactive natural aggregates and nonreactive adhered mortar there is no risk of ASR. On the contrary, if the natural aggregates are highly reactive then the ASR expansion in RAC increases with the increase of RA

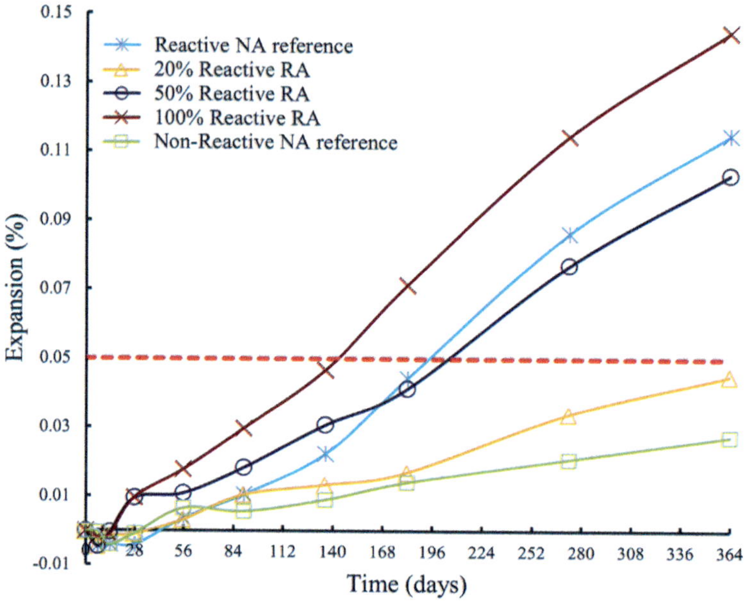

Figure 8.7 Expansion evolution due to ASR (Barreto Santos et al., 2020).

replacement ratio. A replacement level of 20% does not cause a significant increase in the concrete mix reactivity (Barreto Santos et al., 2020).

Fig. 8.7 demonstrates the ASR expansion rate of concrete mixes made with aggregates of different reactivity levels.

Assessment of the potential secondary damage in RAC is limited by the lack of information about the demolished concrete strength, type, and nature of aggregates used. This data is often not available and when coupled with the complex nature of RAC becomes more difficult to understand the ASR mechanism in RAC system (Barreto Santos et al., 2020). Therefore, more research is needed on this particular topic.

The integration of mineral admixtures (SCM) in recycled aggregate concrete mixes and using low alkali cement have been recommended to reduce the expansion due to ASR (Al-Yaqout et al., 2020; Environmental Council of Concrete Organizations, 1999; Guo et al., 2018; Makul et al., 2021).

8.8 Gas penetration

The main gases that affect the service life of concrete structures are carbon dioxide CO_2 and oxygen O_2 through their role in the reinforcement corrosion mechanism. The presence of oxygen close to the reinforcing steel increases the risk of corrosion. The gas penetration rate is related to the permeability of concrete. As discussed in section 8.2, using recycled aggregates in concrete increases its permeability due to the high porosity of such aggregates. Therefore, as the RA replacement ratio increases, the oxygen permeability increases. A recent study (Berredjem et al., 2020) stated that concrete made with full aggregates replacement (fine and coarse aggregates) recorded a maximum increase in gas permeability coefficient reaching 400% higher than mixes made with natural aggregates. The fine recycled aggregate (sand) has more profound influence on concrete vulnerability to aggressive environments. The same study reported a 300% increase in gas permeability coefficient for mixes made with fine RCA and natural coarse aggregate.

Likewise, lowering W/C ratio in the mix or using carbonated RCA reduces significantly the negative effect of RA on the gas permeability of RAC (Olorunsogo & Padayachee, 2002; Xuan et al., 2017).

8.9 Future advances

The presence of the adhered mortar in the RA has led to a more complex structure of concrete made with such recycled aggregates. The resulting RAC is more porous and contains a large number of microcracks and it is featured with low durability properties as compared to conventional concrete. Researchers have investigated several methods to mitigate the lower quality of RA and to enhance the microstructure of the multiple transition zones in order to improve the RAC durability performance. To achieve these goals, two approaches have been identified up to date, the first is to remove the residual mortar and the second is to treat it using different techniques with the objective of making the RA denser and less porous. A comprehensive study about all promising treatment methods can be found in references (Raman & Ramasamy, 2021; Tam et al., 2021a). According to the review studies, methods to remove the residual mortar can be divided into physical and chemical treatments. The physical treatments include mechanical, thermal, electrodynamics, and water treatments. The chemical treatments include acid soaking and chemico-mechanical (hybrid) treatments. The second approach followed to improve the microstructural and mechanical properties of RCA includes carbonation, self-healing, fortification by coating and permeation, sequential mixing, and the addition of mineral admixture (Tam et al., 2021b). The selection of the proper enhancement method depends on many factors, such as cost-effectiveness, environment friendliness, ease and time of processing. Studies proved remarkable improvement in the treated RCA properties when compared to untreated RCA. A substantial lack of research is marked on investigating the effect of different types of chemicals and solutions for coating and permeation. There could be a promising technology for treatment if the CO_2 carbonation method is combined with chemical permeation. The resulting hybrid method can be superior to both the individual methods of carbonation and chemical treatment (Tam et al., 2021b).

More on the enhancement methods of RCA properties can be found in the following references (Andal et al., 2016; Behera et al., 2014; Guo et al., 2018; Kou & Poon, 2012; Li et al., 2021; Liang et al., 2020, 2021, Raman & Ramasamy, 2021; Safiuddin et al., 2013; Shi et al., 2016; Surya et al., 2015; Xuan et al., 2017; Zeng et al., 2020).

More research needs to be conducted on RAC long-term mechanical properties and durability behavior to assess the RAC performance from both material and structural aspects.

References

Al-Yaqout, A., El-Hawary, M., Nouh, K., & Khan, P. B. (2020). Corrosion resistance of recycled aggregate concrete incorporating slag. *ACI Materials Journal, 117*(3), 111–122. https://doi.org/10.14359/51722406

Andal, J., Shehata, M., & Zacarias, P. (2016). Properties of concrete containing recycled concrete aggregate of preserved quality. *Construction and Building Materials, 125*, 842–855. https://doi.org/10.1016/j.conbuildmat.2016.08.110

Anderson, K. W., Uhlmeyer, J. S., & Russell, M. (2009). *Use of recycled concrete aggregate in PCCP*. https://www.wsdot.wa.gov/research/reports/fullreports/726.1.pdf.

Bao, J., Li, S., Zhang, P., Ding, X., Xue, S., Cui, Y., & Zhao, T. (2020). Influence of the incorporation of recycled coarse aggregate on water absorption and chloride penetration into concrete. *Construction and Building Materials, 239*, 117845. https://doi.org/10.1016/j.conbuildmat.2019.117845

Barreto Santos, M., De Brito, J., & Santos Silva, A. (2020). A review on alkali-silica reaction evolution in recycled aggregate concrete. *Materials, 13*(11), 2625. https://doi.org/10.3390/ma13112625

Behera, M., Bhattacharyya, S. K., Minocha, A. K., Deoliya, R., & Maiti, S. (2014). Recycled aggregate from C&D waste & its use in concrete - a breakthrough towards sustainability in construction sector: A review. *Construction and Building Materials, 68*, 501–516. https://doi.org/10.1016/j.conbuildmat.2014.07.003

Berredjem, L., Arabi, N., & Molez, L. (2020). Mechanical and durability properties of concrete based on recycled coarse and fine aggregates produced from demolished concrete. *Construction and Building Materials, 246*, 118421. https://doi.org/10.1016/j.conbuildmat.2020.118421

Bogas, J. A., De Brito, J., & Ramos, D. (2016). Freeze-thaw resistance of concrete produced with fine recycled concrete aggregates. *Journal of Cleaner Production, 115*, 294–306. https://doi.org/10.1016/j.jclepro.2015.12.065

Environmental Council of Concrete Organizations. (1999). *Recycling concrete and masonry*. https://p2infohouse.org/ref/14/13608.pdf.

Etxeberria, M., & Vázquez, E. (2010). Reacción álcali sílice en el hormigón debido al mortero adherido del árido reciclado. *Materiales de Construccion, 60*(297), 47–58. https://doi.org/10.3989/mc.2010.46508

Etxeberria, M., Vázquez, E., Marí, A., & Barra, M. (2007). Influence of amount of recycled coarse aggregates and production process on properties of recycled aggregate concrete. *Cement and Concrete Research, 37*(5), 735–742. https://doi.org/10.1016/j.cemconres.2007.02.002

Evangelista, L., & de Brito, J. (2010). Durability performance of concrete made with fine recycled concrete aggregates. *Cement and Concrete Composites, 32*(1), 9–14. https://doi.org/10.1016/j.cemconcomp.2009.09.005

García-González, J., Rodríguez-Robles, D., Juan-Valdés, A., Pozo, J. M. M.d., & Guerra-Romero, M. I. (2014). Pre-saturation technique of the recycled aggregates: Solution to the water absorption drawback in the recycled concrete manufacture. *Materials, 7*(9), 6224–6236. https://doi.org/10.3390/ma7096224

Gómez-Soberón, J. M. V. (2002). Porosity of recycled concrete with substitution of recycled concrete aggregate: An experimental study. *Cement and Concrete Research, 32*(8), 1301–1311. https://doi.org/10.1016/S0008-8846(02)00795-0

Guo, H., Shi, C., Guan, X., Zhu, J., Ding, Y., Ling, T. C., Zhang, H., & Wang, Y. (2018). Durability of recycled aggregate concrete – a review. *Cement and Concrete Composites, 89*, 251–259. https://doi.org/10.1016/j.cemconcomp.2018.03.008

Hansen, T. C. (2011). *Recycling of demolished concrete and masonry*. https://doi.org/10.1201/9781482267075

Hoffmann, C., Schubert, S., Leemann, A., & Motavalli, M. (2012). Recycled concrete and mixed rubble as aggregates: Influence of variations in composition on the concrete properties and their use as structural material. *Construction and Building Materials, 35*, 701–709. https://doi.org/10.1016/j.conbuildmat.2011.10.007

Hua, J., & Song, S. (2007). Study on mix proportions and durability of recycled aggregate concrete. *Chinese Ready-Mixed Concrete, 5*, 46–51.

Johnson, R., & Shehata, M. H. (2016). The efficacy of accelerated test methods to evaluate alkali silica reactivity of recycled concrete aggregates. *Construction and Building Materials, 112*, 518–528. https://doi.org/10.1016/j.conbuildmat.2016.02.155

Joseph, M., Boehme, L., Cornelly, C., Anseeuw, I., Declercq, W., & Vandewalle, L. (2016). Recycled concrete in weak acidic environments. In *Key engineering materials* (Vol 677, pp. 224–232). Trans Tech Publications Ltd. https://doi.org/10.4028/www.scientific.net/KEM.677.224

Khalil, N. J., & Aouad, G. (2019). *Durability of recycled aggregate concrete*. SP 336 (pp. 91–100). American Concrete Institute, ACI Special Publication 2019. American Concrete Institute https://www.concrete.org/topicsinconcrete/topicdetail/special%20publication.

Khalil, N., Touma, A., Touma, T., & Daher, R. (2016). A study into the relationships between the mechanical properties of recycled concrete aggregates. In *II Int. Conf. Concr. Sustain. - Iccs16* (pp. 1146–1158).

Kim, Y., Hanif, A., Kazmi, S. M. S., Munir, M. J., & Park, C. (2018). Properties enhancement of recycled aggregate concrete through pretreatment of coarse aggregates – comparative assessment of assorted techniques. *Journal of Cleaner Production, 191*, 339–349. https://doi.org/10.1016/j.jclepro.2018.04.192

Kosmatka, S., & Wilson, M. (2016). *Design and control of concrete mixtures*.

Kou, S. C., & Poon, C. S. (2010). Properties of concrete prepared with PVA-impregnated recycled concrete aggregates. *Cement and Concrete Composites, 32*(8), 649–654. https://doi.org/10.1016/j.cemconcomp.2010.05.003

Kou, S. C., & Poon, C. S. (2012). Enhancing the durability properties of concrete prepared with coarse recycled aggregate. *Construction and Building Materials, 35*, 69–76. https://doi.org/10.1016/j.conbuildmat.2012.02.032

Kou, S. C., & Poon, C. S. (2013). Long-term mechanical and durability properties of recycled aggregate concrete prepared with the incorporation of fly ash. *Cement and Concrete Composites, 37*(1), 12–19. https://doi.org/10.1016/j.cemconcomp.2012.12.011

Kwan, W. H., Ramli, M., Kam, K. J., & Sulieman, M. Z. (2012). Influence of the amount of recycled coarse aggregate in concrete design and durability properties. *Construction and Building Materials, 26*(1), 565–573. https://doi.org/10.1016/j.conbuildmat.2011.06.059

Li, X. (2008). Recycling and reuse of waste concrete in China. Part I. Material behaviour of recycled aggregate concrete. *Resources, Conservation and Recycling, 53*(1–2), 36–44. https://doi.org/10.1016/j.resconrec.2008.09.006

Liang, C., Cai, Z., Wu, H., Xiao, J., Zhang, Y., & Ma, Z. (2021). Chloride transport and induced steel corrosion in recycled aggregate concrete: A review. *Construction and Building Materials, 282*, 122547. https://doi.org/10.1016/j.conbuildmat.2021.122547

Liang, C., Pan, B., Ma, Z., He, Z., & Duan, Z. (2020). Utilization of CO2 curing to enhance the properties of recycled aggregate and prepared concrete: A review. *Cement and Concrete Composites, 105*, 103446. https://doi.org/10.1016/j.cemconcomp.2019.103446

Limbachiya, M. C., Leelawat, T., & Dhir, R. K. (2000). Use of recycled concrete aggregate in high-strength concrete. *Materials and Structures/Materiaux et Constructions, 33*(9), 574–580. https://doi.org/10.1007/bf02480538

Limbachiya, M., Meddah, M. S., & Ouchagour, Y. (2012). Use of recycled concrete aggregate in fly-ash concrete. *Construction and Building Materials, 27*(1), 439–449. https://doi.org/10.1016/j.conbuildmat.2011.07.023

Limbachiya, M., Seddik Meddah, M., & Ouchagour, Y. (2012). Performance of portland/silica fume cement concrete produced with recycled concrete aggregate. *ACI Materials Journal, 109*(1), 91–100. https://doi.org/10.14359/51683574

Li, Y., Yang, X., Lou, P., Wang, R., Li, Y., & Si, Z. (2021). Sulfate attack resistance of recycled aggregate concrete with NaOH-solution-treated crumb rubber. *Construction and Building Materials, 287*, 123044. https://doi.org/10.1016/j.conbuildmat.2021.123044

Luan, H., Wu, J., & Pan, J. (2021). Freeze-thaw durability of recycled aggregate concrete: An overview. *Journal Wuhan University of Technology, Materials Science Edition, 36*(1), 58–69. https://doi.org/10.1007/s11595-021-2378-x

Lu, C., Wang, W., Zhou, Q., Wei, S., & Wang, C. (2020). Mechanical behavior degradation of recycled aggregate concrete after simulated acid rain spraying. *Journal of Cleaner Production, 262*, 121237. https://doi.org/10.1016/j.jclepro.2020.121237

Makul, N., Fediuk, R., Amran, M., Zeyad, A. M., Murali, G., Vatin, N., Klyuev, S., Ozbakkaloglu, T., & Vasilev, Y. (2021). Use of recycled concrete aggregates in production of green cement-based concrete composites: A review. *Crystals, 11*(3), 1–35. https://doi.org/10.3390/cryst11030232

Medina, C., Frías, M., Sánchez De Rojas, M. I., Thomas, C., & Polanco, J. A. (2012). Gas permeability in concrete containing recycled ceramic sanitary ware aggregate. *Construction and Building Materials, 37*, 597–605. https://doi.org/10.1016/j.conbuildmat.2012.08.023

Medina, C., Sánchez De Rojas, M. I., Thomas, C., Polanco, J. A., & Frías, M. (2016). Durability of recycled concrete made with recycled ceramic sanitary ware aggregate. Inter-indicator relationships. *Construction and Building Materials, 105*, 480–486. https://doi.org/10.1016/j.conbuildmat.2015.12.176

Neville, A. M. (1981). *Properties of concrete.*

Neville, A. M., & Brooks, J. J. (1987). *Concrete technology.*

Nobuaki, O., Shin-ichi, M., & Wanchai, Y. (2003). Influence of recycled aggregate on interfacial transition zone, strength, chloride penetration and carbonation of concrete. *Journal of Materials in Civil Engineering, 15*, 899–1561. https://doi.org/10.1061/(ASCE)0899-1561(2003)15:5(443)

Olorunsogo, F. T., & Padayachee, N. (2002). Performance of recycled aggregate concrete monitored by durability indexes. *Cement and Concrete Research, 32*(2), 179–185. https://doi.org/10.1016/S0008-8846(01)00653-6

Pedro, D., De Brito, J., & Evangelista, L. (2014). Influence of the use of recycled concrete aggregates from different sources on structural concrete. *Construction and Building Materials, 71*, 141–151. https://doi.org/10.1016/j.conbuildmat.2014.08.030

Pepe, M., Toledo Filho, R. D., Koenders, E. A. B., & Martinelli, E. (2014). Alternative processing procedures for recycled aggregates in structural concrete. *Construction and Building Materials, 69*, 124–132. https://doi.org/10.1016/j.conbuildmat.2014.06.084

Poon, C. S., & Chan, D. (2007). Effects of contaminants on the properties of concrete paving blocks prepared with recycled concrete aggregates. *Construction and Building Materials, 21*(1), 164–175. https://doi.org/10.1016/j.conbuildmat.2005.06.031

Qi, B., Gao, J., Chen, F., & Shen, D. (2017). Evaluation of the damage process of recycled aggregate concrete under sulfate attack and wetting-drying cycles. *Construction and Building Materials, 138*, 254–262. https://doi.org/10.1016/j.conbuildmat.2017.02.022

Raman, J. V. M., & Ramasamy, V. (2021). Various treatment techniques involved to enhance the recycled coarse aggregate in concrete: A review. *Materials Today: Proceedings, 45*, 6356–6363. https://doi.org/10.1016/j.matpr.2020.10.935

Sadati, S., & Khayat, K. H. (2018). Can concrete containing high-volume recycled concrete aggregate be durable? *ACI Materials Journal, 115*(3), 471–480. https://doi.org/10.14359/51702190

Safiuddin, M., Alengaram, U. J., Rahman, M. M., Salam, M. A., & Jumaat, M. Z. (2013). Use of recycled concrete aggregate in concrete: A review. *Journal of Civil Engineering and Management, 19*(6), 796–810. https://doi.org/10.3846/13923730.2013.799093

Santos, W. F., Quattrone, M., John, V. M., & Angulo, S. C. (2017). Roughness, wettability and water absorption of water repellent treated recycled aggregates. *Construction and Building Materials, 146*, 502–513. https://doi.org/10.1016/j.conbuildmat.2017.04.012

Schmidt, T., Lothenbach, B., Romer, M., Neuenschwander, J., & Scrivener, K. (2009). Physical and microstructural aspects of sulfate attack on ordinary and limestone blended Portland cements. *Cement and Concrete Research, 39*(12), 1111–1121. https://doi.org/10.1016/j.cemconres.2009.08.005

Shicong, K., & Poon, C. S. (2006). Compressive strength, pore size distribution and chloride-ion penetration of recycled aggregate concrete incorporating class-F fly ash. *Journal of Wuhan University of Technology-Mater, Materials Science Edition, 21*(4), 130–136. https://doi.org/10.1007/bf02841223

Shi, C., Li, Y., Zhang, J., Li, W., Chong, L., & Xie, Z. (2016). Performance enhancement of recycled concrete aggregate - a review. *Journal of Cleaner Production, 112*, 466–472. https://doi.org/10.1016/j.jclepro.2015.08.057

Silva, R. V., Neves, R., De Brito, J., & Dhir, R. K. (2015). Carbonation behaviour of recycled aggregate concrete. *Cement and Concrete Composites, 62*, 22–32. https://doi.org/10.1016/j.cemconcomp.2015.04.017

Soares, D., De Brito, J., Ferreira, J., & Pacheco, J. (2014). Use of coarse recycled aggregates from precast concrete rejects: Mechanical and durability performance. *Construction and Building Materials, 71*, 263–272. https://doi.org/10.1016/j.conbuildmat.2014.08.034

Surya, M., Rao, V. V. L. K., & Lakshmy, P. (2015). Mechanical, durability, and time-dependent properties of recycled aggregate concrete with fly ash. *ACI Materials Journal, 112*(5), 653–661. https://doi.org/10.14359/51687853

Tam, V. W. Y., & Tam, C. M. (2007). Assessment of durability of recycled aggregate concrete produced by two-stage mixing approach. *Journal of Materials Science, 42*(10), 3592–3602. https://doi.org/10.1007/s10853-006-0379-y

Tam, V. W. Y., Gao, X. F., Tam, C. M., & Chan, C. H. (2008). New approach in measuring water absorption of recycled aggregates. *Construction and Building Materials, 22*(3), 364–369. https://doi.org/10.1016/j.conbuildmat.2006.08.009

Tam, V. W. Y., Kotrayothar, D., & Xiao, J. (2015). Long-term deformation behaviour of recycled aggregate concrete. *Construction and Building Materials, 100*, 262–272. https://doi.org/10.1016/j.conbuildmat.2015.10.013

Tam, V. W. Y., Soomro, M., & Evangelista, A. C. J. (2021). Quality improvement of recycled concrete aggregate by removal of residual mortar: A comprehensive review of approaches adopted. *Construction and Building Materials, 288*. https://doi.org/10.1016/j.conbuildmat.2021.123066

Tam, V. W. Y., Wattage, H., Le, K. N., Buteraa, A., & Soomro, M. (2021). Methods to improve microstructural properties of recycled concrete aggregate: A critical review. *Construction and Building Materials, 270*. https://doi.org/10.1016/j.conbuildmat.2020.121490

Thérèné, F., Keita, E., Naël-Redolfi, J., Boustingorry, P., Bonafous, L., & Roussel, N. (2020). Water absorption of recycled aggregates: Measurements, influence of temperature and practical consequences. *Cement and Concrete Research, 137*, 106196. https://doi.org/10.1016/j.cemconres.2020.106196

Thomas, C., Setién, J., Polanco, J. A., Alaejos, P., & Sánchez De Juan, M. (2013). Durability of recycled aggregate concrete. *Construction and Building Materials, 40*, 1054–1065. https://doi.org/10.1016/j.conbuildmat.2012.11.106

Thomas, J., Thaickavil, N. N., & Wilson, P. M. (2018). Strength and durability of concrete containing recycled concrete aggregates. *Journal of Building Engineering, 19*, 349–365. https://doi.org/10.1016/j.jobe.2018.05.007

Tian, Y., Jiang, J., Wang, S., Yang, T., Qi, L., & Geng, J. (2021). The mechanical and interfacial properties of concrete prepared by recycled aggregates with chloride corrosion media. *Construction and Building Materials, 282*, 122653. https://doi.org/10.1016/j.conbuildmat.2021.122653

Trottier, C., Ziapour, R., Zahedi, A., Sanchez, L., & Locati, F. (2021). Microscopic characterization of alkali-silica reaction (ASR) affected recycled concrete mixtures induced by reactive coarse and fine aggregates. *Cement and Concrete Research, 144*, 106426. https://doi.org/10.1016/j.cemconres.2021.106426

Xiao, Q., Liu, X., Qiu, J., & Li, Y. (2020). Capillary water absorption characteristics of recycled concrete in freeze-thaw environment. *Advances in Materials Science and Engineering*, 1–12. https://doi.org/10.1155/2020/1620914

Xiao, J., Lu, D., & Ying, J. (2013). Durability of recycled aggregate concrete: An overview. *Journal of Advanced Concrete Technology, 11*(12), 347–359. https://doi.org/10.3151/jact.11.347

Xiao, J., Ying, J., & Shen, L. (2012). FEM simulation of chloride diffusion in modeled recycled aggregate concrete. *Construction and Building Materials, 29*, 12–23. https://doi.org/10.1016/j.conbuildmat.2011.08.073

Xuan, D., Zhan, B., & Poon, C. S. (2017). Durability of recycled aggregate concrete prepared with carbonated recycled concrete aggregates. *Cement and Concrete Composites, 84*, 214–221. https://doi.org/10.1016/j.cemconcomp.2017.09.015

Zega, C. J., Coelho Dos Santos, G. S., Villagrán-Zaccardi, Y. A., & Di Maio, A. A. (2016). Performance of recycled concretes exposed to sulphate soil for 10 years. *Construction and Building Materials, 102*, 714–721. https://doi.org/10.1016/j.conbuildmat.2015.11.025

Zeng, W., Zhao, Y., Zheng, H., Poon, C., & sun. (2020). Improvement in corrosion resistance of recycled aggregate concrete by nano silica suspension modification on recycled aggregates. *Cement and Concrete Composites, 106*, 103476. https://doi.org/10.1016/j.cemconcomp.2019.103476

Zhang, H., Ji, T., & Liu, H. (2020). Performance evolution of recycled aggregate concrete (RAC) exposed to external sulfate attacks under full-soaking and dry-wet cycling conditions. *Construction and Building Materials, 248*, 118675. https://doi.org/10.1016/j.conbuildmat.2020.118675

Zhao, Y., Dong, J., Wu, Y., Wang, H., Li, X., & Xu, Q. (2014). Steel corrosion and corrosion-induced cracking in recycled aggregate concrete. *Corrosion Science, 85*, 241–250. https://doi.org/10.1016/j.corsci.2014.04.028

Alternative binders—high volume bauxite red mud alkali activated cements and concretes

9

Pavel V. Krivenko, Oleksandr Kovalchuk and Victoria Zozulynets
V.D.Glukhovsky Scientific Research Institute for Binders and Materials, Kyiv National University of Construction and Architecture, Kyiv, Ukraine

9.1 Introduction

9.1.1 Production and distribution of red mud (bauxite residue)

The world production of aluminum metal has been increasing steadily and this trend is likely to continue for the foreseeable future due to its large number of industrial applications in the transport, construction, packaging, electrical and other sectors (Dentoni et al., 2014; Power et al., 2009).

Alumina is extracted from its primary ore, bauxite, by the Bayer Process, in which it is treated with hot concentrated sodium hydroxide solution at temperatures of up to 240°C and at pressures of 1−6 atm. The alumina is then electrolytically reduced to aluminum metal by the Hall-Heroult process. The production of 1 ton of aluminum metal typically produces 1−2.5 tons of an insoluble highly alkaline and hazardous waste called red mud as a by-product (Deelwal et al., 2014; Dentoni et al., 2014; Achal & Harinder, 2015; Pérez-Villarejo et al., 2012; Power et al., 2009; Yang & Xiao, 2008). Normally, this red mud is stored either as dried slabs or in large ponds ("lagoons") (Power et al., 2009) where it poses considerable hazards to the surrounding environment, which include humans and plants, if it escapes, as occurred at the Ajkai Timfoldgyar alumina plant in Hungary in October 2010. Thus, the issue of the toxicity of red mud and its disposal has become an increasingly important matter, attracting continued research into remediation of this toxic waste by recycling it in to useful products or by devising much safer storage methods. Table 9.1 shows the annual production of red mud in some countries.

Table 9.1 Annual production of red mud (bauxite residue) in some countries.

Country	Amount (Million tons)	References
China	88	National Bureau of Statistics of China (2017)
Austria	30	Sutar (2014)
India	4.71	Hooda et al. (2016); Sridevi et al. (2016)
Greece	0.7	Sutar (2014)

Recycled Concrete. https://doi.org/10.1016/B978-0-323-85210-4.00001-1

In today's China, for example, it is reported that most of the red mud is disposed-off in landfills and only about 10% of it is being recycled and reused (Balomenos et al., 2011). Despite the fact that most of the red mud is disposed-off, its safe disposal is still a major issue in different countries like USA, India, China, Japan, Australia, France and Greece. Storage of red mud in big volumes leads to environmental disasters resulting from high rainfall and/or failure of the containment structure. Also, its small dust like particle size (0.49 μm), high alkalinity and large volumes make its disposal a major challenge, thus posing serious environmental issues such as, groundwater pH change, leakage, overflow, air pollution by its dust and economic issues like substantial land use for disposal and thus substantial economic costs to the alumina producing industry (Deelwal et al., 2014) as it requires extensive land about approximately 1 km^2 per 5 years for One Million tons/year alumina production plant (Tsakiridis et al., 2004) and the land occupied by red mud cannot be used either for construction or farming even when the red mud is already dry (Liu et al., 2014).

In addition to that, safe disposal of red mud costs companies around 2% of the total aluminum price (Tsakiridis et al., 2004). Some sources even indicate 5% of the production value of the aluminum (Pérez-Villarejo et al., 2012) indicating that red mud disposal is both environmentally and economically problematic.

9.1.2 Chemical and mineralogical characteristics of red mud

The knowledge of chemical and mineralogical composition of red mud is of great importance as its application is based upon its unique physico-chemical properties (Achal & Harinder, 2015; Sutar, 2014) as well as its mineralogical properties (Balomenos et al., 2011). Red mud is a complex material and its chemical and mineralogical composition varies widely, depending upon the origin of the bauxite and the alumina production process (Deelwal et al., 2014; Liu et al., 2014; Tsakiridis et al., 2004).

Generally, regardless of production process, the chemical analysis of red mud shows that it contains six major constituents, namely Fe_2O_3, Al_2O_3, SiO_2, TiO_2, Na_2O and CaO (Table 9.2), an array of minor elements namely such as K, Cr, V, Ba, Cu, Mn, Pb, Zn, P, F, S (Deelwal et al., 2014) and rare earth elements Ga, C, Cd, Ni, Mg, Zr, Hf, Nb, U, Th, K, Sr (Tsakiridis et al., 2004). The red mud also contains 2–3 g/L of Na_2O, which results in a high pH value, between 13 and 14.

Mineral composition of red mud mainly depends on the mineral composition of the source material—bauxite. Bauxite is a multiphase ore that contains more than 100 minerals (Cablik, 2007). The major mineral phases of Guizhou red mud include vishnevite, calcite, tiantanite, perovskite, larnite, kaolinite, quartz and a few amorphous materials.

Mineralogical, the main components of red mud from the Bayer process are hematite (Fe_2O_3), boehmite (γ-$Al(OH)_3$, quartz (SiO_2), sodalite ($Na_4Al_3Si_3O_{12}Cl$), gypsum ($CaSO_4 \cdot 2H_2O$) and gibbsite ($Al(OH)_3$) of (Yang & Xiao, 2008) while the dominating mineralogical constituents of red mud from the sintering process are larnite β-$2CaO \cdot SiO_2$ (Liu et al., 2014; Sutar, 2014), calcite ($CaCO_3$), aragonite ($CaCO_3$), hematite (Fe_2O_3), gibbsite ($Al(OH)_3$) and perovskite ($CaTiO_3$) (Liu et al., 2014) of and few amorphous materials.

Table 9.2 Main chemical composition (wt%) of red mud (Bayer's Process) in other countries.

Country	Components										References
	Al_2O_3	SiO_2	Fe_2O_3	CaO	TiO_2	Na_2O	K_2O	MgO	L.O.I.	Other	Total	References
China	27.45	15.86	7.23	21.58	4.93	15.94	2.89	1.44	3.53	–	100.85	Gu et al. (2012)
Spain	20.20	7.50	47.85	6.22	9.91	8.40	0.11	0.33	–	–	100.52	Pascual et al. (2009)
Greece	19.95	6.80	40.80	12.60	5.80	2.70	0.14	0.20	10.54	0.58	100.11	Tsakiridis et al. (2004)
Greece	17.04	7.79	44.34	11.64	5,12	3.17	0.07	0.57	9.77	–	99.51	Pontikes et al. (2007)
Italy	20.00	11.60	35.20	6.70	9.20	7.50	–	0.4	7.30	2.10	100.00	Sglavo et al. (2000)
Bosnia (Birac Alumina)	14.14	11.53	48.50	3.96	5.42	7.50	0.06	0.05	7.25	0.61	99.02	Cablik (2007)

L.O.I., Loss on ignition.

Particle analysis showed that 10% of the red mud grains had a particle size below 1.146 μm, 50% below 5.897 μm, and 90% below 36.663 μm. The red mud had a specific surface area of 1.46 $m^2\ g^{-1}$, with a mean surface diameter of 2.932 μm and a mean volume diameter of 13.224 μm. The particles of red mud were aggregated with multiple fine grains having different constituents.

9.1.3 Previews attempts to utilize red mud

The key to solving red mud disposal issues to the environment is to develop utilization technologies that consume significant quantities of red mud or transform it into a secondary resource (Reddy & Chandra, 2014; Sutar, 2014). In China, the ministry of industry and information had established a plan in 2012 aiming to increase the utilization of red mud from 4% in 2012 to 15% in 2015 and 20% by 2020 (Mukiza et al., 2019). During the past decades, researchers have been working on developing cost-effective ways, which are very useful in the context of environmental concerns for the utilization and disposal of red mud (Hooda et al., 2016) based to its unique physical, chemical properties (Achal & Harinder, 2015; Sutar, 2014) and mineralogical properties (Balomenos et al., 2011). In (Achal & Harinder, 2015), regarded red mud as a valuable material instead of a useless waste since it has many applications (Hua et al., 2017) including: its use in making bricks (Dodoo-Arhin et al., 2013), glass ceramics (Yang et al., 2008), as adsorbent for the removal of heavy metals from aqueous solutions (Pulford et al., 2012; Ribeiro et al., 2011), in metal recovery for example Fe, in making catalysts (Liu et al., 2014). Among the red mud utilization options, making construction and building materials consume a huge quantity of red mud (Liu & Zhang, 2011) and pose much lower risk for implementation. To a great extent, the utilization of red mud in large quantities is especially in road construction (Deelwal et al., 2014; Sridevi et al., 2016).

However, due to the reason of low activity of red mud, its content in finish product is very low. In some cases a preliminary treatment of extensive energy is required (Ke et al., 2015; Liu et al., 2009; Ye et al., 2014). For example, in the preparation of special cements from red mud, the added quantity of red mud is usually less than 5% (Tsakiridis et al., 2004). Previously, many attempts have been made to utilize red mud alone by itself, as raw material using alkali activation.

Using alkali activation, allows considerably higher incorporation rates of red mud in cement and concretes without a decrease in their physico-mechanical characteristics (Alonso et al., 2018; Boskovic et al., 2013; Croymans et al., 2017; Dimas et al., 2009; Gijbels et al., 2020; Hairi et al., 2015; Hajjaji et al., 2013; He et al., 2012; Ke et al., 2014; Komnitsas & Zaharaki, 2009; Krivenko et al., 2017; Kumar & Kumar, 2013; Pan et al., 2003; Patent of USSR 979294, 1982; Patent of Ukraine UA 10286 A, 1996; Rostovskaya, 1994; Vukčević et al., 2013; Ye et al., 2016; Zhang et al., 2010; Zhang et al., 2014).

However, content of red mud in the final product is not high enough. Moreover, the data for long term studies of concrete based on red mud are absent so far, and recommendations are less forthcoming for its use, in conditions of high content of heavy metals and radionuclides in red mud.

The main goal of the current research work was to optimize the formation process of alkali activated cements and concretes in order to allow the incorporation of a large portion of red mud. To achieve this objective, it is important to control the alkaline environment and the presence of oxides, such as CaO, Al_2O_3, SiO_2, Fe_2O_3 and others in the active form, for the synthesis of hydration products-mineral substances in the alkaline material.

9.1.4 Concept for compositional buildup of alkali activated cement

The process of hydration and hardening can be considered as a complex process involving (first stage) destruction-coagulation, (second stage) coagulation-condensation and (third stage) condensation-crystallization (Palomo et al., 2015). It was demonstrated that the quantity of glass phase is a determining factor at the first stage of hydration process and hardening of alkali-activated cement (Palomo et al., 2015). Crystalline high-basic calcium silicates quickly hydrate in alkaline medium promoting acceleration of the third stage because of their hydration products. These calcium silicates can act as crystallo-chemical intensifiers for hardening. Absence of these crystalline phases in the initial material retards the third stage of structure formation process.

The role played by anions originating from alkaline component is a double one and is determined by the anion type. Anions can be divided into two groups: (1) anions that enter into cement within soluble silicates and aluminates and (2) anions that can be found in all remaining alkaline compounds. The anions of the first group are similar to the hydrated primary destruction products of the aluminum-silicon-oxygen framework in the solid phase (SiOH), they serve as a reserve for the hydrated primary destruction products and, naturally, are the most effective anions in terms of hydration and structure formation. Anions of the second group change the properties of the liquid phase, and some participate in the formation of complexes, which extract the destruction products into a solid phase promoting intensification of the hydration processes (Palomo et al., 2015).

Thus, a compositional buildup of high volume red mud alkali activated cements should be based on a proper choice of the optimal ratio of red mud (with additives of glassy structure) to high-basic additives, which quickly hydrate and crystallize in highly alkaline media (Krivenko et al., 2017; Rostovskaya et al., 1994).

9.2 Role of red mud in the microstructure formation of hardened cement paste

Different compositions of red mud, ground granulated blast-furnace slag (ggbs) (glassy additive) and OPC (high-basic calcium containing additive) were mixed with soda ash, sodium metasilicate and soluble sodium silicate as alkaline components in order to produce alkali activated cements. An overview of the compositions of the produced alkali activated cements is given in Table 9.3.

Hydration products of the formulated cements were studied using a set of physical-chemical examination techniques, such as X-ray phase diffractometry, differential thermal analysis (DTA), Thermogravimetry (TG) and Scanning electron microscopy (SEM). X-ray phase diffraction analysis was done using diffractometers DRON-3M and DRON-4-07 with a copper tube at voltage = 30 kV, current = 10–20 mA and angle range $2\theta = 10–60$ at a speed of counter rotation = 2 per minute. Differential thermal and thermogravimetric analyses were carried out using an instrument of F.Paulik, J.Paulik, L.Erdey (MOM Company, Budapest, Hungary). The specimens were heated at 10°C per minute until a temperature of 1000°C was reached. Scanning electron microscopy (SEM) equipped with micro analyzer REMMA-102-02 (resolving power in the regime of secondary electrons is no more than 5 nm, magnification range is ×10—300,000, a range of accelerating voltage 0.2–40 kV was used, maximum excessive pressure in the electron column was kept at $6.7 \cdot 10^{-4}$Pa).

For the determination of the microstructure of hardened cement paste the compositions in Table 9.3 were used. The role of red mud in the microstructure formation was investigated when using (1) soda ash, (2) sodium metasilicate and (3) sodium silicate as alkaline components.

The determination of the phase composition of hydration products of the formulated cements was not straightforward. This can be explained, on one-hand by a multi-component composition of cementitious systems themselves, and on the other, by

Table 9.3 Compositions of alkali activated cements containing red mud.

Composition	Red mud, (% by mass)	ggbs, (% by mass)	OPC (% by mass)
Alkaline component: Soda ash (Na_2CO_3) (5% by mass of ggbs + red mud)			
K1	50	50	–
K2	60	30	10
K3	70	30	–
K4	70	20	10
K5	80	15	5
Alkaline component: Sodium metasilicate (Na_2SiO_3) (5% by mass of ggbs + red mud)			
K6	50	50	–
K7	70	25	5
Alkaline component: Soluble sodium silicate (Ms = 2.8, ρ = 1400 kg/m^3) (soluble sodium silicate/ggbs + red mud = 0.4)			
K8	50	50	–
Alkaline component: Soluble sodium silicate (Ms = 2.8, ρ = 1300 kg/m^3) (soluble sodium silicate/ggbs + red mud = 0.4)			
K9	60	30	10
K10	50	50	–

the fact that each individual component is also a complex system (red mud, slag and cement). In some cases, this resulted in a superposition of characteristic responses making identification of the hydration products more complicated.

The results of X-ray phase diffraction analysis (Fig. 9.1) of the cements K1–K5 (Table 9.2) in which soda ash is used as alkaline component showed that hydration products are represented mainly by tri-calcium aluminates $3CaO\cdot Al_2O_3$ (diffraction characteristics 0.272; 0.191; 0.161, PDF#330251) and calcite $CaCO_3$ (0.368; 0.262; 0.228; 0.209; 0.191; 0.188, PDF#721937). The higher contents of the red mud in the cement composition (K5 and K4) resulted in the formation of zemkorite $Na_2Ca(CO_3)_2$ (0.639; 0.438; 0.425; 0.303, PDF#411440) and gaylussite $Na_2Ca(-CO_3)_2\cdot 5H_2O$ (0.639; 0.270; 0.272; 0.321, PDF#210343). The addition of OPC (up to 10%) (K2, K4, and K5) was found to accelerate mineral formation processes at the early stages of hardening and improves the formation of crystalline structure for the resulted cement stone.

In case of sodium metasilicate use (Fig. 9.1), the phase composition is somewhat different. The following minerals are formed: klinoferrosilite $FeSiO_3$ (0.643; 0.335; 0.321; 0.304; 0.260, PDF#821832) and lawsonite $CaAl_2[Si_2O_7]\ (OH)_2\cdot H_2O$ (0.487; 0.417; 0.368; 0.273, PDF#771994). When going to higher contents of the red mud from 50% to 70% (K6 and K7) the structure formation process decelerates, though it does not stop completely. This means that the red mud takes an active part in cement hydration processes. The formation of klinoferrosilite in the hydration products supports the assumption that hematite contained in large quantities in the red mud takes part and provides the condition of an alkaline medium, in the formation of hydration

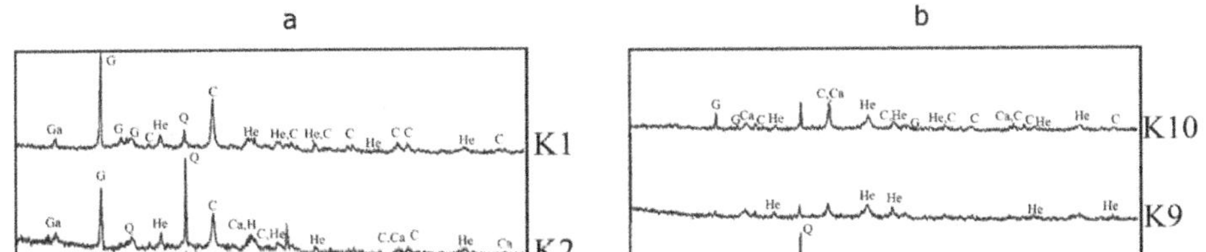

Figure 9.1 X-ray phase diffraction patterns of a cement stone from the alkali activated cements containing red mud and alkaline component: a—soda ash; b—soluble sodium silicate + sodium metasilicate. Designation: He—hematite, Ca—tri-calcium aluminate, C—calcite, Q—quartzite, G—gibbsite, Z—zemkorite, Ga—gaylussite, Cm—caminite, Cli—klinoferrosilite, La—lawsonite.

products. The use of soluble sodium silicate (K8) as alkaline component considerably accelerates the formation of calcium silicate hydrates and is a base of crystallization processes in the cement. Soluble sodium silicate enters actively into interaction with red mud and calcium-containing additive. In the case of compositions with lower density of soluble glass (K9, K10) the prevailing hydration products are found to be tri-calcium aluminates $3CaO \cdot Al_2O_3$ (0.272; 0.191; 0.161, PDF#330251), calcites $CaCO_3$ (0.368; 0.262; 0.228; 0.209; 0.191; 0.188, PDF#411440) and crystals of hematite (PDF#850599).

The obtained results coincide well with the results of DTA. At temperatures between 270°C and 350°C an endothermic effect takes place as a result of loss by gibbsite of OH^- groups. The occurrence of tri-calcium aluminate is confirmed by heat adsorption at 740°C (Fig. 9.2). A wide endothermic band in the DTA curve between 390°C and 700°C gives an indication of the release of crystal water with break of hydrogen bonds.

Electron microscopy studies of the cleavage surface of a cement stone from the alkali activated cement in case of soda ash as alkaline component showed the presence of calcium silicate hydrates, crystals of gaylussite (Ref to: https://www.mindat.org/min-

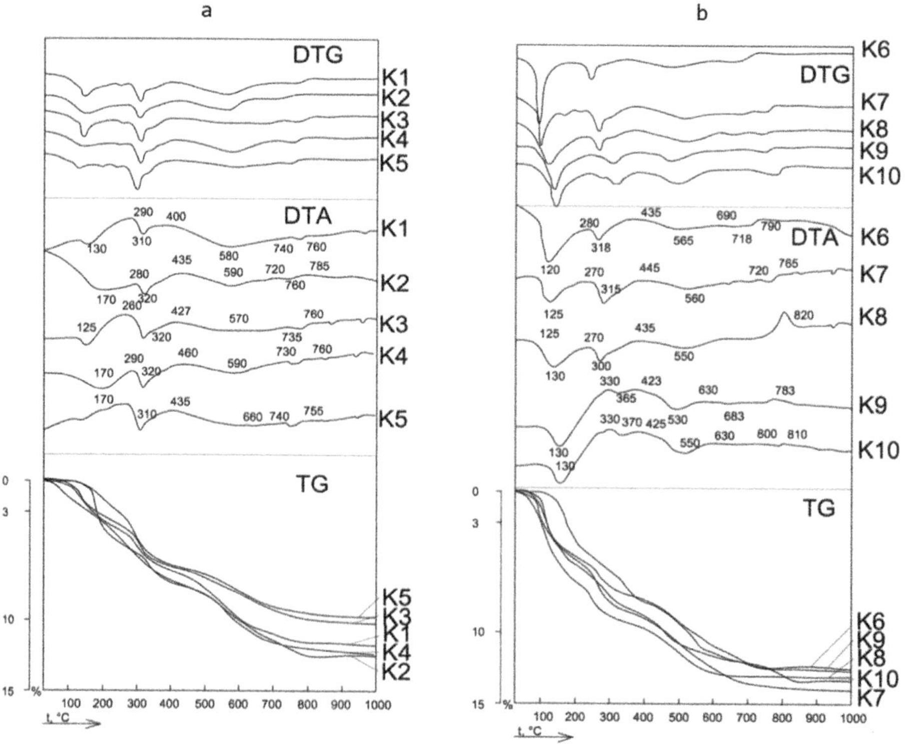

Figure 9.2 DTA curves of a cement stone from alkali activated cement containing red mud and alkaline components a—soda ash; b—soluble sodium silicate + sodium metasilicate.

1662.html) and particles of hematite (Ref to: https://www.mindat.org/min-1856.html) (Fig. 9.3a and c). Photos taken of the cleavage surface case of sodium metasilicate, showed the formation of calcium silicate hydrates (Fig. 9.3b and d, Fig. 9.4).

A cleavage surface featured a dense character without cracks, testifying to a uniform flow of the structure formation processes.

Pore space contains prismatic and fibrous crystals, characteristics of calcium silicate hydrates and alkaline alumino—and ferrosilicate hydrates in sub-microcrystalline state.

In case of soluble sodium silicate as alkaline component, the quantity of calcium silicate hydrates tends to increase considerably, testifying to a more active interaction of the components owing to the use of alkaline components with the higher silicate modulus.

At the same time, the formation of a gel-like phase of alkaline ferro-aluminosilicate hydrates could not be excluded (Dimas et al., 2009).

Figure 9.3 SEM images of the alkali activated cements with incorporated red mud (a)—K3, (b)—K6, (c)—K4, (d)—K7.

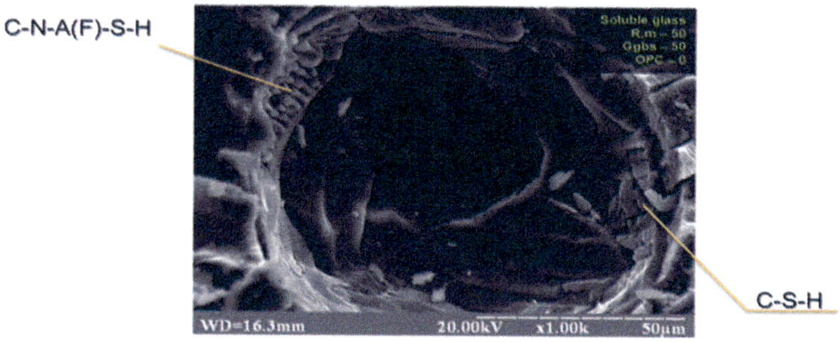

Figure 9.4 SEM images of ne formation in the pore space.

9.3 Compositions and properties of red mud containing alkali activated materials

Red mud containing alkali activated materials can be divided into three groups, namely:

(a) alkali activated red mud based cements and pastes;
(b) alkali activated concretes for common use;
(c) alkali activated pressed materials on red mud bases.

These materials differ in mix design, and red mud plays different roles in the structure formation properties—from cement constituent to the aggregate. The highest load of red mud in the composition is possible in case when it is used as both, cement constituent and aggregate.

Table 9.4 Composition and properties of red mud based alkali activated cements on red mud bases.

	Cement composition, % by mass				Strength (MPa) after hardening (days)					
					2		7		28	
No	GGBS	Red mud	Alkaline component	Flow (mm)	R_c	R_f	R_c	R_f	R_c	R_f
R1	50	50	5	102	12.5	2.3	19.7	4.7	29.3	5.4
R2	45	55	5	152	10.6	1.9	23.5	4.9	27.8	5.4
R3	40	60	7	136	4.1	1.4	20.6	4.2	26.3	6.1
R5	40	60	7	132	15.0	1.9	25.0	3.5	35.3	6.2
R6	40	60	8	140	15.3	3.1	23.1	4.4	36.0	4.6
R7	50	50	7	152	24.0	3.8	28.1	4.3	39.4	5.4
R8	45	55	7	150	19.7	3.0	26.9	4.2	40.3	5.8

9.3.1 Red mud based alkali activated cements and pastes

Alkali activated cements and pastes could be used as for the common application, so far as matrixes for immobilization of heavy metals and hazardous materials. Examples of red mud containing pastes and cement mortars are given in Table 9.4.

It was observed that optimal content of red mud in the cement composition is 55%–60% by mass. Higher content of red mud significantly reduces strength properties and setting times of the cement. In case of use soluble glass as an alkaline component even using 60% of red mud in the cement composition it is possible to obtain proper strength properties. Regularities of strength variation depending on soluble glass properties are shown on Fig. 9.5.

Regulation of soluble glass density makes it possible not only to regulate setting times, but also to change the viscosity of the system, improving workability. Use of soluble glass with different density in the system with 60% of red mud content opens up the possibility to obtain cement stone with compressive strength at 7 days up to 30 MPa, and up to 50 MPa at 28 days.

On the other hand, modern trends in the construction need to be used "dry" cement systems (with dry alkaline components). Such approach simplifies alkali activated cement use and also makes it possible to decrease shrinkage deformations. Compositions of alkali activated cements with dry alkaline component using red mud and their properties are given in Table 9.4. The mixed dry alkaline component (sodium carbonate and metasilicate) was used in these systems.

Determination of shrinkage was done for different types of alkaline components. As a reference material, was taken slag alkali activated cement with dry mix of alkalis (sodium carbonate and sodium metasilicate). Composition of alkali activated cements is shown in Table 9.5.

As it seems from the test results, shrinkage of basic composition (slag alkali activated cement) at 28 days age is 1.1 mm/m. At the same time, alkali activated cements based on red mud basis 1.6 mm/m for system on dry alkaline component and

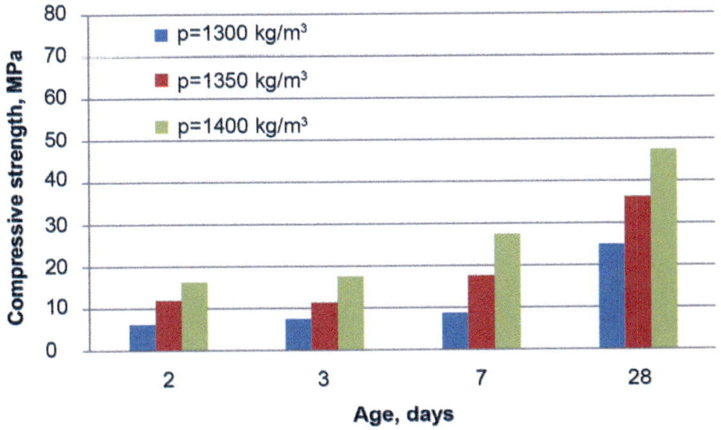

Figure 9.5 Variation of compressive strength depending from the soluble glass density.

Table 9.5 Composition of alkali activated cements to determine shrinkage deformations.

No	Cement composition (%)				Flow (mm)
	Red mud	GGBS	OPC	Alkaline component	
1	60	30	10	Soluble glass (P = 1400 kg/m^3, M_b = 2.8)	115
2	55	45	–	Dry mix (7%)	115
3	–	100	–	Dry mix (5%)	115

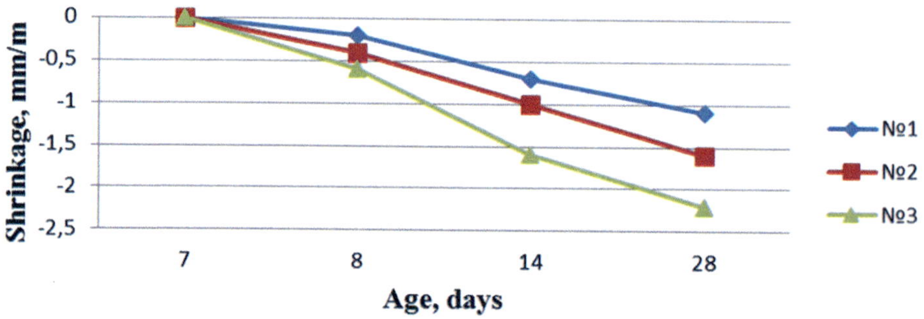

Figure 9.6 Shrinkage of alkali activated cement on red mud basis.

2.2 mm/m for the cement on soluble glass basis. This is probably caused by extraction of water from gel micro pores and capillary pores, resulting in increasing of surface tension forces, leading to the shrinkage. Test results are shown in Fig. 9.6.

Alkali activated red mud based cements and pastes are also known as a good matrix for immobilization of toxic and hazardous metals (Alonso et al., 2018). Thus, Pb immobilization rate is over 99.7% and Ni immobilization rate is over 99.9%, making the material an ideal matrix for heavy metals immobilization and utilization.

9.3.2 *Red mud based alkali activated concretes*

Incorporation of red mud into alkali activated cement compositions, even in quantities reaching 60% by mass of the cement, will only lead to 14.5% by mass of red mud in the final product—concrete, under the assumption of a cement content of 500 kg/m^3.

In order to increase the quantity of the incorporated bauxite residue, the red mud in a state "as produced" can be used as fine aggregates in making the alkali activated cement concretes. Increasing of red mud content in the concrete mix to almost 40% by mass of concrete constituents, leads to some problems at early stages of hardening. However, optimization of the aggregate content made it possible to have a good strength (over 20 MPa) only after 1 day. Results of the concrete test are given in Table 9.6.

Table 9.6 Characteristics of high volume red mud alkali activated concretes using soluble glass (water glass) (quantities of red mud—reaching up to 38.6% by mass of dry constituents).

		Compressive strength, (MPa), age		
No	**Slump (mm)**	**1 day**	**7 days**	**28 days**
C1	170	13.72	43.57	63.5
C2	110	22.89	49.81	70.8

Substitution of fine aggregate—sand—to up to 38.6% by mass of red mud allows producing rather high-slump concretes with high strength reaching 70.8 MPa after 28 days (Table 9.6). Incorporation of even higher mass percentages of red mud in concrete mixtures can create technical difficulties because of its (red muds) high water demand as a result of the high fineness characteristic of red mud. The water content does not affect the concrete strength characteristics; however, it influences the performance properties of the resulted concretes leading to a lower freeze-thaw- and corrosion resistance.

It should be mentioned that alkaline component in such concrete mixes was represented by soluble glass (water glass). However, use of soluble glass for common application materials is not favorable from the technical and economical points of view. Much interesting is to use alkaline component in dry state in a mix containing other concrete constituents and only with water added to it.

The studies (provided in reference section) had shown that about 40% red mud content in the concrete mix when used as dry alkaline component do not provide compressive strength higher than 20 MPa. Reducing the content of red mud makes it possible to obtain concretes C25 and C35. Results of the tests are shown in Fig. 9.7.

Red mud could be used in concrete in two variations—as a part of cement or as an aggregate. Alkali activated concrete containing red mud as a part of cement mix design and its properties are given in Table 9.7.

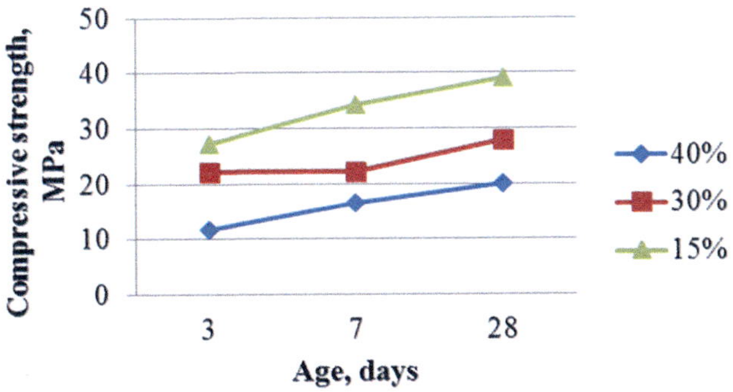

Figure 9.7 Variation of concrete compressive strength depending from red mud content.

Table 9.7 Concrete mix design and properties using red mud as a part of cement.

	Mix composition of concrete, kg/m^3										Compressive strength, MPa, after, days		
	Cement composition, kg/m^3						Aggregates, (kg/m^3)						
No.	GGBS	Red mud	Alkaline component		OPC	LST	Sand, 0.63–1.2 mm	Stone,5–20 mm	W/C, L/S	Slump (cm)	3	7	28
CR1	332	–	Soda ash, 7	Sodium metasilicate,11	–	–	850	1150	0.34	2	22.0	27.6	34.2
CR2	105	210	Soluble glass(P = 1400 kg/m^3, Ms = 2.8)		35	–	850	1150	0.37	3	20.1	23.3	34.0
CR3	148	181	Soda ash, 3.5	Sodium metasilicate, 17.5	–	2,62	850	1150	0.35	4	21.0	22.1	32.4

Table 9.8 Concrete mix design and properties using red mud as an aggregate.

	Mix composition											
	Cement, kg/m3				Aggregate, kg/m3							
No	GGBS	OPC	Soda ash	Sodium metasilicate	Red mud	Sand	Granite dropouts	Fr. 5–20 mm	Slump (cm)	Compressive strength after steam curing (MPa)	Total (integral) pore volume, %	Frost resistance
1B	400	11.5	23.1	46.2	885	187	187	374	5	17.2	18.3	–
2B	390	11.3	22.7	45.5	655	239	239	478	6	44.5	18.0	–
3B	370	4.1	16.4	16.4	210	370	370	740	9	30.5	15.1	F75
4B	414	4.8	19.2	38.5	318	330	330	660	7	22.3	18.8	F100
5B	388	16.0	17.6	35.2	861	136	136	272	9	31.8	18.6	–

On the other hand, use of red mud as an aggregate makes it possible to utilize higher quantities of red mud (up to 885 kg/m^3 of concrete). Concrete mix design and properties are given in Table 9.8.

According to above mentioned study, it is possible to increase concrete compressive strength by using poly-fractional aggregates, namely: red mud, sand, granite dropouts, aggregates fractions 5−10 mm and 10−20 mm.

9.3.3 Red mud based alkali activated pressed materials

To tackle the issue of water demand, a possibility to utilize red mud in quantities in excess of 35% (by mass) is to use it in concrete products manufactured by pressing (moisture content is 8%−16% by mass). By means of pressing, large percentages up to 90% by mass of red mud, can be used when considering them for applications in the road base. Results of the study are given in Table 9.9.

As is demonstrated in Table 9.9 (adapted from Croymans, 2017), even after seven days of hardening in normal conditions, the strength of the concrete was higher than 3.5 MPa, which is required for compressive strength of concrete road bases. Testing for water resistance showed that the specimen C1, after 7 days of hardening in normal conditions and following 2 days of saturation with water, had a softening coefficient of 0.88 while specimen C4 had a softening coefficient of 1.02. In this way, it is demonstrated that the materials are water resistant and from the technical point of view they could be used as a basis for road construction even when red mud incorporation is 90% by mass.

9.3.3.1 Porosity and specific surface area

The isotherms of N_2 adsorption/desorption for each alkali-activated mortar sample can be found in Fig. 9.8. All isotherms showed type IIb characteristics with an H3 hysteresis loop (P1, P2, P3, C1, C2, C3, C4) or an H4 hysteresis loop (P4, P5). According to the literature, this means that the produced samples have mainly meso- and macro pores, in addition to fewer slit-shaped micro-pores. The high-slump-casting technique (samples C1−C4) resulted generally in a higher amount of N_2 adsorbed at the maximum relative pressure, while the semidry pressing technique (samples P1−P5) gave larger hysteresis loops, irrespective of the sample's mix design.

Fig. 9.9 presents the specific surface area of the samples, calculated by the BET method and include specific surface area of both micropores plus meso- and macropores. The micropore surface area was obtained from the T-plot analysis method, whereupon the specific surface area of meso- and macropores was calculated. A major influence of both the amount of red mud incorporated and water content is kept under observation. The specific surface area generally decreases with decreasing amount of red mud and water, except for sample C4. Regarding the content of red mud, the specific surface area was higher when the high-slump-casting technique was applied, e.g., samples P4 and C2 both incorporating 50 wt% of red mud, while the specific surface area of sample C2 was approximately 3 times higher. The latter is expected to be caused by a higher water content in the mix design of sample C2, compared with

Table 9.9 Sample's mix design (in wt%) (100% corresponds to dry mass) and properties.

Sample	Cementitious binder	Bauxite residue	Sand	H_2O	Compressive strength (MPa)
P1	10	90	—	17	17.2
P2	15	85	—	17	44.5
P3	25	75	—	16	30.5
P4	25	50	25	14	22.3
P5	25	40	35	12	31.8
C1	25	60	15	32.5	5.10
C2	25	50	25	25.5	9.15
C3	25	40	35	23.5	12.5
C4	25	30	45	11.2	4.09

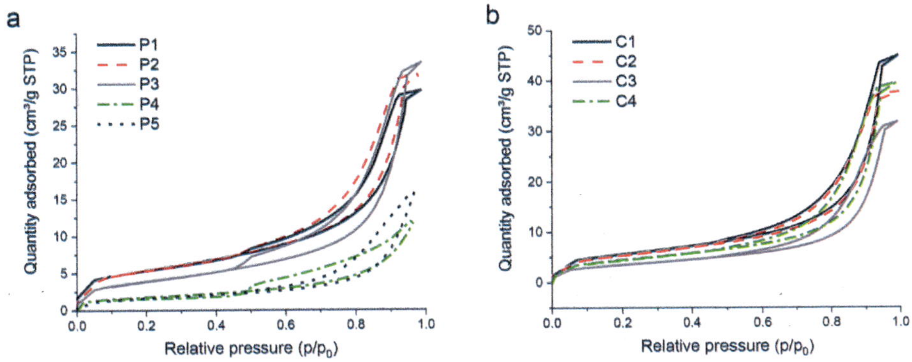

Figure 9.8 N2 adsorption/desorption isotherms of (a) samples P1−P5 produced by the semidry pressing technique and (b) samples C1−C4 produced by the high-slump-casting technique.

sample P4. The relatively high specific surface area of sample C4 is believed to be the consequence of lower flowability of the mortar, since this sample contained the highest incorporated amount of sand and the lowest water content.

The mesopore size distribution is shown in Fig. 9.10, obtained from the desorption data based on the BJH method. It has to be noted that in the application of this method the upper limit of P/P_0 is approximately 0.97. Consequently, the valid range of pore size is lower than 30 nm. For all samples, a bimodal profile was observed, with a maximum between 3 and 4 nm ("smallest mesopores:") and one between 10 and 15 nm ("larger mesopores"). In the case of semidry pressing, the "smallest mesopores" were the most abundant, and a higher amount of red mud and water incorporation resulted in a shift of these "smallest mesopores" to larger pore diameters and a decrease in their amount. Sample P4 showed lowest amount of "larger mesopores," while this amount was largest for samples P1−P3. When the high-slump-casting technique was applied, the "larger mesopores" became the most abundant ones. Irrespective of the amount of red mud or water incorporated, the amount of "larger mesopores" was

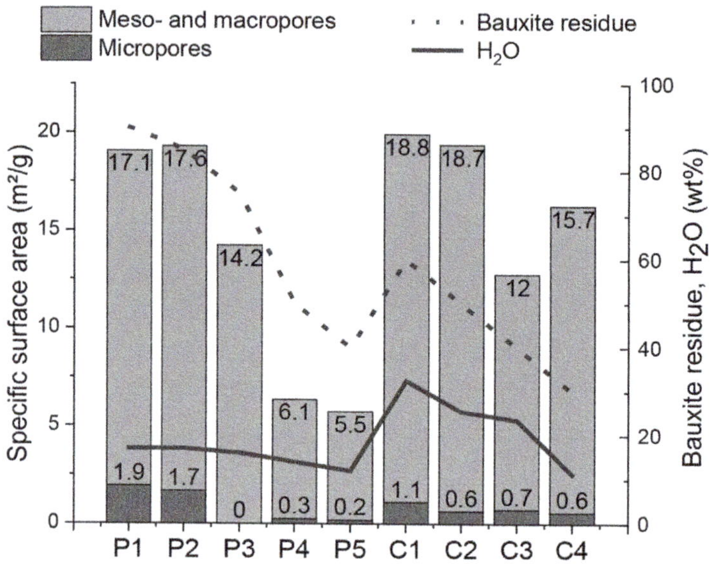

Figure 9.9 Specific surface area of micropores (<2 nm) and meso- and macropores (2–300 nm), and amount of bauxite residue and free water incorporated.

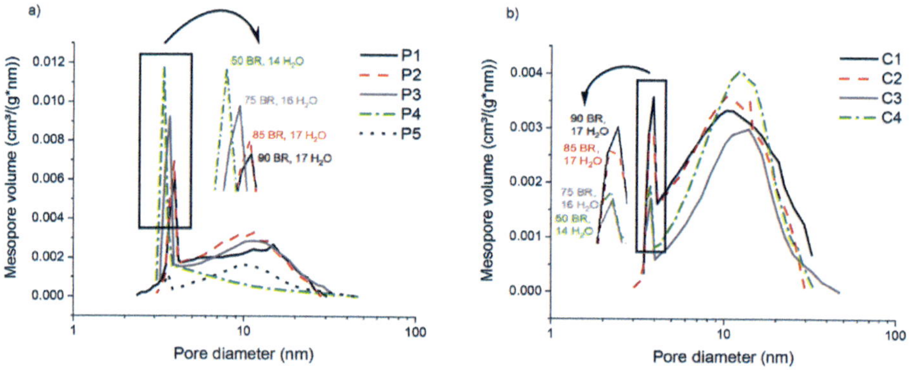

Figure 9.10 Mesopore size distribution of (a) samples P1–P5 produced by the semidry pressing technique and (b) samples C1–C4 produced by the high-slump-casting technique.

comparable for samples C1–C4, while a higher amount of red mud and water incorporation resulted—in contrast to samples P1–P5—in an increase of "smallest mesopores."

The micropore volume, derived from the T-plot analysis, is found in Fig. 9.10. Generally, the micropore volume decreases with decreasing amount of red mud and water incorporated, in line with the trend observed for the specific surface area. The cumulative mesopore volume is shown in Fig. 9.11, where the relation with sample's mix design is less clear. The cumulative mesopore volume is highest for samples

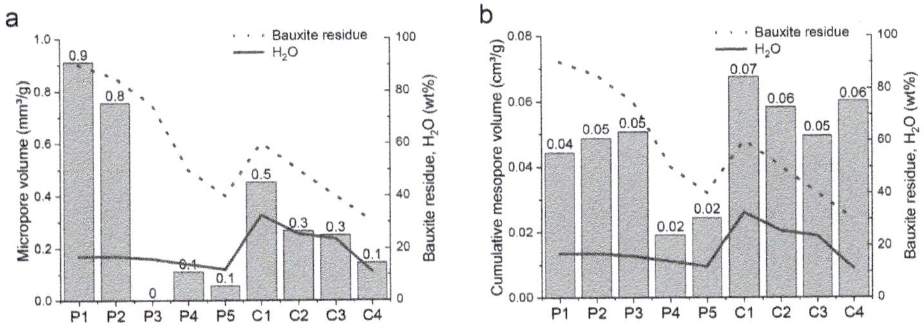

Figure 9.11 (a) Micropore volume and (b) cumulative mesopore volume; and amount of bauxite residue and free water incorporated.

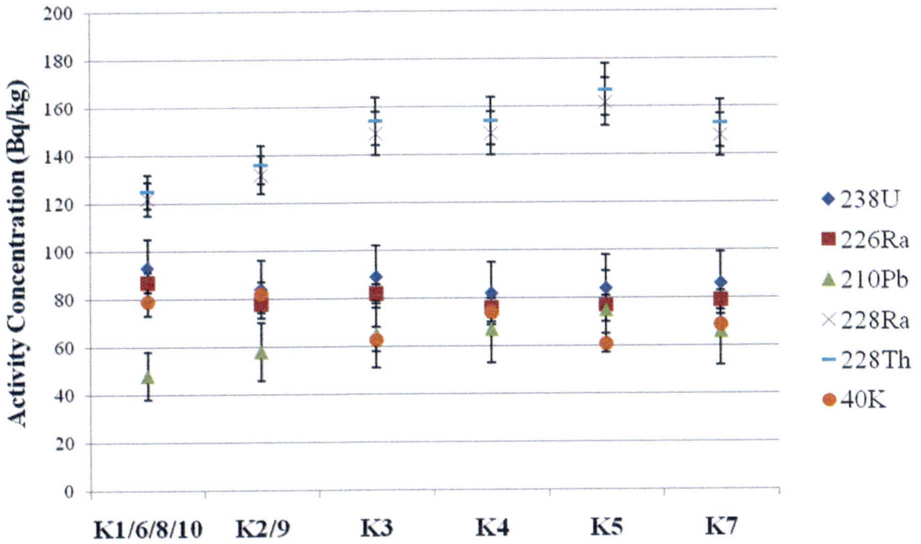

Figure 9.12 Activity concentrations of the produced cements that were calculated on the basis of the measured activity concentrations of the constituents considering the composition in Table 9.3.

produced by the high-slump-casting technique, while this volume gets reduced when semidry pressing was employed.

9.3.4 Radiological characterization

In Fig. 9.12 the activity concentrations are given for the newly produced cements (composition in Table 9.3), calculated from the constituents. For all the cements the activity concentrations of the considered radionuclides of the ^{238}U and ^{232}Th day

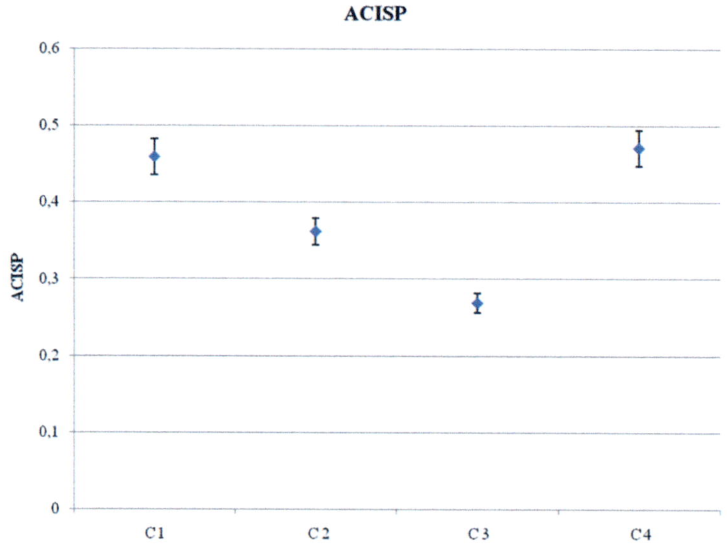

Figure 9.13 ACISP calculated for the produced concretes on the basis of the measured activity concentrations of the constituents considering the composition in Table 9.7.

series and ^{40}K are well below the exemption/clearance levels specified by the EU-BSS of 1 kBq/kg and 10 kBq/kg.

Radiological evaluation of alkali activated concretes for road construction (pressed specimens) are given in Fig. 9.13.

The public and occupation exposure linked to the use of the prepared concrete mixtures (Table 9.9) for road base was verified. Fig. 9.5 shows that all calculated ACI_{SP}-values were below 1. So the external gamma dose for public exposure is lower than the 0.1 mSv/y dose threshold level proposed by Markkanen. The assessment of occupational exposure for workers—via the RP-122 part II "Road construction" scenario—was calculated for samples in Table 9.4. The calculated total doses to the road construction workers are 0.22 0.02; 0.18 0.02; 0.13 0.01 and 0.23 0.063 mSv/y

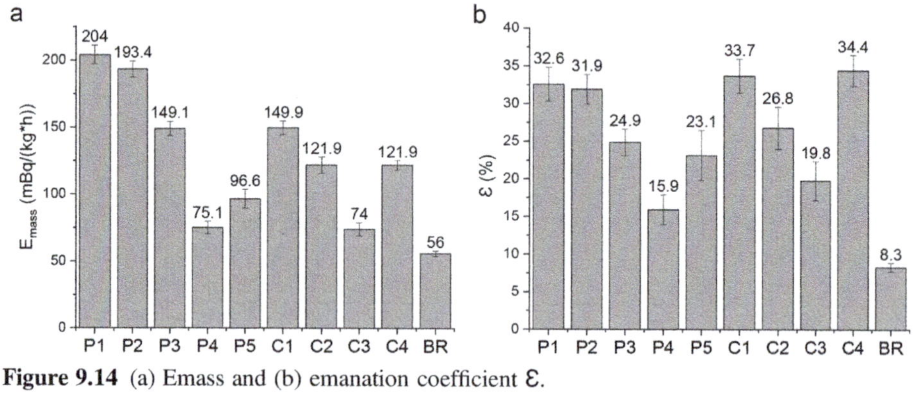

Figure 9.14 (a) Emass and (b) emanation coefficient Ɛ.

respectively, for C1, C2, C3 and C4. These values are below the 0.3 mSv/y dose threshold level proposed by RP-122 (European Commission, Radiation Protection 122 Practical Use of the Concepts of Clearance and Exemption Part II, Application of the Concepts of Exemption and Clearance to Natural Radiation Sources, 2001).

9.3.4.1 Radon exhalation and emanation

The Emass rates of mortar samples and the bauxite residue (BR) are shown in Fig. 9.14 Due to their variable ^{226}Ra concentration, it is more straightforward to evaluate radon release based on the emanation coefficient.

The amount of bauxite residue and water incorporated were the influencing factors for porosity and specific surface of alkali-activated mortar samples. In this section, it is evaluated how the porosity and specific surface area at their turn are influencing the E_{mass} and the emanation coefficient Ɛ.

From Fig. 9.15, it can be concluded that, generally, higher cumulative mesopore volumes resulted in higher exhalation rates. Samples P1 and P2 were characterized with the highest volume of micropores (0.9 mm^3/g and 0.8 mm^3/g, respectively), which additionally enhanced their exhalation capacity. Correlation ($R^2 = 0.6$) was found between the specific surface area of the samples and their emanation coefficient Ɛ. The correlation is not high, likely induced by differences in pore size distribution meaning that also the pore size (and shape) partly determines the emanation. Despite, measurements of the specific surface area of porous materials containing naturally occurring radionuclides (and more specific ^{226}Ra), can provide useful information about their emanation. Compable correlation was also found in literature on alkali-activated materials incorporating phosphogypsum.

Since measurements were executed on dry materials under equal laboratory circumstances, and assuming a homogeneous ^{226}Ra distribution, the amount of emanated radon is dependent on the particle size and porosity. The bauxite residue shows lowest

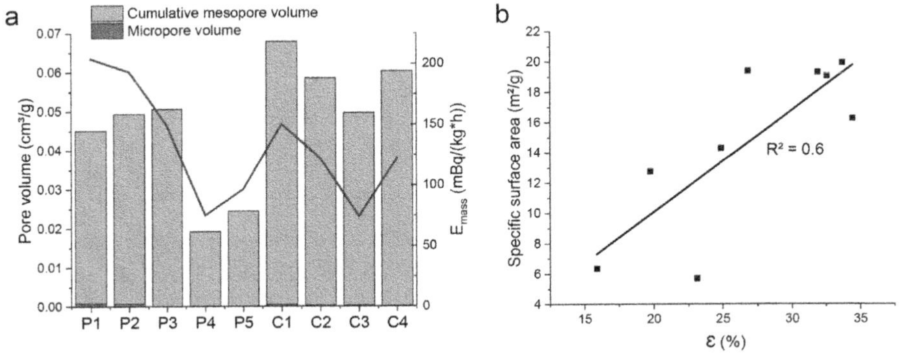

Figure 9.15 (a) Pore volume versus Emass and (b) emanation coefficient Ɛ versus specific surface area.

emanation, compared to alkali-activated mortar samples. Since the radon atom size is 0.38 nm, while the particle sizes of bauxite residue ranged from 0.05 to 350 μm, most emanated radon stayed inside the particles. Upon alkali activation, a part of the particles get dissolved, and consequently more radon can escape. A large variability in ε for the alkali-activated mortar samples can be observed from Fig. 9.14. It is anticipated that the porosity and specific surface area are inherently determining the release of radon, influenced by sample's mix design and casting technique.

9.4 Conclusions

The red mud is Al rich industrial by-product, able to be active component of cements and concretes compositions. However, it contains a lot of alkalis, heavy and hazardous metals, limiting its use in Ordinary Portland cement materials. Moreover, because of alumina production technology it has very high specific surface (1000 ... 1500 m^2/kg, Blaine), demanding a lot of water and reducing the properties of final product.

Use of red mud in alkali activated materials makes it possible to obtain cements with red mud content up to 60% by mass, meeting the requirements of cement Grades 32.5 and 42.5 at the same time. Also, red mud could be used as an aggregate in common application concretes, where red mud content could be 15 ... 40% by mass (350 ... 900 kg/m^3) for concrete classes C15 ... C40. For pressed materials red mud content could reach 90% by mass of the material, meeting the specified requirements for road base.

Various tests had showed that red mud containing alkali activated materials are ecologically clean, safe and environmentally friendly from the point of view of radiological safety (radioactivity, emanation of radon), and also could serve as a matrix for the immobilization/stabilization of heavy and hazardous metals, showing immobilization rates, close to 100%.

Thus in alkali activated materials, industrial wastes and by-products e.g., red mud loading could be up to the 95% by mass of material, creating a great source for recycling high amounts of red mud in the concrete production.

References

Achal, G., & Harinder, Y. (2015). Study of red mud as an alternative building material for interlocking block manufacturing in constructionIndustry. *International Journal of Materials Science and Engineering, 3*(4), 295–300. https://doi.org/10.17706/ijmse.2015.3.4.295-300. http://www.ijmse.net/uploadfile/2016/0113/20160113041604844.pdf

Alonso, M. M., Pasko, A., Gascó, C., Suarez, J. A., Kovalchuk, O., Krivenko, P., & Puertas, F. (2018). Radioactivity and Pb and Ni immobilization in SCM-bearing alkali-activated matrices. *Construction and Building Materials, 159*, 745–754. https://doi.org/10.1016/j.conbuildmat.2017.11.119

Balomenos, E., Gianopoulou, I., Panias, D., Paspaliaris, I., Perry, K., & Boufounos, D. (2011). Efficient and complete exploitation of the bauxite residue (red mud) produced in the Bayer Process. In, *3. Proceedings - European metallurgical conference* (pp. 745–758). EMC, 2011.

Boskovic, I., Vukcevic, M., Krgovic, M., Ivanovic, M., & Zejak, R. (2013). The influence of raw mixture and activators characteristics on red-mud based geopolymers. *Research Journal of Chemistry and Environment, 17*, 34–40.

Cablik, V. (2007). Characterization and applications of red mud from bauxite processing. *Gospodarka Surowcami Mineralnymi, 23*, 27–38.

Croymans, T., Schroeyers, W., Krivenko, P., Kovalchuk, O., Pasko, A., Hult, M., Marissens, G., Lutter, G., & Schreurs, S. (2017). Radiological characterization and evaluation of high volume bauxite residue alkali activated concretes. *Journal of Environmental Radioactivity, 168*, 21–29. https://doi.org/10.1016/j.jenvrad.2016.08.013

Deelwal, K., Dharavath, K., & Kulshreshtha, M. (2014). Evaluation of characteristic properties of red mud for possible use as a geotechnical material in civil construction. *International Journal of Advanced Engineering and Technology, 7*, 1053–1059.

Dentoni, V., Grosso, B., & Massacci, G. (2014). Environmental sustainability of the alumina industry in Western Europe. *Sustainability (Switzerland), 6*(12), 9477–9493. https://doi.org/10.3390/su6129477

Dimas, D. D., Giannopoulou, I. P., & Panias, D. (2009). Utilization of alumina red mud for synthesis of inorganic polymeric materials. *Mineral Processing and Extractive Metallurgy Review, 30*(3), 211–239. https://doi.org/10.1080/08827500802498199

Dodoo-Arhin, D., Konadu, D. S., Annan, E., Buabeng, F. P., & Yaya, A. (2013). Fabrication and characterisation of Ghanaian bauxite red mud-clay composite bricks for construction applications. *American Journal of Materials Science, 3*, 110–119. https://doi.org/10.5923/j.materials.20130305.02

European Commission. (2001). *Radiation protection 122 practical use of the concepts of clearance and exemption Part II, Application of the concepts of exemption and clearance to natural radiation sources.* https://ec.europa.eu/energy/sites/default/files/documents/122_part2.pdf

Gijbels, K., Krivenko, P., Kovalchuk, O., Pasko, A., Schreurs, S., Pontikes, Y., & Schroeyers, W. (2020). The influence of porosity on radon emanation in alkali-activated mortars containing high volume bauxite residue. *Construction and Building Materials, 230*. https://doi.org/10.1016/j.conbuildmat.2019.116982

Gu, H., Wang, N., & Liu, S. (2012). Radiological restrictions of using red mud as building material additive. *Waste Management & Research, 30*(9). https://doi.org/10.1177/0734242X12451308

Hairi, S. N. M., Jameson, G. N. L., Rogers, J. J., & MacKenzie, K. J. D. (2015). Synthesis and properties of inorganic polymers (geopolymers) derived from Bayer process residue (red mud) and bauxite. *Journal of Materials Science, 50*(23), 7713–7724. https://doi.org/10.1007/s10853-015-9338-9

Hajjaji, W., Andrejkovičová, S., Zanelli, C., Alshaaer, M., Dondi, M., Labrincha, J. A., & Rocha, F. (2013). Composition and technological properties of geopolymers based on metakaolin and red mud. *Materials and Design, 52*, 648–654. https://doi.org/10.1016/j.matdes.2013.05.058

He, J., Yu, Y., & Zhang, G. (2012). The strength and microstructure of two geopolymers derived from metakaolin and red mud-fly ash admixture: a comparative study. *Construction and building materials, 30*, 80–91.

He, J., Zhang, J., Yu, Y., & Zhang, G. (2012a). The strength and microstructure of two geopolymers derived from metakaolin and red mud-fly ash admixture: A comparative study. *Construction and Building Materials, 30*, 80–91. https://doi.org/10.1016/j.conbuildmat.2011.12.011

Hooda, A., Malik, P., & Sangwan, V. (2016). Research paper on gauging engineering properties of red mud-fly ash-marble dust amalgamation. *International Journal of All Research Education and Scientific Methods, 4*, 252–258.

Hua, Y., Heal, K. V., & Friesl-Hanl, W. (2017). The use of red mud as an immobiliser for metal/metalloid-contaminated soil: A review. *Journal of Hazardous Materials, 325*, 17–30. https://doi.org/10.1016/j.jhazmat.2016.11.073

Ke, X., Bernal, S. A., Ye, N., Provis, J. L., & Yang, J. (2015). One-part geopolymers based on thermally treated red Mud/NaOH blends. *Journal of the American Ceramic Society, 98*(1), 5–11. https://doi.org/10.1111/jace.13231

Ke, X., Ye, N., Bernal, S. A., Provis, J. L., & Yang, J. (2014). Preparation of one-pai geopolymer from thermal alkali activated bauxite red mud. In *Proc. Second Inl Conf. On advances in chemically-activated materials*, 204–211.

Komnitsas, K., & Zaharaki, D. (2009). Utilisation of low-calcium slags to improve the strength and durability of geopolymers. In *Geopolymers: Structures, processing, properties and industrial applications* (pp. 343–375). Elsevier Ltd. https://doi.org/10.1533/9781845696382.2.343

Krivenko, P., Kovalchuk, O., Pasko, A., Croymans, T., Hult, M., Lutter, G., Vandevenne, N., Schreurs, S., & Schroeyers, W. (2017). Development of alkali activated cements and concrete mixture design with high volumes of red mud. *Construction and Building Materials, 151*, 819–826. https://doi.org/10.1016/j.conbuildmat.2017.06.031

Kumar, A., & Kumar, S. (2013). Development of paving blocks from synergistic use of red mud and fly ash using geopolymerization. *Construction and Building Materials, 38*, 865–871. https://doi.org/10.1016/j.conbuildmat.2012.09.013

Liu, W., Chen, X., Li, W., Yu, Y., & Yan, K. (2014). Environmental assessment, management and utilization of red mud in China. *Journal of Cleaner Production, 84*(1), 606–610. https://doi.org/10.1016/j.jclepro.2014.06.080

Liu, W., Yang, J., & Xiao, B. (2009). Review on treatment and utilization of bauxite residues in China. *International Journal of Mineral Processing, 93*(3–4), 220–231. https://doi.org/10.1016/j.minpro.2009.08.005

Liu, X., & Zhang, N. (2011). Utilization of red mud in cement production: A review. *Waste Management and Research, 29*(10), 1053–1063. https://doi.org/10.1177/0734242X11407653

Mukiza, E., Zhang, L. L., Liu, X., & Zhang, N. (2019). Utilization of red mud in road base and subgrade materials: A review. *Resources, Conservation and Recycling, 141*, 187–199. https://doi.org/10.1016/j.resconrec.2018.10.031

National Bureau of Statistics of China. (2017). Statistical Communiqué of the People's Republic of China on the 2016 National Economic and Social Development. http://www.stats.gov.cn/english/PressRelease/201702/t20170228_1467503.html

Palomo, Á., Kavalerova, E., Fernández-Jiménez, A., Krivenko, P., García-Lodeiro, I., & Maltseva, O. (2015). *A review on alkaline activation: New analytical perspectives.*

Pan, Z., Li, D., Yu, J., & Yang, N. (2003). Properties and microstructure of the hardened alkali-activated red mud-slag cementitious material. *Cement and Concrete Research, 33*(9), 1437–1441. https://doi.org/10.1016/S0008-8846(03)00093-0

Patent of USSR 979294, (1982), Krivenko P., Pushkareva E., Rumunova G. 2977103/29-33, 07.12.1982.

Patent of Ukraine UA 10286 A, (1996), Krivenko P., Petropavlovsky O., Rostovskaya G., C 04 B7/06, 7/06, 25.12.1996.

Pascual, J., Corpas, F., Lopez-Beceiro, J., Benttez-Guerrero, M., & Artiaga, R. (2009). Thermal characterization of a Spanish red mud. *Journal of Thermal Analysis and Calorimetry, 96*(2), 407–412.

Pérez-Villarejo, L., Corpas-Iglesias, F. A., Martínez-Martínez, S., Artiaga, R., & Pascual-Cosp, J. (2012). Manufacturing new ceramic materials from clay and red mud derived from the aluminium industry. *Construction and Building Materials, 35*, 656–665. https://doi.org/10.1016/j.conbuildmat.2012.04.133

Pontikes, Y., Nikolopoulos, P., & Angelopouios, G. (2007). Thermal behaviour of clay mixtures with bauxite residue for the production of heavy-clay ceramics. *Journal of the European Ceramic Society, 27*, 1645–1649.

Power, G., Gräfe, M., & Klauber, C. (2009). *Review of current bauxite residue management, disposal and storage: Practices, engineering and science.*

Pulford, I. D., Hargreaves, J. S. J., Ďurišová, J., Kramulova, B., Girard, C., Balakrishnan, M., Batra, V. S., & Rico, J. L. (2012). Carbonised red mud - a new water treatment product made from a waste material. *Journal of Environmental Management, 100*, 59–64. https://doi.org/10.1016/j.jenvman.2011.11.016

Reddy, N. G., & Chandra, K. S. (2014). Characterization and comprehensive utilization of red mud - an overview. *IJSRD - International Journal for Scientific Research and Development, 2*(1), 670–673.

Ribeiro, D. V., Labrincha, J. A., & Morelli, M. R. (2011). Potential use of natural red mud as pozzolan for Portland cement. *Materials Research, 14*(1), 60–66. https://doi.org/10.1590/S1516-14392011005000001

Rostovskaya, G. S. (1994). Alkaline binders based on bauxite residues. *First Int.Conference on Alkaline Cements and Concretes, 1*, 329–346.

Sglavo, V., Campostrini, R., Maurina, S., Carturan, G., Monagheddu, M., Budroni, G., & Cocco, G. (2000). Bauxite red mud in the ceramic industry. *Journal of the European Ceramic Society, 20*, 235–244.

Sridevi, G., Sanjeet, S., & Subhrajeet, S. (2016). *Stabilization of expansive soil with red mud and lime*. Indian Geotech.

Sutar, H. (2014). Progress of red mud utilization: An overview. *American Chemical Science Journal*, 255–279. https://doi.org/10.9734/ACSJ/2014/7258

Tsakiridis, P. E., Agatzini-Leonardou, S., & Oustadakis, P. (2004). Red mud addition in the raw meal for the production of Portland cement clinker. *Journal of Hazardous Materials, 116*(1–2), 103–110. https://doi.org/10.1016/j.jhazmat.2004.08.002

Vukčević, M., Turović, D., Krgović, M., Bošković, I., Ivanović, M., & Zejak, R. (2013). Utilization of geopolymerization for obtaining construction materials based on red mud. *Materiali in Tehnologije, 47*(1), 99–104. http://mit.imt.si/Revija/izvodi/mit131/vukcevic.pdf

Yang, J., & Xiao, B. (2008). Development of unsintered construction materials from red mud wastes produced in the sintering alumina process. *Construction and Building Materials, 22*(12), 2299–2307. https://doi.org/10.1016/j.conbuildmat.2007.10.005

Yang, J., Zhang, D., Hou, J., He, B., & Xiao, B. (2008). Preparation of glass-ceramics from red mud in the aluminium industries. *Ceramics International, 34*(1), 125–130. https://doi.org/10.1016/j.ceramint.2006.08.013

Ye, N., Yang, J., Ke, X., Zhu, J., Li, Y., Xiang, C., Wang, H., Li, L., & Xiao, B. (2014). Synthesis and characterization of geopolymer from bayer red mud with thermal pretreatment. *Journal of the American Ceramic Society, 97*(5), 1652–1660. https://doi.org/10.1111/jace.12840

Ye, N., Yang, J., Liang, S., Hu, Y., Hu, J., Xiao, B., & Huang, Q. (2016). Synthesis and strength optimization of one-part geopolymer based on red mud. *Construction and Building Materials, 111*, 317–325. https://doi.org/10.1016/j.conbuildmat.2016.02.099

Zhang, M., El-Korchi, T., Zhang, G., Liang, J., & Tao, M. (2014). Synthesis factors affecting mechanical properties, microstructure, and chemical composition of red mud-fly ash based geopolymers. *Fuel, 134*, 315–325. https://doi.org/10.1016/j.fuel.2014.05.058

Zhang, G., He, J., & Gambrell, R. P. (2010). Synthesis, characterization, and mechanical properties of red mud-based geopolymers. *Transportation Research Record, 2167*, 1–9. https://doi.org/10.3141/2167-01

Life cycle assessment for structural and non-structural concrete

10

Karoline Figueiredo and Assed Haddad
Programa de Engenharia Ambiental (PEA), Universidade Federal do Rio de Janeiro (UFRJ), Rio de Janeiro, Brazil

10.1 Introduction

As the most used construction material worldwide and one of the most commonly used structural materials, concrete has been increasingly investigated regarding environmental and sustainability assessments. Compared to other construction materials, concrete can have a reduced environmental impact per kilogram (Kleijer et al., 2017). However, the massive volume of concrete and cement used in construction makes the concrete industry responsible for a large portion of global CO_2 emissions and other significant environmental impacts. In addition to carbon emissions, concrete production raises other concerns, such as the waste associated with this production, the large consumption of water, and natural resource depletion.

Some possible solutions have been discussed in the literature and applied in several countries, such as the use of recycled aggregates from construction and demolition waste. However, despite the critical role that these recycled aggregates play in the sustainability of the built environment, it is clear that, in general, the compressive strength, the flexural strength, and splitting tensile strength of concrete that uses these aggregates tend to be lower (Wang et al., 2021). Therefore, it is observed that the use of recycled aggregates occurs mainly in non-structural concretes, but the interest in green concretes for structural use has grown and become more realistic in recent years.

Unfortunately, even with the use of recycled aggregates in several countries, the concrete production process remains one of the biggest consumers of virgin raw material. Furthermore, the most significant impact lies in the production of Portland cement, an indispensable ingredient in concrete. Cement, whose consumption is close to two billion tons per year, generates a large volume of rock extraction and earth movement and directly influences global warming and the greenhouse effect. However, in order to assess the environmental impacts generated by concrete, it is not enough to assess energy use and greenhouse gas emissions of the production process. Instead, a holistic assessment of the impacts generated is necessary, considering all stages of the concrete life cycle: acquisition of raw materials, concrete production, concrete use in construction, operation and maintenance, repair, demolition, recycling, and waste disposal.

Recycled Concrete. https://doi.org/10.1016/B978-0-323-85210-4.00007-2

In this context, the Life Cycle Assessment (LCA) methodology seems to be an adequate tool to assess the potential environmental impacts of structural and non-structural concretes. According to the standard ISO 14040, the Environmental Life Cycle Assessment (E-LCA) methodology, usually referred to just as Life Cycle Assessment (LCA), is the compilation of inputs, outputs, and potential environmental impacts of a product system throughout its life cycle (ISO, 2006). This methodology is widely applied in the construction industry to assess the environmental impacts of construction materials (Saade et al., 2020) and is very useful for improving the decision-making process about concrete and its aggregates (Kurda et al., 2018).

A four-phase framework represents the LCA, namely: (i) Goal and Scope definition; (ii) Life Cycle Inventory (LCI); (iii) Life Cycle Impact Assessment (LCIA); and (iv) Interpretation. Briefly, it can be said that the first phase involves defining the goal and the scope by determining different aspects of the study, such as the functional unit, the system boundary (i.e., cradle-to-grave, cradle-to-gate, gate-to-gate, gate-to-grave), and the assumptions and limitations to be considered (Figueiredo et al., 2021). In turn, the LCI phase represents the compilation and quantification of inputs and outputs for the chosen functional unit regarding the chosen system boundary (Crawford et al., 2018), and the LCIA phase consists of evaluating the magnitude and significance of the potential environmental impacts (Santos et al., 2021). Lastly, the interpretation phase comprises identifying and assessing all the information from the previous stages so that the results can be communicated to interested parties and that they can be used to aid decision-making (Hollberg et al., 2021). Each of these phases will be further detailed and explained in the following sections.

Although the LCA methodology is already standardized and widely discussed in several industries, it is clear that its application in the construction sector still needs further advances. This is also very true when referring specifically to the concrete industry. For example, a recent landmark study conducted by the International Institute for Sustainable Development (IISD) has shown that LCA is the best approach to measuring the carbon emissions of construction products at each stage of their life cycle (Stiebert et al., 2019). However, the report also discusses the need for more data, transparency, and robust LCA standards for the built environment. Therefore, this chapter intends to help professionals linked to the concrete industry make use of this powerful tool, fully understanding the concepts and opportunities generated from the LCA application.

The application of LCA for concrete is interesting for several target audiences. The environmental analyses are critical for policymakers to adopt conscious public policies and for ecological standards developers to create realistic standards for sustainable buildings. Besides, these analyses are also very useful for builders, civil engineers, and architects. With a reliable database and representative results of their regions and projects, professionals can minimize the environmental impacts of their constructions more easily. Furthermore, with the dissemination of this information and the growing awareness of the global population about the importance of sustainable development, end users are increasingly interested in obtaining green buildings and can benefit from LCA studies for concrete.

It is important to stress that the LCA is not a decision-maker per se but rather a tool to provide useful information to interested parties using a standardized and transparent approach (Corominas et al., 2020). For the LCA to be a handy tool to minimize the environmental impacts of structural and non-structural concretes, it is essential that the professional knows how to correctly choose which approach to use according to the objective of the study, what level of detail is required in each case, what software and databases to use and what functional unit and system boundaries should be more representative. Therefore, these concepts will be deepened in the following sections so that the reader can use them as a reference for future studies.

10.2 LCA phases and related concepts

The LCA methodology has its structure standardized by the ISO 14040 series. Until 2006, this series consisted of the following standards: ISO 14040, which dealt with general principles and guidelines; ISO 14041, aimed at the phases of defining the objective and scope and analyzing the life cycle inventory; ISO 14042, regarding the life cycle impact assessment phase; and ISO 14043, focused on life cycle interpretation. As of 2006, the technical standards were compressed into only two standards, which are currently responsible for guiding LCA practitioners worldwide: ISO 14040 and ISO 14044. The first deals with the principles and structure of LCA (ISO, 2006), while the second addresses its requirements and guidelines (ISO - International Organization for Standardization, 2006).

There are also international technical reports and a technical specification published by ISO, which can be used in conjunction with the ISO 14040 series to guide LCA applications: ISO/TR 14047 (ISO 2012a,b), ISO/TS 14048 (ISO, 2002), and ISO/TR 14049 (ISO, 2012a,b). The two technical reports aim to provide examples to illustrate the current practice of life cycle inventory analysis (LCI) and life cycle impact assessment (LCIA) according to ISO 14044. In turn, the technical specification ISO/TS 14048 provides the requirements and a structure for a data documentation format to be used in LCA studies.

This section will be divided into four main parts related to the four phases of the LCA framework and will be based on these ISO standards and reports. The order of the concepts to be explained was chosen to build with the reader a comprehensive discussion from the theoretical aspects to a practical application of LCA for structural and non-structural concretes.

10.2.1 Goal and scope

The first phase of LCA corresponds to the goal and scope definition. The aspects defined at this point depend on the subject and the intended use of the LCA study. It can be said that the objective of applying the LCA is linked to one of two main aspects: the practitioner hopes to identify opportunities to improve the environmental performance of a given product, or it is expected to make an environmental comparison between different products that fulfill equivalent functions. In both cases, LCA has proven to be a valuable tool to benefit the decision-making process about products and services.

Regarding concrete analysis, the goal of each study can vary enormously: it can be choosing the best concrete mixture proportions for a specific purpose, defining the carbon footprint for concrete structures during a given stage of a building life cycle, or the definition of how the concrete production process could be improved to reduce a specific impact category, among several other possibilities. According to ISO 14040 standard, the study's goal has to communicate the intended LCA application and the target audience (ISO, 2006). This is essential for the LCA to help the future decision-making process, as knowing the intended audience ensures that the professional will choose the most coherent impact categories. With the objective well defined, it is necessary to define the scope of the analysis.

For the scope definition, the ISO 14040 standard suggests that some crucial aspects be determined for the complete realization of the study. First, the practitioner must define the product system to be evaluated with its respective functions. The product system corresponds to the product and its upstream and downstream processes, including components manufacturing, distribution, product use, and final disposal. Besides, it is essential to consider all transportation and energy used during the process because this information has to be part of the product system. Fig. 10.1 presents an example of the processes that could be considered during a concrete LCA analysis.

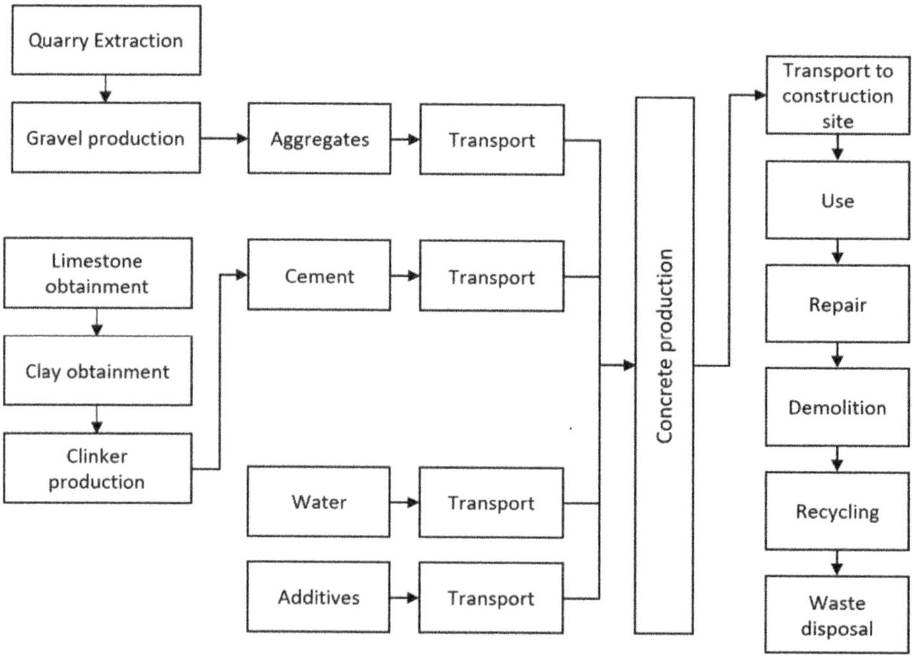

Figure 10.1 Processes associated with the concrete life cycle.

The practitioner must define the level of detail required to determine the processes that constitute the product's life cycle to be analyzed. There is no correct answer to this definition; the more detailed the processes, the more realistic the analysis, but the more labor-intensive the data collection and its evaluation. The level of detail chosen and the assumptions adopted must be clearly communicated so that this does not interfere with interpreting the results at the end of the study.

Another example of a product system is shown in Fig. 10.2. In this second example, the processes necessary for the construction of a reinforced concrete building are considered. Therefore, the processes that involve not only the production of concrete but also steel and wood formwork are exposed. Note that recycling processes were not considered in this case. The formwork could be reused for other constructions and, after the building's demolition, the concrete could be used as recycled aggregates. This level of detail is at the discretion of the LCA practitioner, who must make all decisions in an informed manner.

A final example is given, now regarding the concrete recycling process. In this case, the professional considers that after the building is demolished, there will be an on-site sorting of the construction and demolition (C&D) waste, with subsequent processing of the recycled concrete. Considering that it will not be possible to recycle or reuse all the waste generated, a number of materials will be transported to landfills. This product system is represented in Fig. 10.3.

Then, it has to be determined the functional unit and the reference flows. The functional unit refers to a quantified description of the product that serves as the reference

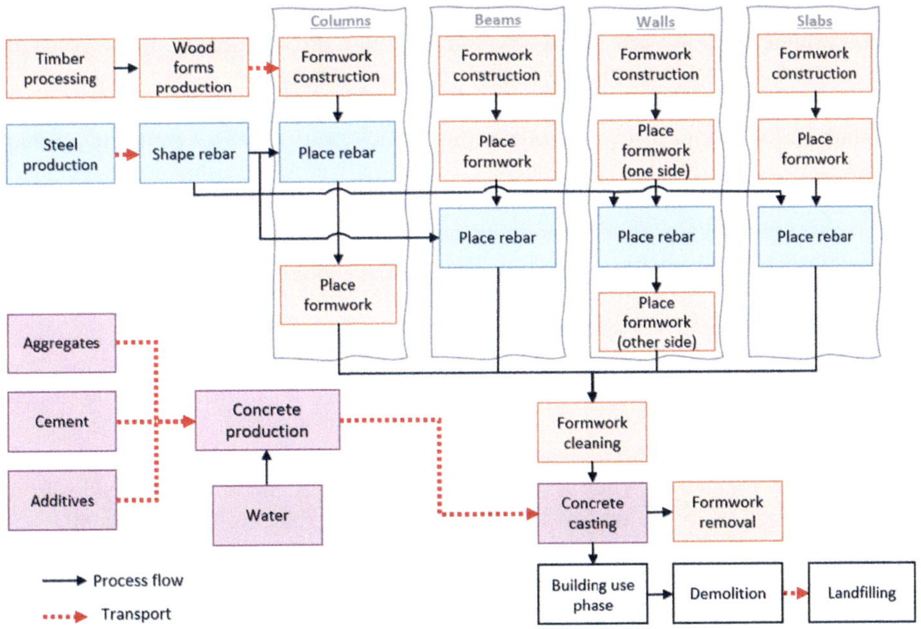

Figure 10.2 Processes associated with the construction of a building using reinforced concrete.

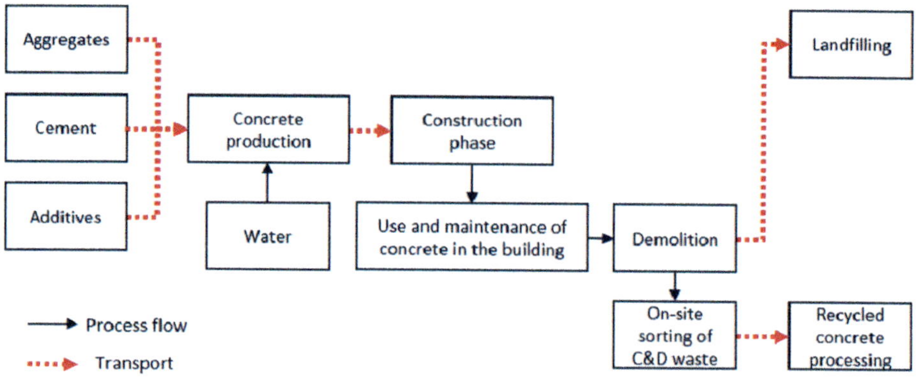

Figure 10.3 Life cycle of concrete, considering the recycling stage.

basis, while the system boundary determines which elementary processes should be included in the study. Regarding the concrete analysis, the functional unit could be one cubic meter (1 m^3) of concrete with a given compressive strength, for example. In this case, the environmental impacts will be calculated concerning this volume of concrete with this specification. Instead, the practitioner could decide to use one square meter (1 m^2) of construction or 1 kilometer (1 km) of bridge as the functional unit. This decision needs to make sense for the study and be consistent with the objectives already defined. On the other hand, the reference flow represents the amount of product necessary to fulfill the established function. For example, if the professionals choose one square meter for ready-mixed concrete as the functional unit, they can analyze different reference flows that fulfill the same function: different ready-mixed concrete elements and products to make one square meter of concrete.

From these first definitions, it remains for the practitioner to determine the system boundary in the study. It means to determine which unit processes from the product system shall be included in the analysis. Ideally, all processes should be included. Nevertheless, this decision is often neither possible nor practical due to data and cost constraints. Again, this decision needs to be made consciously and must be communicated to all interested parties. The system boundary definition needs to be done to answer the questions that interest the target audience completely.

A possible example of a system boundary would be considering only the processes associated with the concrete production without going through the construction, use, and end-of-life phases. This type of system boundary is known as cradle-to-gate because the assessment is based on a partial product life cycle from resource extraction (cradle) to the factory gate (i.e., before being transported to the construction site). It is represented in Fig. 10.4. When the analysis contemplates the entire life cycle, from resource extraction to the end-of-life phase, it is said that the system boundary is cradle-to-grave. It is represented in Fig. 10.5.

In most of the cement and concrete LCAs found in the literature, a cradle-to-gate approach is usually considered, including in the system boundary only the main processes related to producing these materials. For example, regarding cement production,

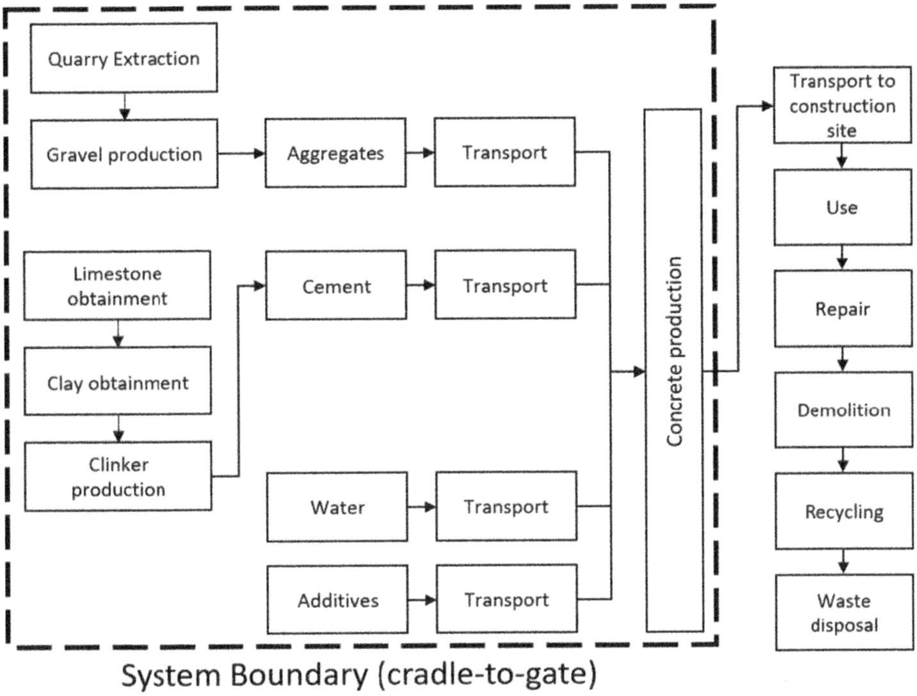

Figure 10.4 Example of a cradle-to-gate system boundary for a concrete LCA.

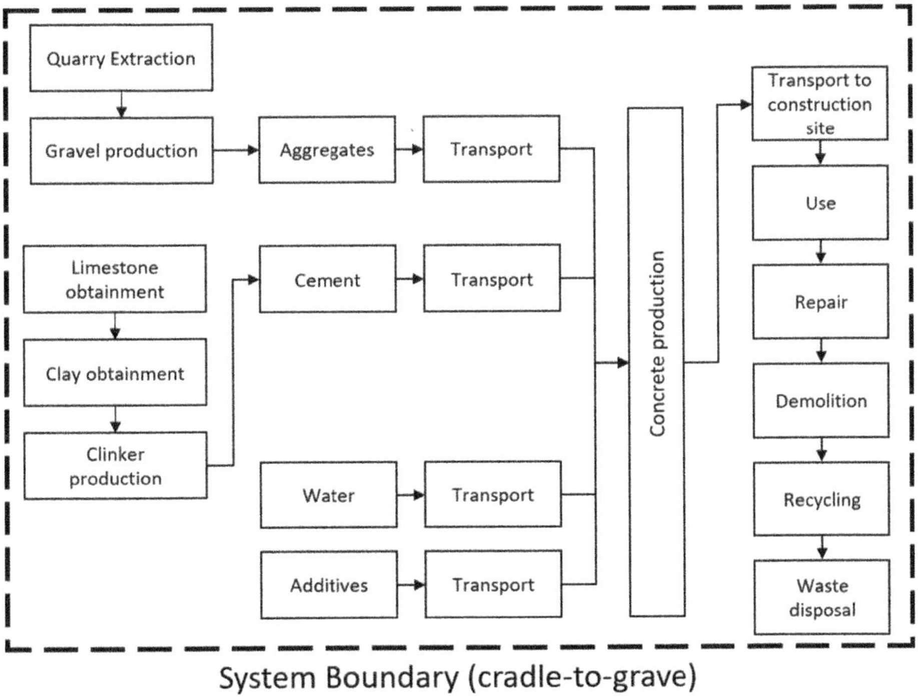

Figure 10.5 Example of a cradle-to-grave system boundary for a concrete LCA.

many studies consider the following processes: extraction of raw materials, raw material preparation, pyroprocessing, clinker cooling, finish milling, packaging handling, and transportation (Petek Gursel et al., 2014).

It is noteworthy that it is up to the practitioner to define which parts of the system will be left out, and often this decision is made based on quantitative cut-off criteria. It means that processes whose cumulative contribution in terms of total mass or energy is less than some certain percentage defined by the practitioner must be excluded from the system. However, this method can lead to distortion of LCA results, impairing the interpretation of the study. Even though the environmental impacts can be considered small concerning the studied functional unit, the impacts generated by these neglected processes can represent a large number when global production volumes are considered. To avoid this, it is recommended to consider each process's environmental relevance and conduct sensitivity analyses to increase the study's reliability.

During the process of defining the scope, the practitioner may perceive that some allocation method will be needed. This is because some processes are responsible for producing more than one product, and in this case, it may be necessary to divide the environmental impacts of this process among its products. Therefore, allocation in LCA refers to the partition of the input or output flows between the product system under study and other product systems. The ISO 14044 standard establishes some general considerations and procedures regarding allocation. For example, allocation procedures for reuse and recycling should consider as a basis for allocation some physical property (i.e., mass), economic value, or the number of subsequent uses of the recycled material (ISO - International Organization for Standardization, 2006).

Although it is preferable to avoid the use of impact allocation, it is sometimes impossible or not realistic to decide not to use it, as is the case of utilizing by-products as aggregates to improve the concrete's characteristics and make it more suitable for a specific use. In this particular case, the practitioner needs to analyze whether or not it would be more appropriate to allocate a part of the environmental impacts of these products to the concrete production (van den Heede & de Belie, 2012). In order to define an adequate allocation procedure, it is important to be able to clearly distinguish which are the primary production processes, responsible for producing the main product and the by-products, and the secondary processes, which aim to treat the by-product to become suitable for its use as a concrete component (Chen et al., 2010).

Finally, it is essential that the practitioner report all limitations and assumptions to be adopted in the study so that the scope of the LCA is clear and well defined. The ISO 14040 standard also suggests that the impact assessment and interpretation methods should be defined at this phase, as well as the impact categories to be used. Then, with the objective and scope determined, it proceeds to the inventory phase.

10.2.2 Life cycle inventory (LCI)

The second phase of LCA corresponds to data collection and calculation procedures to quantify a product system's relevant inputs and outputs. This process is considered iterative because as data is collected, knowledge about the system increases, and changes may occur. These changes may be linked to new requirements or limitations of the study, or even changes in data collection procedures so that the study objectives can still be met.

The inventory can be considered as the basis of LCA. All essential data related to the production chain are quantified at this phase so that analyses can occur in the following phases. Consequently, it is considered that most of the limitations found in LCA studies occur at this stage of data collection. Often, the practitioner is faced with the unavailability of specific and reliable sources of information to carry out the study. This is because it is not always possible to conduct the entire study using only primary data obtained from direct measurements. It may also be necessary to use secondary data.

Different data types can be used to build the life cycle inventory, such as on-site data, data obtained from the literature, international databases, or data provided by third parties, such as companies, government agencies, trade associations, and analysis laboratories. Data quality requirements are determined by the geographical, temporal, and technological coverage adopted for the study. The ISO 14044 standard states that, in practice, all data used in the study may include a mixture of measured, calculated, or estimated data (ISO - International Organization for Standardization, 2006). For example, in a published study on the application of LCA to compare the impacts of recycled and ordinary concretes, the authors found that some company-specific processes resemble those modeled by an international database, but raw data information is more detailed (Kleijer et al., 2017). Therefore, in this study, all essential data relating to materials and processes were obtained directly by the company, but the professionals used background environmental data from the international database.

10.3 Databases available

There are more than 40 national and international databases currently available. These databases vary in terms of the amount of data and territorial coverage. Some of the most important ones are highlighted in Fig. 10.6.

Unfortunately, not all of the available databases have information on building materials (Martínez-Rocamora et al., 2016). Among the most used databases to support life cycle analysis in the construction sector, Ecoinvent, GaBi, Athena, and U.S. Life Cycle Inventory Database can be mentioned. These four databases provide information on concrete and cement but with different degrees of detail and coverage. The advantage of the Ecoinvent and GaBi databases is that they are representative of different countries. The Ecoinvent database is one of the most comprehensive international LCI databases, with 18.000 LCI datasets in many areas such as energy supply, agriculture, transport, construction materials, wood, and waste treatment (Wernet et al., 2016). Users can consult the data online, download the data directly, or use the database in an LCA-based software. In turn, GaBi Databases offer over 15,000 datasets based on primary data collection worldwide in different areas such as agriculture, building and construction, chemicals, and energy (Sphera). Also, GaBi presents regionalized water and land use data that can be used with the practitioner's regionalized data.

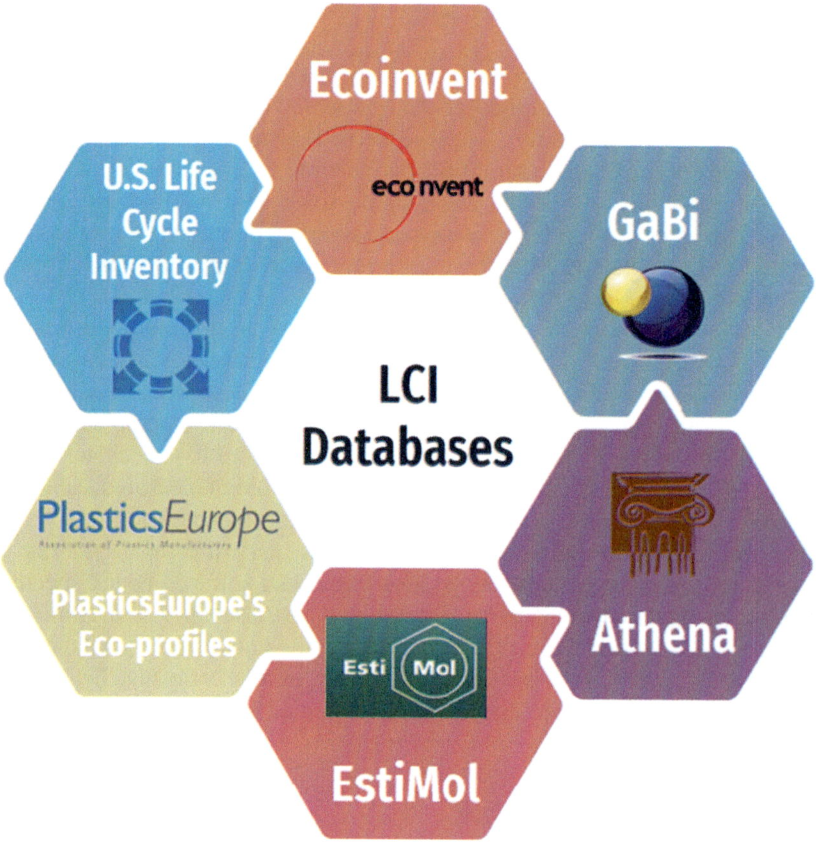

Figure 10.6 LCI databases available.

On the other hand, the Athena and U.S. Life Cycle Inventory databases present data from specific countries. Regarding the Athena database, the information offered to the users includes data for construction materials, energy, transport, construction and demolition processes, maintenance, repairing, and waste disposal (Martínez-Rocamora et al., 2016). However, the information is specifically about the manufacturing process in Canada and the United States. Finally, the U.S. Life Cycle Inventory database considers the input and output flow of materials specifically about the United States. This database focuses on metals, wood materials, and plastics but also presents information about cement and concrete product manufacturing (NREL).

10.4 LCA-based software tools

Data research can be time-consuming and, therefore, many professionals resort to some software based on the LCA methodology. Some companies offer a combination

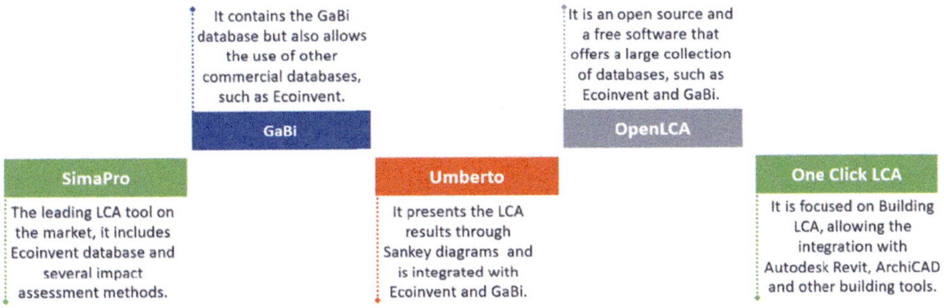

Figure 10.7 Examples of LCA software tools.

of an LCI database and a tool for assessing environmental impacts. As an example of this, we have GaBi Solutions, offering a powerful LCA engine and an international LCI database. Other companies offer computational tools based on the LCA framework that accept the insertion of different LCI databases. Thanks to agreements between the providers of these software tools and the institutions that make different LCI databases, different LCA-based programs can act as resellers for these databases. Fig. 10.7 shows the most used LCA-based software tools worldwide.

Studies show that sometimes differences are identified in the LCA results, depending on which software and database are used (Herrmann & Moltesen, 2015). This issue needs to be increasingly debated among software developers and all LCA professionals. This concern proves the importance of careful analysis and interpretation by LCA practitioners during the whole process.

10.5 Life cycle impact assessment (LCIA)

In the LCIA phase, the elementary flows and all data collected in the inventory phase are converted into environmental impacts. It is from the LCIA results that it will be possible to conduct the interpretation of the study. Therefore, the ISO 14040 and ISO 14044 standards suggest some steps to be followed so that the impact assessment can take place reliably: selection, classification, characterization, normalization, grouping, and weighting. These standards also determine that three of these steps are mandatory (i.e., selection, classification, and characterization), while the remaining three are optional. These steps are represented in Fig. 10.8.

Regarding the mandatory parts of an LCIA, the first step is to select impact categories, indicators, and characterization models. This decision has to be made following the goal defined at the beginning of the study. For most LCA studies for concrete, the practitioners choose impact categories widely discussed in the literature, as is the case of recent studies that considered global warming potential, acidification potential, and ozone depletion potential (Liu et al., 2021; Ntimugura et al., 2021). However, this decision is up to the practitioner and must satisfy the objective and scope defined for the LCA. If this is not possible through existing impact categories, new categories are

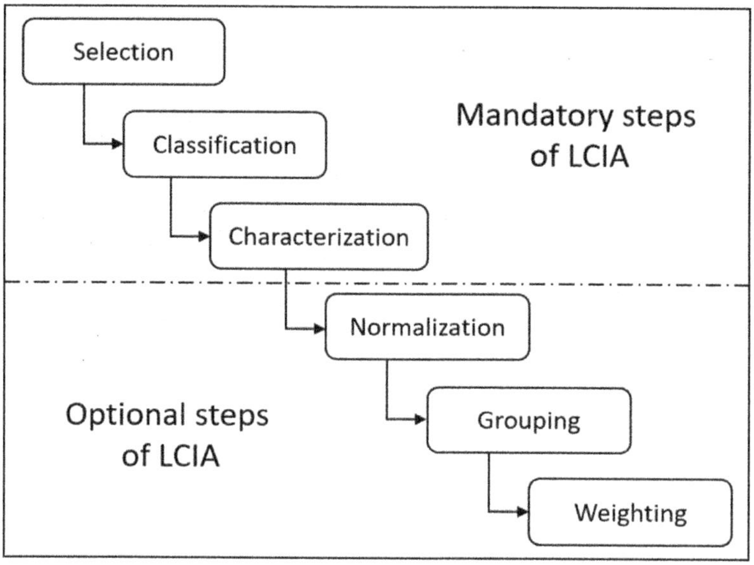

Figure 10.8 Steps of the LCIA phase.

defined, making it necessary to create new characterization models for those indicators.

The second mandatory step is classification, which corresponds to the assignment of LCI results to the selected impact categories. Finally, the third mandatory step refers to the characterization, which indicates calculating the category indicator results by multiplying the amount found in LCI with a characterization factor. The result of this calculation is a numerical result that represents the indicator (ISO - International Organization for Standardization, 2006). For this, the practitioner uses categorization models based on a specific method of impact assessment.

The main difference between the existed impact assessment methods is related to which stage of the cause-effect chain will be examined to calculate the impact. In this way, methods are classified into midpoint or endpoint. A midpoint method analyzes the impacts that occur earlier along the cause-effect chain. On the other hand, an endpoint method analyzes the environmental impact at the end of this chain, being more focused on the damages that occur with the production of the analyzed product. Therefore, it can be said that midpoint methods quantify problem-oriented impacts, while endpoint methods quantify damage-oriented impacts associated with the analyzed process.

For example, regarding concrete analysis, the professional could use a midpoint method, choosing impact categories such as ozone depletion, water scarcity, and climate change. On the other hand, if it makes more sense to use an endpoint method according to the study's objective and the target audience, the professional can consider impact categories related to human health, ecosystem health, or resource availability. It is noteworthy that the calculation of endpoint impact categories is usually associated with greater uncertainties when compared to midpoint indicators (Wu & Su, 2020). In addition, the practitioner may decide to assess midpoint and endpoint impacts in the same study as long as this makes sense for the analysis.

There are several LCIA methods available, each corresponding to a set of impact categories. These methods can contain a midpoint approach, an endpoint approach, or both. The practitioner must choose the method that best suits their needs for each new LCA study. For some widely used impact categories, many LCIA methods use the same units. For example, all midpoint methods consider CO_2-equivalents (CO_2eq) as the unit for global warming potential, representing the number of metric tons of CO_2 emissions with the same global warming potential as one metric ton of another greenhouse gas. However, the use of standardized units is not valid for all existing impact categories. It is crucial, thus, to deeply understand how the chosen method performs its characterization calculations and what each unit means. Some of the most used LCIA methods around the world are shown in Fig. 10.9.

Another important point about choosing which characterization method to use is that this choice is directly linked to the study's uncertainties. Commonly, a professional chooses one method indiscriminately, without comparing results from different methods. Besides, this choice is generally left to the discretion of the user in an LCA software. Thus, inappropriate methods for a specific situation can be chosen, directly influencing the analysis and interpretation of data. An example of this would be using an LCIA method with inappropriate global warming time intervals in a study with a specific time scope (Chen et al., 2021). These decisions need to be made consciously by the practitioner, always considering the objective and scope outlined at the beginning of the analysis.

After characterizing the impacts, the professional can choose to follow the optional LCIA steps: normalization, grouping, and weighting. The normalization step is beneficial in LCA studies because the results found for the indicators are often reported in different metrics, making the interpretation phase more difficult. In this way, the normalization process helps the analysis by presenting the relative magnitudes of

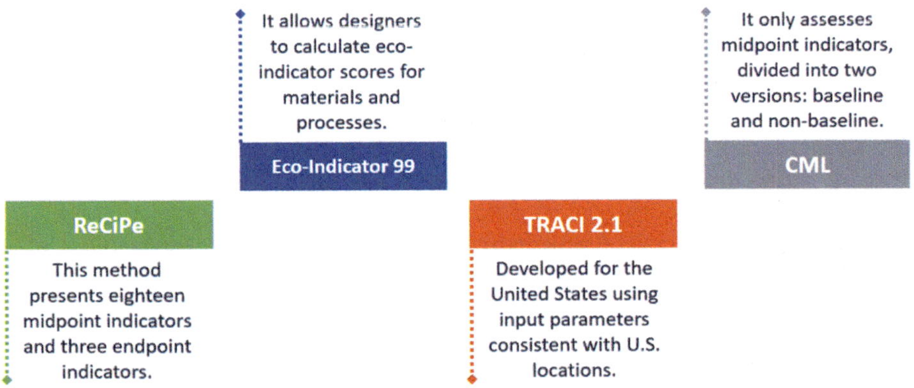

Figure 10.9 Different LCIA methods used worldwide.

the impact indicators, all on the same scale. Different approaches can be used in this step, and they are divided into internal and external normalization approaches (Laurent & Hauschild, 2015). Internal normalization brings non-commensurate impact indicator scores into a standard metric, while the external one uses an external reference system and generally at a larger scale.

With the normalized results, the practitioner can still decide to follow the other two optional steps. The grouping step refers to assembling impact categories into one or more groups. The impact categories analyzed can, for example, be grouped according to a given hierarchy that makes sense for the target audience of that study. Finally, the LCA practitioner may decide to weigh the results. The weighting step corresponds to apply weighting factors that represent the importance assigned to each impact category. With that, the LCIA phase ends.

10.6 Interpretation

Interpretation is the last phase of LCA, where the questions posed in the goal definition are answered. The results obtained throughout the study cannot be interpreted anyhow; the life cycle interpretation is a systematic technique to identify, quantify, check and evaluate information from LCI and LCIA phases (ISO, 2006). As a result of this phase, the practitioner can determine the level of confidence in the study's results and accurately communicate them (Cao, 2017).

To start the interpretation process, the practitioner must keep in mind that it is essential to understand the accuracy of the results, ensuring they meet the LCA's goal. For this, the professional has to evaluate the sensitivity of the data used, assess the study's completeness and consistency, and draw conclusions and recommendations (Hernandez et al., 2018). The vast importance of this phase is related to the possibility of transforming the LCI and LCIA results into comparable and comprehensible ones (Vieira et al., 2016).

A fundamental topic to emphasize is that, although the interpretation corresponds to the last phase of the LCA, this phase can accompany all the previous ones. The professional can start to perform the life-cycle interpretation from the goal and scope definition phase. An example of this is when the definition of the functional unit and the system boundary is not so intuitive. In this situation, the practitioner needs to check and evaluate the proposals before the continuation of LCA. Another recurrent situation is analyzing the quality of data collected at the end of the LCI phase before continuing the study. This would also be another example of when the interpretation phase occurs concurrently with the other phases of the study.

For this reason, the ISO 14040 standard presents a conceptual framework of LCA in which the interpretation phase is directly linked to the other three phases of the methodology. Based on the image found in ISO 14040 standard, Fig. 10.10 is presented in this work to facilitate the reader's understanding and to condense all the knowledge discussed so far.

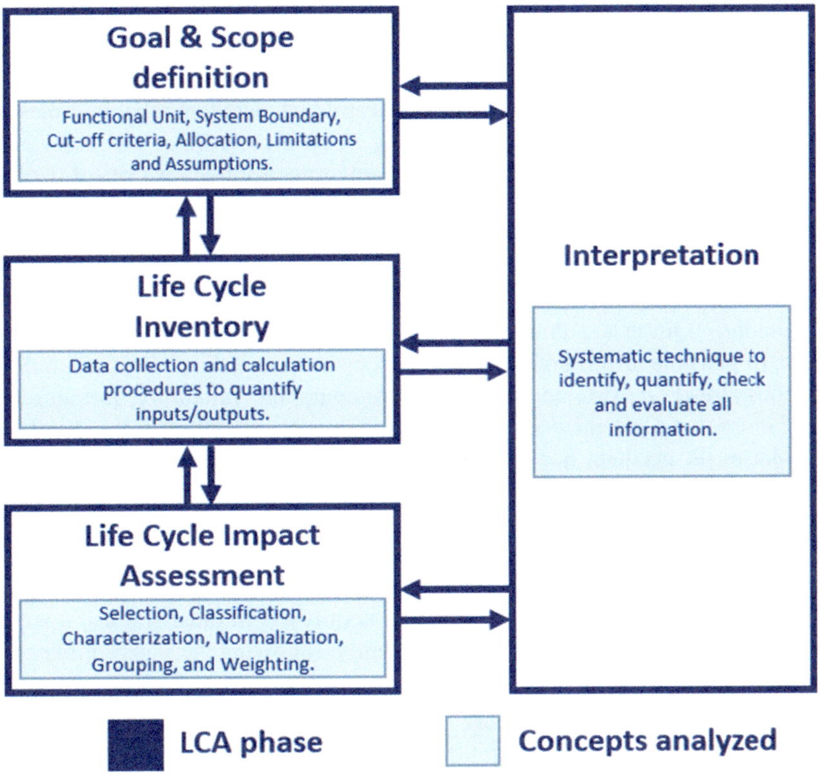

Figure 10.10 LCA framework.

10.7 Key aspects to consider during a concrete-LCA practice

After fully understanding how the application of LCA works and the concepts that must be considered at each stage of the study, it is essential to discuss the most significant challenges found in concrete LCA studies. Therefore, this section is divided into five parts to present different limitations found in concrete LCAs, in addition to discussing concepts that have been associated with the LCA application in the literature and that can benefit the decision-making process in the concrete industry.

10.8 Data reliability

The data uncertainty is seen as a huge problem during the interpretation phase of an LCA. Although it is crucial to quantify this uncertainty throughout the study, the practitioner should always look for ways to manage and reduce it. In this context, the insufficiency in the available data can be highlighted as a substantial limitation in applying LCA methodology for concrete.

It is undeniable that for a robust environmental analysis of structural and non-structural concretes, it is necessary to consider all energy consumption and emission generation issues involving all existing processes in producing and using these materials. However, it is clear that the LCA studies still lack data that reliably represent the technological and regional variations existing in each case. Since inventory analysis often varies from location to location, improving data accuracy is critical (Colangelo et al., 2018).

An attempt to assess and minimize uncertainties during the LCI, responsible for collecting and evaluating data, is observed in the literature. However, LCA experts claim that the LCIA phase contributes more significantly to the overall uncertainty of an LCA outcome (Qin et al., 2020). LCA practitioners and researchers should always be aware of possible uncertainties and make decisions accordingly. Although different approaches already exist to adopt regional or geographic variability, further development of more complete methods for characterizing uncertainties during the LCI and LCIA phases is needed. For instance, a recent paper compared commonly-used LCIA methods and showed that the uncertainty generated by using different methods could be significant (Chen et al., 2021). The authors claim that being aware of LCIA uncertainties is essential but often ignored during LCA studies.

Another interesting point is that the procedure suggested in the ISO 14044 standard can become a problem if it is not understood iteratively. In older studies, it is noticed that LCI and LCIA were developed independently, following the steps stipulated in the ISO standard: (1) data collection related to the functional unit, (2) cumulative LCI results computation, (3) impact assessment conduction. However, an iterative understanding of the LCA framework helps improve the reliability and the usefulness of LCA results (Anonymous, 2016).

10.9 Concrete constituents

Concrete mix design refers to the process of preparing concrete in which different ingredients are used so that the mix achieves the strength, durability, and other properties necessary for a specific function of the concrete structure. Determining the concrete mix design is one of the challenging issues in the construction sector, and this subject has been an object of investigation in recent years (Golafshani & Behnood, 2019; Rumman et al., 2020; Sassani et al., 2017).

Materials with totally different properties can be used in this process. Therefore, all ingredients must also be considered concerning the environmental impacts they could potentially generate. It is observed in the literature that the majority of LCA studies regarding the concrete analysis consider the primary constituents of concrete (i.e., cement, water, and aggregates) but neglect the potential impacts caused by other concrete constituents, such as admixtures, pigments, fibers, polymers, and reinforcement. Some of the materials that can be considered in a concrete mix design are represented in Fig. 10.11.

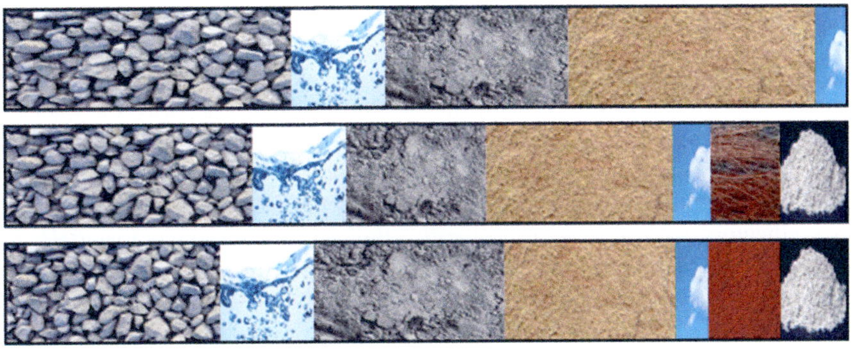

Figure 10.11 Representation of different concrete constituents proportions.

The consideration of only primary constituents of concrete during an environmental analysis can be explained by the small quantities of these constituents in concrete, often representing less than 1% of the total mass. Nonetheless, as discussed before, cut-off criteria are often adopted without due caution. Although these materials represent a low percentage of the total mass of concrete, it is immature to consider the impacts of these materials to be negligible without further quantitative assessment (Petek Gursel et al., 2014). Besides, several chemical substances used in concrete admixtures contain toxicological properties, and it has to be a concern among professionals. LCA applications for concrete need to be increasingly comprehensive and representative, but to do so, an investment in creating reliable and comprehensive databases for the concrete industry is necessary.

10.10 Circular economy

The circular economy is a concept directly linked to sustainability. Its application results in energy savings, reduced resource consumption, reduced waste, and new job generation (Stahel, 2016). The circular economy considers production processes to outline how to reuse, repair, and recycle items, thus increasing sustainability. This economy consists of a continuous positive development cycle to preserve and value natural capital, optimize resource production, and minimize systemic risks (Moreau et al., 2017).

The circular economy can happen in an open or closed loop. Applying a closed-loop circular-economy principle means to reuse the material in an identical function to what it had before (Lonca et al., 2020). In contrast, an open-loop circular economy principle is related to the use of waste to manufacture different materials (Huysveld et al., 2019), which can happen for different reasons, such as the change in the physical and chemical properties of the material. Both approaches bring many benefits to society, as thanks to the reuse and recycling of materials, fewer materials and production processes are needed to create new products. This change minimizes the consumption of fossil fuels and consequently the emission of greenhouse gases (Durán-Romero et al., 2020).

In 2020, the Life Cycle Initiative, a public-private, multi-stakeholder partnership hosted by UN Environment, commented through a position paper on integrating LCA and circular economy concepts (Peña et al., 2020). The Initiative claims that the LCA methodology can provide technical support to circular economy decision-makers, enabling these professionals to assess impact offsets on various environmental impact indicators, such as water use, energy, climate change, and raw materials. Thus, new research from different industries has been published based on integrating these concepts, and the concrete and cement industries have also sought to benefit from this integration.

Some waste materials are considered low-value resources as they require long treatments to meet the quality standards of a production chain. In this case, many of these materials are taken directly to landfills, not being reused. This is the mixed waste glass case, consisting of a mixture of different types of glass containing small impurities. In many countries, this kind of waste is not useful in a closed-loop model (i.e., they are not used to produce new glasses).

From this finding, a recent study presented the application of an open-loop circular-economy principle in the concrete LCA, using glass powder as supplementary cementitious material (Deschamps et al., 2018). The authors show that glass powder has pozzolanic properties, which guarantees the generation of hydrated calcium silicate when the glass chemically reacts with the products of cement hydration. In this way, the glass powder can guarantee an increase in the compressive strength of concrete. Then, the researchers compared the usual concrete production model and a glass-powder concrete production process model and found that this use can reduce the overall environmental impact of the production and use of concrete.

Other recent publications integrated LCA and the circular economy concept for a concrete analysis. For example, a study regarding the use of eco-cement pastes elaborated with recycled concrete powder from construction and demolition wastes evaluated LCA's environmental impacts (Caneda-martínez et al., 2021). The authors found that the overall environmental impact could be minimized in the same proportion of the replacement rate, proving the advantages of applying the circular economy concept. Another recent study assessed greenhouse gas emissions in the life cycle of wood bio-concretes production using recycled wood shavings to achieve the circular economy strategy (Caldas et al., 2021). The authors evaluated the life-cycle impacts of two mixtures of wood bio-concretes with different contents of wood shavings and concluded that wood waste could be beneficial for the production of low carbon materials in the concrete industry.

For the LCA methodology to help in minimizing environmental impacts, it must necessarily be accompanied by behavioral changes and other concepts related to sustainability, such as the circular economy. This integration is already a reality in the research field on concrete, but it still needs to be applied on a large scale globally. This requires transparency of assumptions, data reliability, and critical interpretation of results (Peña et al., 2020) so that the LCA practitioner can make decisions in a supported way.

10.11 Dynamic LCA framework

LCA is proven to be a great tool to benefit the decision-making process of different projects. However, when applied within the construction industry, it is necessary to consider the complexity of a building compared to other products or processes as an object of analysis. Beyond concrete, a building is composed of several materials, systems, and components, in addition to having a long lifespan. Therefore, data of different natures are evaluated jointly regarding the Building LCA, and the analysis becomes extensive and consequently more complex.

Much effort is required to collect all the data needed to perform an LCA analysis. In order to simplify the study, assumptions are often made that do not represent the reality of the building. A clear example of this is the quantification of energy consumption. It is known that the energy consumption is not constant during the processes, and this data needs to be evaluated as a function of several factors. Energy consumption is a time-dependent parameter, but this information is not considered in many analyses published in the literature.

In this context, an innovative approach emerged to improve the LCA applications: the dynamic LCA framework. A dynamic LCA refers to considering the temporalization of background and foreground systems into the LCA application, proposed to improve the LCA accuracy by addressing the inconsistency of temporal assessment (Levasseur et al., 2010). This recent approach includes the dynamic LCI and dynamic LCIA, allowing the evaluation of different temporal parameters involved in the processes (Pigné et al., 2020).

As implemented in conventional LCA, the use of a fixed time horizon deprives practitioners of essential information (Shimako et al., 2018). Since the operational phase of a building is extensive, it significantly influences the accuracy of evaluation results. Therefore, some articles already consider time-dependent parameters for the LCA application in the construction sector, such as technological progress, the variation in occupancy behavior, and dynamic weighting factors (Su et al., 2017). Over time, more in-depth studies on this new approach aimed at the construction sector will emerge. It is expected, then, that the concrete industry will benefit from this, as the dynamic approach makes the analysis more realistic and more representative.

10.12 Life cycle thinking

It is undeniable that working to minimize the environmental impacts generated by the concrete industry is urgent and imperative. As mentioned before, concrete production is responsible for large consumption of water, abundant depletion of natural resources, and a massive emission of greenhouse gases. Besides, as it is a construction material used on a large scale worldwide, any positive change made during the production of concrete results in enormous gains for the environment. However, it is always worth emphasizing the importance of adopting a life cycle thinking approach, referring to providing information to interested parties to make the best decisions regarding the life cycles of products (UNEP, 2013). It is done when we apply the LCA methodology considering not only the concrete production stage but its entire life cycle, from the extraction of raw materials to the end-of-life stage.

Nevertheless, when thinking about sustainable development, a common misconception is to consider only the environmental concerns of a product or service since sustainability is related to a triple-bottom-line framework composed of environmental, social, and economic factors. It is essential to reduce a product's resource use and emissions to the environment, but this goal must be accompanied by improving the socio-economic performance of the product. In this context, the Life Cycle Sustainability Assessment (LCSA) methodology emerged. This is a more comprehensive methodology based on the life cycle thinking approach, and it can be applied to the concrete industry. LCSA framework considers that all phases in a product's life cycle cause environmental impacts and socio-economic consequences, and to achieve sustainability, all these issues need to be assessed (UNEP, 2012).

LCSA is a methodology resulting from the sum and harmonization of three other methodologies based on life cycle thinking: (i) Life Cycle Assessment (LCA), related to the environmental pillar of sustainability (Nwodo & Anumba, 2019); (ii) Social Life Cycle Assessment (S-LCA), representing the social pillar (Ramos Huarachi et al., 2020); and (iii) Life Cycle Costing (LCC), referring the economic pillar (Illankoon & Lu, 2019). This idea is represented in Fig. 10.12.

The LCSA methodology can be very advantageous for the concrete and cement industry, although its application is linked to more significant difficulties in data collection and impact analysis. This is because the inventory contains information of different natures, and the analysis usually involves different stakeholders. The application of this more comprehensive methodology in the concrete industry is still in its infancy, but there are already studies published in the literature about it.

A paper from 2013 presented an ongoing LCSA study regarding concrete recycling (Hu et al., 2013). Although the authors discuss necessary steps for the operationalization of the LCSA, presenting some limitations found during the study, the article does

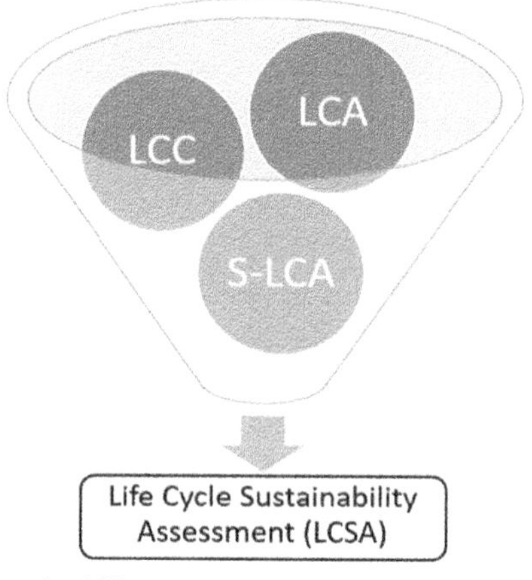

Figure 10.12 Life cycle sustainability assessment concept.

not present a practical application of the methodology. Another article used the LCSA methodology to compare the impacts generated by two typical structural systems: the wood frame and concrete frame buildings (Hossaini et al., 2015). The authors developed a case study considering environmental, social, and economic impact categories. They considered, for example, fossil fuel depletion, global warming potential, and acidification potential, among other environmental categories; occupant comfort, safety, seismic and fire resistance, and affordability as the social impact categories; and the overall cost of the building project as the economic indicator. The authors highlighted that the main challenge of the study was making a comparison between different criteria because there is no agreed method to determine the relative importance of each category.

A more recent paper presented the LCSA application of fly ash concrete structures using a proposed three-dimensional coordinate diagram in order to combine the different results (i.e., environmental, social, and economic units) into a single sustainable value (Wang et al., 2017). The authors considered one cubic meter of fly ash concrete as the functional unit and analyzed environmental, social, and economic impact categories, such as climate change, living conditions, and the cost of energy consumed in concrete preparation. Based on a case study, the authors could demonstrate that replacing cement with fly ash can reduce social, environmental, and economic impacts.

Indeed, the harmonization of the three techniques (LCA, S-LCA, and LCC) is considered a challenge for LCSA by different authors (Fauzi et al., 2019; Llatas et al., 2020). Much still needs to be discussed in the literature, especially in the construction sector, which is lagging behind other industries in using the life cycle thinking approach. This directly affects the cement and concrete industries. However, it is crucial to bear in mind that this holistic thinking in pursuit of sustainability is essential and that the concrete industry has a lot to contribute in this regard for the environment and society.

10.13 Summary - environmental impacts and their management

Arguably, the environmental impacts of concrete exceed CO_2 emissions and climate change and may also be linked to acid rain (Mahdikhani et al., 2018), health risks from high local concentrations of cement dust (Owonikoko et al., 2021), and drinking water supply depletion (Miller et al., 2018). Besides, the energy consumption during cement production is a substantial environmental concern since it is one of the most energy-intensive processes. Although these environmental problems are worrying, concrete has many advantages as a building material, and it is not possible to imagine a significant minimization of its use worldwide in the near future. Therefore, all professionals associated with the concrete industry are responsible for managing these impacts in the best way possible.

Several exploratory studies in the literature try to substitute ordinary Portland cement with other less damaged materials to the environment. For example, a recent paper analyzed the slag waste incorporation in high early strength concrete as a cement replacement (Kim et al., 2018). The authors found that it was possible to reduce the carbon footprint of the resulting precast concrete, maintaining the hydration characteristics and mechanical strength of concrete. Another paper found that blast furnace slag and fly ash, two different wastes, can be used as substitutes in portland cement to produce concrete with high strength (Bayraktar, 2019). With this replacement, the number of fine aggregates could be reduced, and the pozzolan thinness could be increased.

Additionally, the use of recycled aggregates to replace natural aggregates in concrete has proven to be a great solution not only for the environment but also for saving on construction costs. Some recent studies of concrete LCA present a comparison between natural aggregates and recycled aggregates, proving the advantages obtained through this replacement (Estanqueiro et al., 2018; Shan et al., 2017; Yazdanbakhsh et al., 2018). Although some papers showed that the mechanical properties of recycled aggregate concrete could sometimes be inferior compared to natural aggregate concrete, limiting the application to structural concrete (Wang et al., 2021), other studies already apply recycled aggregates for structural applications (Knoeri et al., 2013).

Another point that deserves due attention is dust emission. A single cement plant is responsible for a large amount of dust, and it is the reason for complaints from residents around cement plants (Adeyanju & Okeke, 2019). Unfortunately, the effects are immense and affect human beings, vegetation, and the aquatic environment. There are some solutions to deal with dust emissions. One of them is related to the use of cement kiln dust into concrete. Cement kiln dust is a fine powdery material generated as a by-product of Portland cement manufacturing and can be used to create blended concretes (Kunal & Rajor, 2012). This solution helps to minimize the amount of material landfilled and the amount of clinker required in cement production.

Through improved and consistent data collection and monitoring methods, it will be possible to accurately measure the environmental impacts caused during the concrete life cycle. In this context, the LCA application is an excellent way to improve the management of the environmental impacts and identify improvements through the concrete value chain. We hope that LCA methodology and other life-cycle thinking approaches will be increasingly applied in the concrete industry, both in the research field and in real projects. Finally, we emphasize the importance of government investments in creating comprehensive databases on concrete and its aggregates and public policies to encourage the use of recycled aggregates and other solutions to minimize environmental impacts.

Acknowledgment

This study was financed in part by the Coordenação de Aperfeiçoamento de Pessoal de Nível Superior - Brasil (CAPES) – Finance Code 001.

References

Adeyanju, E., & Okeke, C. A. (2019). Exposure effect to cement dust pollution: A mini review. *SN Applied Sciences, 1*, 1–17. https://doi.org/10.1007/s42452-019-1583-0

UNEP/SETAC life cycle initiative, global guidance for life cycle impact assessment indicators.(2016). *Vol. 1*. Paris www.lifecycleinitiative.org.

Bayraktar, O. Y. (2019). The possibility of fly ash and blast furnace slag disposal by using these environmental wastes as substitutes in portland cement. *Environmental Monitoring and Assessment, 191*, 1–12. https://doi.org/10.1007/s10661-019-7741-4

Caldas, L. R., Saraiva, A. B., Lucena, A. F. P., da Gloria, M. Y., Santos, A. S., & Filho, R. D. T. (2021). Building materials in a circular economy: The case of wood waste as CO2-sink in bio concrete. *Resources, Conservation and Recycling, 166*, 105346. https://doi.org/10.1016/j.resconrec.2020.105346

Caneda-martínez, L., Monasterio, M., Moreno-juez, J., Martínez-ramírez, S., García, R., & Frías, M. (2021). Behaviour and properties of eco-cement pastes elaborated with recycled concrete powder from construction and demolition wastes. *Materials, 14*, 1–19. https://doi.org/10.3390/ma14051299

Cao, C. (2017). Sustainability and life assessment of high strength natural fibre composites in construction. In *Advanced high strength natural fibre composites in construction* (pp. 529–544). Elsevier Inc. https://doi.org/10.1016/B978-0-08-100411-1.00021-2

Chen, C., Habert, G., Bouzidi, Y., Jullien, A., & Ventura, A. (2010). LCA allocation procedure used as an incitative method for waste recycling: An application to mineral additions in concrete. *Resources, Conservation and Recycling, 54*, 1231–1240. https://doi.org/10.1016/j.resconrec.2010.04.001

Chen, X., Matthews, H. S., & Griffin, W. M. (2021). Uncertainty caused by life cycle impact assessment methods: Case studies in process-based LCI databases. *Resources, Conservation and Recycling, 172*, 105678. https://doi.org/10.1016/j.resconrec.2021.105678

Colangelo, F., Forcina, A., Farina, I., & Petrillo, A. (2018). Life Cycle Assessment (LCA) of different kinds of concrete containing waste for sustainable construction. *Buildings, 8*, 70. https://doi.org/10.3390/buildings8050070

Corominas, L., Byrne, D. M., Guest, J. S., Hospido, A., Roux, P., Shaw, A., & Short, M. D. (2020). The application of life cycle assessment (LCA) to wastewater treatment: A best practice guide and critical review. *Water Research, 184*, 116058. https://doi.org/10.1016/j.watres.2020.116058

Crawford, R. H., Bontinck, P. A., Stephan, A., Wiedmann, T., & Yu, M. (2018). Hybrid life cycle inventory methods – a review. *Journal of Cleaner Production, 172*, 1273–1288. https://doi.org/10.1016/j.jclepro.2017.10.176

Deschamps, J., Simon, B., Tagnit-Hamou, A., & Amor, B. (2018). Is open-loop recycling the lowest preference in a circular economy? Answering through LCA of glass powder in concrete. *Journal of Cleaner Production, 185*, 14–22. https://doi.org/10.1016/j.jclepro.2018.03.021

Durán-Romero, G., López, A. M., Beliaeva, T., Ferasso, M., Garonne, C., & Jones, P. (2020). Bridging the gap between circular economy and climate change mitigation policies through eco-innovations and Quintuple Helix Model. *Technological Forecasting and Social Change, 160*, 120246. https://doi.org/10.1016/j.techfore.2020.120246

Estanqueiro, B., Dinis Silvestre, J., de Brito, J., & Duarte Pinheiro, M. (2018). Environmental life cycle assessment of coarse natural and recycled aggregates for concrete. *European*

Journal of Environmental and Civil Engineering, 22, 429–449. https://doi.org/10.1080/19648189.2016.1197161

Fauzi, R. T., Lavoie, P., Sorelli, L., Heidari, M. D., & Amor, B. (2019). Exploring the current challenges and opportunities of life cycle sustainability assessment. *Sustainability, 11*, 1–17. https://doi.org/10.3390/su11030636

Figueiredo, K., Pierott, R., Hammad, A. W. A., & Haddad, A. (2021). Sustainable material choice for construction projects: A life cycle sustainability assessment framework based on BIM and fuzzy-AHP. *Building and Environment, 196*. https://doi.org/10.1016/j.buildenv.2021.107805

Golafshani, E. M., & Behnood, A. (2019). Estimating the optimal mix design of silica fume concrete using biogeography-based programming. *Cement and Concrete Composites, 96*, 95–105. https://doi.org/10.1016/j.cemconcomp.2018.11.005

van den Heede, P., & de Belie, N. (2012). Environmental impact and life cycle assessment (LCA) of traditional and "green" concretes: Literature review and theoretical calculations. *Cement and Concrete Composites, 34*, 431–442. https://doi.org/10.1016/j.cemconcomp.2012.01.004

Hernandez, P., Oregi, X., Longo, S., & Cellura, M. (2018). Life-cycle assessment of buildings. In *Handbook of energy efficiency in buildings: A life cycle approach* (pp. 207–261). Elsevier. https://doi.org/10.1016/B978-0-12-812817-6.00010-3

Herrmann, I. T., & Moltesen, A. (2015). Does it matter which life cycle assessment (LCA) tool you choose? - a comparative assessment of SimaPro and GaBi. *Journal of Cleaner Production, 86*, 163–169. https://doi.org/10.1016/j.jclepro.2014.08.004

Hollberg, A., Kiss, B., Röck, M., Soust-Verdaguer, B., Wiberg, A. H., Lasvaux, S., Galimshina, A., & Habert, G. (2021). Review of visualising LCA results in the design process of buildings. *Building and Environment, 190*, 107530. https://doi.org/10.1016/j.buildenv.2020.107530

Hossaini, N., Reza, B., Akhtar, S., Sadiq, R., & Hewage, K. (2015). AHP based life cycle sustainability assessment (LCSA) framework: A case study of six storey wood frame and concrete frame buildings in vancouver. *Journal of Environmental Planning and Management, 58*, 1217–1241. https://doi.org/10.1080/09640568.2014.920704

Hu, M., Kleijn, R., Bozhilova-Kisheva, K. P., & di Maio, F. (2013). An approach to LCSA: The case of concrete recycling. *International Journal of Life Cycle Assessment, 18*, 1793–1803. https://doi.org/10.1007/s11367-013-0599-8

Huysveld, S., Hubo, S., Ragaert, K., & Dewulf, J. (2019). Advancing circular economy benefit indicators and application on open-loop recycling of mixed and contaminated plastic waste fractions. *Journal of Cleaner Production, 211*, 1–13. https://doi.org/10.1016/j.jclepro.2018.11.110

Illankoon, I. M. C. S., & Lu, W. (2019). Optimising choices of 'building services' for green building: Interdependence and life cycle costing. *Building and Environment, 161*. https://doi.org/10.1016/j.buildenv.2019.106247

ISO - International Organization for Standardization. (2006). *ISO 14044:2006 - environmental management — life cycle assessment — requirements and guidelines*.

ISO. (2002). *ISO/TS 14048:2002 - environmental management — life cycle assessment — data documentation format*.

ISO. (2006). *ISO 14040:2006 - environmental management — life cycle assessment — principles and framework*.

ISO. (2012a). *ISO/TR 14047:2012 - environmental management — life cycle assessment — illustrative examples on how to apply ISO 14044 to impact assessment situations*.

ISO. (2012b). *ISO/TR 14049:2012 - environmental management — life cycle assessment — illustrative examples on how to apply ISO 14044 to goal and scope definition and inventory analysis*.

Kim, Y., Hanif, A., Usman, M., Munir, M. J., Kazmi, S. M. S., & Kim, S. (2018). Slag waste incorporation in high early strength concrete as cement replacement: Environmental impact and influence on hydration & durability attributes. *Journal of Cleaner Production, 172*, 3056–3065. https://doi.org/10.1016/j.jclepro.2017.11.105

Kleijer, A. L., Lasvaux, S., Citherlet, S., & Viviani, M. (2017). Product-specific Life Cycle Assessment of ready mix concrete: Comparison between a recycled and an ordinary concrete. *Resources, Conservation and Recycling, 122*, 210–218. https://doi.org/10.1016/j.resconrec.2017.02.004

Knoeri, C., Sanyé-Mengual, E., & Althaus, H. J. (2013). Comparative LCA of recycled and conventional concrete for structural applications. *International Journal of Life Cycle Assessment, 18*, 909–918. https://doi.org/10.1007/s11367-012-0544-2

Kunal, R. Siddique, & Rajor, A. (2012). Use of cement kiln dust in cement concrete and its leachate characteristics. *Resources, Conservation and Recycling, 61*, 59–68. https://doi.org/10.1016/j.resconrec.2012.01.006

Kurda, R., Silvestre, J. D., & de Brito, J. (2018). Life cycle assessment of concrete made with high volume of recycled concrete aggregates and fly ash. *Resources, Conservation and Recycling, 139*, 407–417. https://doi.org/10.1016/j.resconrec.2018.07.004

Laurent, A., & Hauschild, M. Z. (2015). Normalisation. In M. Hauschild, & M. Huijbregts (Eds.), *Life cycle impact assessment* (pp. 271–300). Springer. https://doi.org/10.1007/978-94-017-9744-3_14

Levasseur, A., Lesage, P., Margni, M., Deschěnes, L., & Samson, R. (2010). Considering time in LCA: Dynamic LCA and its application to global warming impact assessments. *Environmental Science and Technology, 44*, 3169–3174. https://doi.org/10.1021/es9030003

Liu, X., Lai, G., Guan, J., Qian, S., Wang, Z., Cui, S., Gao, F., Jiao, Y., & Tao, R. (2021). Technical optimization and life cycle assessment of environment-friendly superplasticizer for concrete engineering. *Chemosphere, 281*, 130955. https://doi.org/10.1016/j.chemosphere.2021.130955

Llatas, C., Soust-Verdaguer, B., & Passer, A. (2020). Implementing life cycle sustainability assessment during design stages in building information modelling: From systematic literature review to a methodological approach. *Building and Environment, 182*. https://doi.org/10.1016/j.buildenv.2020.107164

Lonca, G., Lesage, P., Majeau-Bettez, G., Bernard, S., & Margni, M. (2020). Assessing scaling effects of circular economy strategies: A case study on plastic bottle closed-loop recycling in the USA PET market. *Resources, Conservation and Recycling, 162*, 105013. https://doi.org/10.1016/j.resconrec.2020.105013

Mahdikhani, M., Bamshad, O., & Fallah Shirvani, M. (2018). Mechanical properties and durability of concrete specimens containing nano silica in sulfuric acid rain condition. *Construction and Building Materials, 167*, 929–935. https://doi.org/10.1016/j.conbuildmat.2018.01.137

Martínez-Rocamora, A., Solís-Guzmán, J., & Marrero, M. (2016). LCA databases focused on construction materials: A review. *Renewable and Sustainable Energy Reviews, 58*, 565–573. https://doi.org/10.1016/j.rser.2015.12.243

Miller, S. A., Horvath, A., & Monteiro, P. J. M. (2018). Impacts of booming concrete production on water resources worldwide. *Nature Sustainability, 1*, 69–76. https://doi.org/10.1038/s41893-017-0009-5

Moreau, V., Sahakian, M., van Griethuysen, P., & Vuille, F. (2017). Coming full circle: Why social and institutional dimensions matter for the circular economy. *Journal of Industrial Ecology, 21*, 497–506. https://doi.org/10.1111/jiec.12598

NREL, U.S. Life Cycle Inventory Database, (n.d.). https://www.nrel.gov/lci/(accessed June 23, 2021).
Ntimugura, F., Vinai, R., Harper, A. B., & Walker, P. (2021). Environmental performance of miscanthus-lime lightweight concrete using life cycle assessment: Application in external wall assemblies. *Sustainable Materials and Technologies, 28*. https://doi.org/10.1016/j.susmat.2021.e00253
Nwodo, M. N., & Anumba, C. J. (2019). A review of life cycle assessment of buildings using a systematic approach. *Building and Environment, 162*. https://doi.org/10.1016/j.buildenv.2019.106290
Owonikoko, M. W., Emikpe, B. O., & Olaleye, S. B. (2021). Standardized experimental model for cement dust exposure; tissue heavy metal bioaccumulation and pulmonary pathological changes in rats. *Toxicology Reports, 8*, 1169–1178. https://doi.org/10.1016/j.toxrep.2021.06.001
Peña, C., Civit, B., Gallego-Schmid, A., Druckman, A., Caldeira-Pires, A., Weidema, B., Mieras, E., Wang, F., Fava, J., Milà I Canals, L., Cordella, M., Arbuckle, P., Valdivia, S., Fallaha, S., & Motta, W. (2020). *Position paper of the life cycle initiative - using life cycle assessment to achieve a circular economy.*
Petek Gursel, A., Masanet, E., Horvath, A., & Stadel, A. (2014). Life-cycle inventory analysis of concrete production: A critical review. *Cement and Concrete Composites, 51*, 38–48. https://doi.org/10.1016/j.cemconcomp.2014.03.005
Pigné, Y., Gutiérrez, T. N., Gibon, T., Schaubroeck, T., Popovici, E., Shimako, A. H., Benetto, E., & Tiruta-Barna, L. (2020). A tool to operationalize dynamic LCA, including time differentiation on the complete background database. *International Journal of Life Cycle Assessment, 25*, 267–279. https://doi.org/10.1007/s11367-019-01696-6
Qin, Y., Cucurachi, S., & Suh, S. (2020). Perceived uncertainties of characterization in LCA: A survey. *International Journal of Life Cycle Assessment, 25*, 1846–1858. https://doi.org/10.1007/s11367-020-01787-9
Ramos Huarachi, D. A., Piekarski, C. M., Puglieri, F. N., & de Francisco, A. C. (2020). Past and future of social life cycle assessment: Historical evolution and research trends. *Journal of Cleaner Production, 264*. https://doi.org/10.1016/j.jclepro.2020.121506
Rumman, R., Bari, M. S., Manzur, T., Kamal, M. R., & Noor, M. A. (2020). A durable concrete mix design approach using combined aggregate gradation bands and rice husk ash based blended cement. *Journal of Building Engineering, 30*, 101303. https://doi.org/10.1016/j.jobe.2020.101303
Saade, M. R. M., Guest, G., & Amor, B. (2020). Comparative whole building LCAs: How far are our expectations from the documented evidence? *Building and Environment, 167*. https://doi.org/10.1016/j.buildenv.2019.106449
Santos, T., Almeida, J., Silvestre, J. D., & Faria, P. (2021). Life cycle assessment of mortars: A review on technical potential and drawbacks. *Construction and Building Materials, 288*, 123069. https://doi.org/10.1016/j.conbuildmat.2021.123069
Sassani, A., Ceylan, H., Kim, S., Gopalakrishnan, K., Arabzadeh, A., & Taylor, P. C. (2017). Influence of mix design variables on engineering properties of carbon fiber-modified electrically conductive concrete. *Construction and Building Materials, 152*, 168–181. https://doi.org/10.1016/j.conbuildmat.2017.06.172
Shan, X., Zhou, J., Chang, V. W. C., & Yang, E. H. (2017). Life cycle assessment of adoption of local recycled aggregates and green concrete in Singapore perspective. *Journal of Cleaner Production, 164*, 918–926. https://doi.org/10.1016/j.jclepro.2017.07.015
Shimako, A. H., Tiruta-Barna, L., Bisinella de Faria, A. B., Ahmadi, A., & Spérandio, M. (2018). Sensitivity analysis of temporal parameters in a dynamic LCA framework. *Science of the Total Environment, 624*, 1250–1262. https://doi.org/10.1016/j.scitotenv.2017.12.220

Sphera, GaBi Solutions - Databases, (n.d.). https://gabi.sphera.com/databases (accessed June 23, 2021).

Stahel, W. R. (2016). The circular economy. *Nature, 531*, 435–438. https://doi.org/10.1038/531435a

Stiebert, S., Echeverría, D., Gass, P., & Kitson, L. (2019). *Emission Omissions: Carbon accounting gaps in the built environment.* www.iisd.org. (Accessed 21 June 2021).

Su, S., Li, X., Zhu, Y., & Lin, B. (2017). Dynamic LCA framework for environmental impact assessment of buildings. *Energy and Buildings, 149*, 310–320. https://doi.org/10.1016/j.enbuild.2017.05.042

UNEP. (2012). *Towards a life cycle sustainability assessment.*

UNEP. (2013). *Greening the economy through life cycle thinking: Ten years of the UNEP/SETAC life cycle initiative.* https://doi.org/10.1248/cpb.37.3229

Vieira, D. R., Calmon, J. L., & Coelho, F. Z. (2016). Life cycle assessment (LCA) applied to the manufacturing of common and ecological concrete: A review. *Construction and Building Materials, 124*, 656–666. https://doi.org/10.1016/j.conbuildmat.2016.07.125

Wang, J. J., Wang, Y. F., Sun, Y. W., Tingley, D. D., & Zhang, Y. R. (2017). Life cycle sustainability assessment of fly ash concrete structures. *Renewable and Sustainable Energy Reviews, 80*, 1162–1174. https://doi.org/10.1016/j.rser.2017.05.232

Wang, B., Yan, L., Fu, Q., & Kasal, B. (2021). A comprehensive review on recycled aggregate and recycled aggregate concrete. *Resources, Conservation and Recycling, 171*, 105565. https://doi.org/10.1016/j.resconrec.2021.105565

Wernet, G., Bauer, C., Steubing, B., Reinhard, J., Moreno-Ruiz, E., & Weidema, B. (2016). The ecoinvent database version 3 (part I): Overview and methodology. *International Journal of Life Cycle Assessment, 21*, 1218–1230. https://doi.org/10.1007/s11367-016-1087-8

Wu, Y., & Su, D. (2020). Review of life cycle impact assessment (LCIA) methods and inventory databases. In *Sustainable product development* (pp. 39–55). Springer International Publishing. https://doi.org/10.1007/978-3-030-39149-2_3

Yazdanbakhsh, A., Bank, L. C., Baez, T., & Wernick, I. (2018). Comparative LCA of concrete with natural and recycled coarse aggregate in the New York City area. *International Journal of Life Cycle Assessment, 23*, 1163–1173. https://doi.org/10.1007/s11367-017-1360-5

Asset management of recycled concrete applications

11

Sheyla Serra [1], *Carlos Soares* [2] *and Luiz Sarti Junior* [1]
[1]Universidade Federal de São Carlos (UFSCar), São Carlos, São Paulo, Brazil; [2]Universidade Federal Fluminense (UFF), Niterói, Rio de Janeiro, Brazil

11.1 Introduction

Concrete is one of the most used materials in the modern world, and losses inevitably occur during its use, whether in the stages of manufacturing, transport, use, and selective demolition. These steps generate by-products that can be reused or recycled and can be used to compose new products. The by-products can rise to Recycled Concrete Aggregates (RCA) when the residues meet specific characteristics and situations. RCA and derivative products can be classified as "assets" of an organization. Therefore, it can enjoy the benefits of asset management, which can be understood as a set of best management practices for the physical and intangible assets that make up a company's assets to obtain sustainable results.

The chapter is structured in four sections about the central issues related to Asset management of recycled concrete applications. First, in section 11.1 "Asset Management Systems and Standards," the essential systems and standards aimed at managing assets: PAS 55, Series ISO 55000, GFMAM Asset Management Landscape, and Asset Management Implementation Plan (Six Whats).

In section 11.2, "Decision-support tools for asset management," is discuss three commonly used tools in asset management processes: Life Cycle Assessment (LCA), Life Cycle Costing (LCC), and SWOT (Strengths, Weaknesses, Opportunities, and Threats) Analysis. The decision-support tools improve the efficiency and effectiveness of the decision-making process of asset managers, helping to reduce its complexity.

Section 11.3, "Contextual Factors for the Asset Management System," addresses three critical contexts that influence the RCA management process: political, technological, and organizational. The political context is related to the degree of importance of the civil construction sector in the economy and the implementation of strategies that encourage sectoral actions and the circular economy. The technological context considers the characteristics present in Construction and Deconstruction Waste (CDW) and the technical conditions for its better management. The organizational context seeks to identify the primary environmental certification standards, the main stakeholders in managing CDW and RCA, and the potential relationship forms.

Recycled Concrete. https://doi.org/10.1016/B978-0-323-85210-4.00008-4

This discussion is presented with the international context and history of initiatives and practices used by the RCA and CDW in general.

Finally, it presents essential issues for designing a recycled concrete asset management system, aiming to provide insights to fulfill their role in building, operating, and improving asset management, focusing on managing RCA.

11.2 Asset Management Systems and Standards

According to International Organization for Standardization (ISO) 55000 (ISO, 2014, p. 19), an asset is an "item, thing or entity that has potential or actual value to an organization." The value of an asset is highly dependent on the balance between costs, opportunities, and risks, which requires a management system that optimizes the relationship of these three variables, enabling the assets to fulfill their function, and that collaborate to the potential of assets is appropriately used and maintained. Assets that fail to fulfill their functions imply substantial losses, such as, for example, loss of revenue due to production stoppage or noncompliance, and intangibles, such as loss of stakeholder confidence.

Asset management has been approached from different points of view; however, two widely accepted concepts. The first is proposed from the British Standards Institution (BSI, 2008), which defines asset management as "systematic and coordinated activities and practices through which an organization optimally and sustainably manages its assets and asset systems, their associated performance, risks and expenditures over their life cycles to achieve its organizational strategic plan." The second point is present from ISO 55000 (ISO, 2014, p. 19) that defines asset management as "the coordinated activity of an organization to realize value from assets".

Asset management must incorporate the specificities of the industrial sector in which the organization is inserted. According to Alfatih et al. (2015), engineering asset management should guide the effective use of all physical engineering assets within an organization, covering the stages of operation, maintenance, upgrade, and disposal, that is, the entire lifecycle. Next, the principal asset management systems identified will be discussed.

11.2.1 Publicly Available Specification (PAS) 55

Publicly Available Specification 55 (PAS 55) is a set of standard specifications for physical asset management developed by the Institute of Asset Management (IAM), in collaboration with the British Standards Institution (BSI, 2008), and assistance from various organizations and individuals. It has been structured according to the British Standards ISO format. PAS 55 provides a broad view of managing assets throughout their lifecycle. Although its scope focuses mainly on managing physical assets, human, information, financial and intangible assets are also considered when influencing the optimized management of physical assets.

PAS 55 is formed by PAS 55-1 "Specification for the optimized management of physical assets," which contains the requirements specifications for best practices, and PAS 55-2 "Guidelines for the application of PAS 55-1" guide the application of the procedure. In preparing the standard, two main strategies were adopted aiming at compatibility with related management systems: a) standardize asset management in the form of specifications and break down information into essential requirements; b) use the Plan-Do-Check-Act (PDCA) continuous improvement cycle framework. Fig. 11.1 presents the structure of the PAS management system and your relationship with in to the PDCA cycle.

11.2.2 Series ISO 55000 asset management

The wide acceptance of PAS 55 by the industrial sector led the International Standards Organization (ISO) to adopt this standard as a reference for the preparation of ISO 55000. It is also intended for managing physical assets in particular, but it can also be used for managing other types of assets. This scope expansion is the main difference between PAS 55. It consists of three documents:

- ISO 55000 Asset management—Overview, principles, and terminology: which provides an overview of asset management, its principles and terminology, and the expected benefits of adopting asset management;
- ISO 55001 Asset management—Management systems—Requirements: which presents the specification of the requirements for the establishment, implementation, maintenance and improvement of an asset management system";
- ISO 55002 Asset management—Management systems—Guidelines for the application of ISO 55000: which provides guidelines for applying an asset management system following the requirements of ISO 55001.

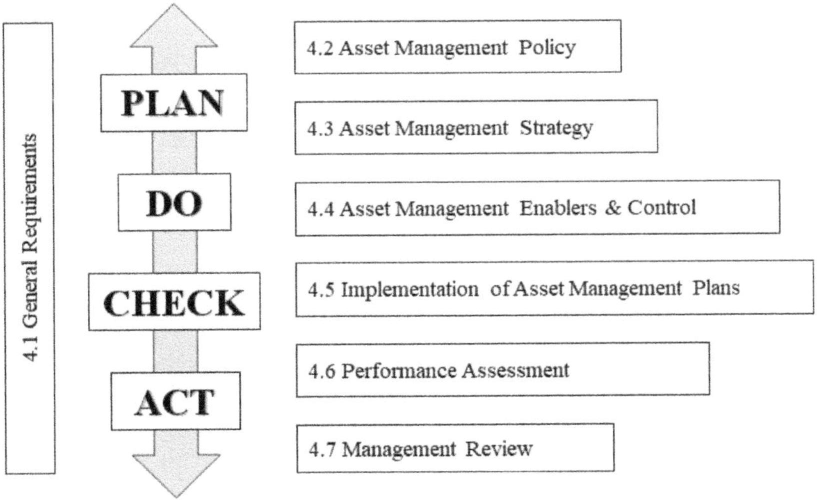

Figure 11.1 Structure of PAS 55-1 management system with the PDCA cycle.

Fig. 11.2 shows the key elements of asset management that must be considered according to the ISO 55000 series.

11.2.3 GFMAM Asset Management Landscape

Global Forum on Maintenance and Asset Management (GFMAM) was created in 2010 with representatives from different countries (https://gfmam.org/) and shared advances, knowledge, maintenance, and asset management standards. The Asset Management Landscape (GFMAM, 2014) is one of the group's products and is characterized as a framework that presents a common global approach to Asset Management. It is a conceptual model, in line with the ISO 55000 series, which presents 39 Asset Management Subjects and Principles and a framework for describing best practices, maturity, and standards. Table 11.1 shows the Asset Management Landscape subjects.

11.2.4 Asset management implementation plan (six whats)

The Asset Management Implementation Plan is a multi-stage plan for implementing asset management proposed by Vanier (2001) to structure the collection of data and information required by the decision-making process of the strategic plans of an organization's fixed assets. However, the main focus is the infrastructure of cities. It consists of six asset management levels of implementation based on Six "Whats" as shown in Table 11.2.

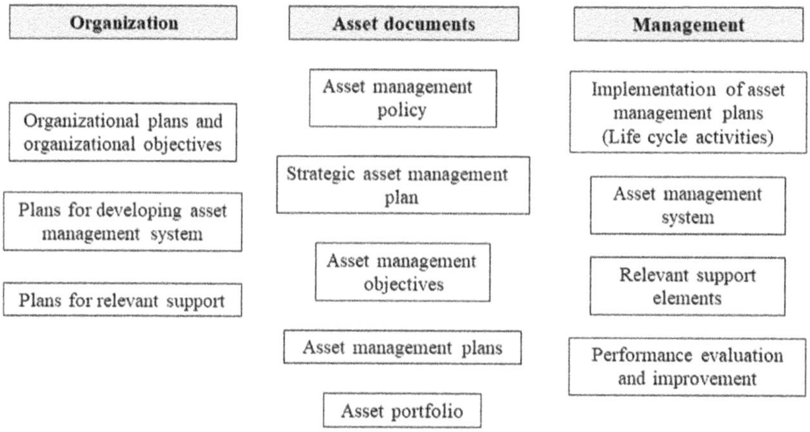

Figure 11.2 The key elements of an asset management system according to the ISO 55000.

Table 11.1 Asset Management Landscape subjects.

Landscape subject group	Landscape second edition subject
Strategy and planning	Asset management policy
	Asset management strategy and objectives
	Demand analysis
	Strategic planning
	Asset management planning
Asset management decision making	Capital investment decision-making
	Operations and maintenance decision-making
	Life cycle value realization
	Resourcing strategy
	Shutdowns and outage strategy
Life cycle delivery	Technical standards and legislation
	Asset creation and acquisition
	Systems engineering
	Configuration management
	Maintenance delivery
	Reliability engineering
	Asset operations
	Resource management
	Shutdown and outage management
	Fault and incident management
	Asset decommissioning and disposal
Asset information	Asset information strategy
	Asset information standard
	Asset information systems
	Data and information management
Organization and people	Procurement and supply chain management
	Asset management leadership
	Organizational structure
	Organizational culture
	Competence management
Risk and review	Risk assessment and management
	Contingency planning and resilience analysis
	Sustainable development
	Management of change
	Asset performance and health monitoring
	Asset management system monitoring
	Management review, audit and assurance
	Asset costing and valuation
	Stakeholder engagement

Reproduced from Global Forum on Maintenance and Asset Management (GFMAM). (2014). *The asset management Landscape* (2nd. edn.). https://gfmam.org/webform/download-the-am-landscape-v2.

Table 11.2 Levels of asset management implementation plan.

Asset management level	Action
What do you own?	Managers must identify the extent and composition of their portfolios.
What is it worth?	Managers must assign realistic values to each asset in their portfolio, considering: (a) The "book value" of the asset; (b) the historical cost calculated in present-day dollars; (c) the current asset replacement value; (d) the prescribed value of the current asset for the user; (e) the cost avoided as a result of having control of an asset; (f) the market value.
What is deferred maintenance?	Referred maintenance refers to the postponement of maintenance activities, aiming to realign budget resources, saving costs, or adjusting budget funding levels. Answering this question enables additional metrics for the allocation of maintenance funds.
What is its condition?	It is an extension of the "what is it worth?" collaborates in prioritizing maintenance, repair, and renovation, by identifying the asset's performance level.
What is the remaining service life?	Managers should establish the remaining useful life to calculate life cycle costs to develop alternative maintenance, repair, and renewal strategies.
What do you fix first?	Managers should prioritize their maintenance expenses, considering the results of the previous five levels.

Adapted from Vanier, D. J. (2001). Why industry needs asset management tools. *Journal of Computing in Civil Engineering*, *15*(1). https://doi.org/10.1061/(ASCE)0887-3801(2001)15:1(35).

11.3 Decision-support tools for asset management

The decision-support tools improve the efficiency and effectiveness of the decision-making process of asset managers, helping to reduce its complexity. Next, three commonly used tools in asset management processes will be discussed: LCA, LCC, and SWOT Analysis.

11.3.1 Life Cycle Assessment (LCA)

The LCA is a vital decision support tool, which enables the use of systems thinking to assess impacts and compare the environmental performance of assets throughout their life cycle. The asset lifecycle is the time interval between identifying the need for an asset and the deactivation of the asset or any associated liabilities (BSI, 2008).

The LCA is beneficial for structuring the compilation and assessment of assets' inputs, outputs, and environmental impacts. ISO 14,040 (ISO, 2006) structures the LCA into four phases: (a) the goal and scope definition phase, (b) the inventory analysis phase, (c) the impact assessment phase, and (d) the interpretation phase. Table 11.3 summarizes the phases of LCA based on the ISO 14040 series.

11.3.2 Life Cycle Costing (LCC)

The Life Cycle Costing, also called whole-life costing, is a tool used to estimate the costs of an asset throughout its life cycle, taking into account the aspects of planning, acquisition, operation, maintenance, and disposal. It makes it possible to evaluate the asset based on all relevant economic consequences over its useful life. LCC can be used, for example, to assess the feasibility and requirements of existing and new assets, estimate project or investment requirements, identify cost drivers, and develop value-oriented decision support systems. The costs usually considered include the initial investment, the investments required over the useful life, and the annual operating and maintenance costs.

According to El-Akruti et al. (2013), the integration of LCC models in the asset management decision process, according to Blanchard's model, can be carried out according to the following steps: define system requirements, specify the system life-cycle and identify activities by phase, develop a cost breakdown structure, identify input data requirements, establish costs for each category similar to Cost Breakdown Structure (CBS), select a cost model for analysis and evaluation, develop a cost profile and summary, identify high-cost contributors and cause-effect relationships, conduct

Table 11.3 The four phases of LCA based on ISO 14040 series.

Goal and scope definition	Inventory analysis	Impact assessment	Interpretation
Purpose Scope Level of detail System boundaries Assumptions and limitations Data quality requirements	Inventory of input/output data Flow diagram of the processes Data collection methodology Relevant data sources Data collection Modeling and analysis Reporting of results.	Relevant impact categories Classification and characterization of the environmental impacts Analysis using category indicators	Interpretation of results and usability Conclusions about completeness, significance, limitations, and consistency with goals and scope

Adapted from ISO 14040 (International Organization for Standardization (ISO). (2006). *ISO 14040: Environmental management—life cycle assessment—principles and framework*. https://www.iso.org/standard/37456.html)

sensitivity analysis, identify priorities for problem resolution, identify additional alternatives, evaluate feasible alternatives and select a preferred approach.

11.3.3 SWOT analysis

The SWOT analysis is a management tool that aims to build strategic planning, demonstrating the organization's internal and external environments. The acronym SWOT comes from the words Strengths, Weaknesses, Opportunities, and Threats. Strengths include internal capabilities, resources, and positive situational factors that can help the company achieve its goals; Weaknesses include internal limitations and negative factors that can interfere with its performance. Opportunities are favorable items, such as factors or trends in the external environment that the company can exploit for its benefit. Finally, threats are external factors or unfavorable trends that can present performance challenges.

An example of the use of SWOT analysis in waste management can be found in the study by Yuan (2013), who applied the tool to understand the construction waste management (CWM) model in China. The author used data obtained from government reports, regulations related to waste management, literature reviews, and group meetings with professionals working in the area. There are seven critical strategies for addressing CWM: Establishing a mechanism for determining the responsibility of various departments involved; Promulgating detailed CWM regulations; Investigating the amount of construction waste generated and planning waste facilities properly; Implementing CWM throughout the life cycle of construction projects; Implementing a pilot program of applying recycled construction materials; Establishing a construction waste research institute; and Raising CWM awareness via training and promotion activities.

11.4 Contextual factors for the asset management system

The asset management system can be generically organized considering the main elements that guide the stages of system definition and the main issues that must be raised to support decision making. For the Federal Highway Administration (FHWA, 2017), the fundamental components of the asset management system are goals and policies, asset inventory, condition assessment and performance modeling, alternatives evaluation and program optimization, budget allocations, short and long-range plans, program implementation, and performance monitoring. Some of the fundamental questions to be answered are: "What is included in our inventory of assets?", "What is the value of our assets?", "What resources are available? What is the budget level? What is the projected level of future funding?". For each indicated component, tools or strategies can be established that provide essential information. Each company will define its system based on its decision variables, such as policies, goals, types of assets, budgets, procedures, and business practices. For example, in the case of selective demolition, Debacker et al. (2017) recommend that the stakeholders involved reflect on the following guidelines:

- That a predemolition audit be carried out to guide the decision process with alternatives such as whether there will be a possibility of better screening of CDW at the site, to identify the fraction of waste and components with high added value;
- That the planning of machines and personnel be carried out, to have a less idle operation of the demolition and sorting machines;
- That local crushing and recycling plants be implemented, whenever possible;
- That the modes of transport be studied to remove the fraction of waste and materials for recycling, what is the destination of nonrecyclable or reusable waste, what is the potential volume of transport in a single trip; and
- Those vehicles, machinery, and heavy trucks with fossil fuels are avoided, if possible.

Thus, several questions need to be answered. First, it is essential to understand how these variables of the asset management system relate to the internal and external context of the organization, depending on the type of waste to be considered.

In general, this chapter also proposes to present essential aspects for the implementation of a generic RCA asset management system, following the following structure:

- Political context: understand the importance of civil construction in the country's socioeconomic context. Check the central policies and incentives for sustainable management of civil construction waste in general. Analyze whether the economic environment contributes to promoting the circular economy of CDW in the construction industry.
- Technological context: understand how the CDW generation process takes place and what are its main characteristics; know the local conditions, the relevant legislation, and the availability of equipment and spaces for the installation of the recycling center; compare possible transport alternatives;
- Organizational context: know the contexts for adopting certifiable environmental standards; the existence of associations or programs that encourage the use of RCA; and the characteristics of CDW management systems in general.

11.4.1 Political context

11.4.1.1 Recycled concrete in the context of the construction industry

The construction industry is characterized by a high demand for natural resources, particularly large volumes of natural aggregate. Granular materials can be composed of sand, gravel, crushed stone, recycled aggregates, and manufactured according to the European Aggregates Association (Union Européenne des Producteurs de Granulats—UEPG, 2020). Primary aggregates are produced from natural sources, extracted from quarries, sand, and gravel extraction sites, and, in some countries, dredging from the sea. Secondary aggregates include recycled, reused, and manufactured aggregates from industrial processes.

According to UEPG (2020), in 2018, the European Union consumed 3.07 billion tons of aggregates. The production of secondary aggregates, comprising recycled and reused aggregates, was 327,000 tons, representing 10.6% of the total. However, in the national context, there is a significant difference. About 27% of secondary aggregates are used in France, followed by the United Kingdom with 26%, the Netherlands and Malta with 24% each, and Belgium with 21%.

Thus, some consumer countries have an excellent incentive to stimulate the consumption of secondary aggregates, such as RCA. The RE4 project—REuse and REcycling of Construction and Deconstruction Waste (http://www.re4.eu/), for example, aims to develop fully prefabricated and energy-efficient construction solutions of components containing up to 65% in weight of CDW derived materials and structures demolitions (Whittaker et al., 2019).

Furthermore, the construction sector plays an essential and impacting role in the economy of any country, contributing to the Gross Domestic Product (GDP) and generation of millions of jobs, mainly in micro and small businesses (UK Government, 2013).

Among the main improvement strategies for the sector, adopting sustainability practices, creating business opportunities in terms of competitiveness and sustainable development in a national and regional context can be mentioned (European Commission, 2012). Different approaches are used, such as the development of new materials made with CDW, recycling or recovery of CDW, and management systems that consider the systemic view of the business, such as asset management.

For Heinrich and Lang (2019), the mineral aggregate from demolition can be used to produce recycled concrete, which is a form of high-level recycling, since the relevant technical standards and specifications are considered.

11.4.1.2 Policies and incentives for recycling waste in general

Government policies that encourage the creation of alternatives and business models and promote sustainability are essential for disseminating and implementing best practices and new management approaches by stakeholders. In addition, it is also necessary to generate knowledge and prepare technical recommendations that indicate the use of alternatives found with more excellent safety, such as recycling or reusing CDW and the use of RCA.

For Kelly and McMaster (2017), a key motivation for encouraging the recycling of CDW was the increase in costs and restrictions associated with their disposal in waste landfills. Thus, a series of relevant initiatives and regulations were developed that encouraged entrepreneurs to act within the sustainability aspects. In Europe, programs for "Waste and recycling" environmental policy can be highlighted:

- The Waste Framework Directive (2008/98/EC): sets out the basic concepts, definitions, and principles of waste management as the "polluter pays principle" or the "waste hierarchy") (https://eur-lex.europa.eu/legal-content/EN/TXT/?uri=celex%3A32008L0098);
- Landfill Directive (99/31/EC): sets out guidelines to prevent or reduce as much as possible adverse effects on the environment (pollution of water, groundwater, soil, and air) and on the global environment (including the effect greenhouse) and the risks to human health throughout the life cycle of the waste landfill (https://eur-lex.europa.eu/legal-content/EN/TXT/?uri=CELEX%3A31999L0031);

- Green Public Procurement (GPP): is a process by which public authorities acquire goods, services and works with reduced environmental impact throughout their life cycle. As a result, European public authorities started to follow sustainability principles in public procurement, creating a new market for the use of recycled aggregates and the consequent minimization of landfill waste.

The Level(s) program was created in 2018 to help encourage sustainability in the construction industry in Europe, which is a new approach to assess the sustainability performance of buildings throughout their lifecycle, reducing their impact on the built environment. As a result, several indicators were created that can be adopted at different stages of the construction process and by different stakeholders (European Commission, 2021). In addition, the Level(s) program aims to rethink technology standards and organizational practices and establish new types of contracts across the construction industry value chain.

The Level(s) assessment levels provide an overview of best practices. They are divided into three thematic areas: "Resource use and environmental performance," "Health and comfort", and "Cost, value, and risk". In addition, it has six macro-objectives, among which "2. Resource-efficient and circular material life cycles," which have four performance indicators: 2.1 Bill of quantities, materials and lifespans, 2.2 Construction and Demolition waste, 2.3 Design for adaptability and renovation and 2.4 Design for deconstruction. Indicators highlight what types of information to collect and show how key points can be assessed.

For the European Commission (2021), Level(s) Indicator 2.2 aims to encourage professionals to systematically plan the reuse and recovery of materials through segregated collection during construction, renovation and demolition activities.

Usually, these wastes represent a considerable cost, mainly those resulting from the loss of material, the intensive use of workers for separation, the need for equipment for collection at the construction site, the expenses involved in transport, and the waste management processes. Therefore, in Indicator 2.2 the program hopes to strengthen the dialogue between construction professionals to agree on sustainability measures for CDW.

A significant project in the field of recycled concrete was the French National Project Recybéton, created in 2012 to promote the application of recycled aggregate to replace natural aggregate or as a raw material in cement production (https://www.pnrecybeton.fr/en/).

The project sought to study the ecosystem of recycling stakeholders and how the decision-making process about recycling should take place. The following aspects were analyzed: the legal environment, regulations, norms, and public incentives, links with the circular economy, the availability of the resource in terms of distance to the concrete production site, type of products, and quality of the RCA (Larrard and Colina, 2019).

11.4.1.3 Circular economy in construction

Various programs were created to circulate the economy, encourage the reuse of building components and materials after selective demolition, and stimulate the sale of these products. However, recycled concrete is still considered a niche product, and absorption is deficient from primary sources despite significant economic aspects. For the authors, policies and standards must be provided to accept recycled materials in the circular economy better. In general, there is an incentive for waste in general and concrete to be re-incorporated into the construction process (Luscuere et al., 2019).

The certification processes were created to favor recognizing a product as a sustainable "asset" and verify the material conditional on being sold. One example is the "Building as Material Bank" (BAMB) Project, created in 2015 to foster systemic change in the construction sector, focusing on sustainability. Through a tracking mechanism, principles and values were defined for materials, products, and systems to reuse the circular economy. In addition, more than 400 Materials Passports were cataloged, including products, components, or materials to facilitate future reuse at the construction, system, product, and material levels (Luscuere, 2017).

Also created was the European Circular Construction Alliance (ECCA), a project that aims to adopt the principles of the circular economy throughout the lifecycle of buildings. It has considered the phases from the extraction of construction materials, design and production, construction, operation/maintenance, retrofit, even demolition and material recycling (End Of Life—EOL). Circular construction is based on grouping activities and actors in the construction industry interested in similar themes, such as waste management (ECCA, 2017).

The Environmental Product Declarations (EPD) seal was also created, which is an environmental product declaration so that the products sold meet sustainability standards. EPD provides information about the environmental performance of products throughout their life, resulting in a series of advantages, both for producers and the entire production chain, which includes end-users (International EPD System, 2021). Furthermore, a Construction Product Regulation (CPR) establishes harmonized rules for the sale of construction products in the European market (European Commission, 2011).

11.4.2 Technological context

11.4.2.1 CDW generation

Recycled concrete can originate from waste in different situations and with very different technological characteristics, with demolition and losses during the execution of services, manufacturing, and transport of ready-mixed concrete, and precast concrete production.

The reuse of RCA produced by demolishing old buildings has become an option for preserving the environment by reusing inert materials, reducing natural materials, and disposal in landfills. Several countries have been looking for alternatives to the use of RCA and are evolving in their regulations, best practices, and research (Makul et al., 2021).

However, recycling waste generated through demolition is not always recommended. According to Bergmans et al. (2017), a predemolition waste inventory must be carried out to determine the quality and selectivity of potential waste. Demolition and renovation assessment audits should identify the potential for material that can be reused or recycled from deconstruction and removal. Material with risks or hazardous substances and how to reduce conflicts with the region where the construction is located must also be identified (European Commission, 2018).

Losses can also occur during operations with ready-mixed or fresh concrete. This loss usually happens with plants producing ready-mixed concrete, usually transported by concrete mixer trucks. Concrete loss can occur due to excessive transport time, failure to meet the quality criteria observed during material receipt, or concrete leftovers after a delivery (Kazaz & Ulubeyli, 2016). For this situation, it is more common for ready-mixed concrete (RMC) suppliers to provide a sedimentation deposit for unloading the material until it reaches a certain age that allows its reuse, recycling, or disposal. Another solution was identified by Tam and Tam (2007), who verified the use of concrete reclaimer technology within the concrete batching plant to reuse fresh concrete. Based on a test, the authors found that out of the total 2400 tons, around 1000 tons (42%) could be recycled as aggregate in a new concreting.

After transporting the concrete in a concrete mixer truck, the fresh concrete must be washed, and the aggregates can be removed mechanically with the addition of 1000 L of water for reuse in new mixtures (Vefago & Avellaneda, 2010). However, this water cannot be discarded into the environment without prior treatment due to many solid particles and its extremely high pH.

Sandrolini and Franzoni (2001) comment that the practice of recycling water in premixed concrete plants is made possible with the use of sedimentation basins, meeting adequate performance criteria. The sediment can be later disposed of in authorized landfills or recycled.

Finally, there is the actual situation of the concrete from the precast concrete components due to problems in handling the parts during the manufacturing, stripping, storage, transport, and assembly processes. However, considering the quality of concrete and the low contamination with other materials, such a recycling study shows interest from the manufacturers. Thus, associations of precast concrete manufacturers have joined together to disseminate good practices and seek strategies for reusing this high-quality waste (British Precast, 2013).

11.4.2.2 CDW characteristics

To be considered technically suitable for reuse, the CDW from deconstruction or disassembly processes must be considered inert. That is, it cannot cause any adverse environmental effects or harm human health. For International Union of Laboratories and Experts in Construction Materials, Systems, and Structures (RILEM), in addition to not having potentially harmful substances that are self-combusting or present biological hazards (RILEM, 1994).

Recycled concrete and masonry aggregates are classified for aggregates with a diameter of 4 mm or greater, according to RILEM (1994). Type I is classified as aggregate made from masonry rubble; type II aggregates refer to concrete rubble, and type III are aggregates containing concrete rubble and masonry.

In the case of EOL building constructions and demolitions in general, there is a considerable mixture of concrete waste with other materials, requiring classification and separation for reuse or recycling. In addition, it is typical for wood, steel, and mineral residues, such as concrete, to be adhered to with different coatings, such as paint, ceramic, or plaster. Braymand et al. (2017b) verified that RCA has primary or hardened mortar, influencing its behavior differently from natural aggregates.

For Bergmans et al. (2017), there is a lack of low-cost techniques for deconstruction and selective removal of "complex" materials, as selective removal still requires much human effort. In addition, there is often not enough knowledge about the potential presence of hazardous substances (Polychlorinated biphenyls (PCB) in containing transformer, mercury-containing lamps, and asbestos). Finally, the separation of gypsum and organic insulation materials (such as expanded polystyrene and polyurethane foam) from the stony fraction also makes high-grade recycling difficult for these authors.

Through the Recybéton Project, separation techniques based on mechanical procedures and chemical and physical principles were also studied. According to Larrard and Colina (2019), the methods tested in the laboratory were based on: mechanical wear, fragmentation, sandblasting, high-pressure waterjet, microwaves, thermal and chemical treatments. The authors noted that neither method was entirely satisfactory, as the aggregates were never completely clean or damaged. The most promising methods identified were the combined thermo-mechanical processes. However, this solution has only been studied at the laboratory level and not reproduced on an industrial scale.

11.4.2.3 Recycling plant on-site

For Goodship (2010), the use of RCA is particularly beneficial when it occurs in the exact location as the previous construction, mainly when it comprises a high proportion of concrete to be recycled, coming from "concrete-rich" components such as structure, slabs, or pavement. In this way, in processing the RCA in the construction site, collecting and transporting the material will be reduced. The infrastructure for the on-site recycling process needs space for segregation, processing, and classification of the recycled material. The equipment used is usually sieves, primary crushing or crushing equipment, conveyors, mobile bulk handling equipment, and a washing plant (Goodship, 2010).

Basic CDW processing involves the introduction of processing equipment such as crushers and sieves. For Kelly and McMaster (2017), demolition without waste separation induces recycled material for low-value applications, such as landfills. For these authors, it is possible to use wet processing equipment with washing technology that uses physical and water-based methods to separate the recycled material.

It may also be necessary to use equipment and segregation techniques differentiated according to the RCA. The recycling plant and equipment adoption must be consistent with customer and project requirements. Different contamination levels, fines content, and variability of input materials (excavation waste, railroad ballast, etc.) require an individual approach for each recycling plant. Plants can include different phases of waste processing, such as feeding, presorting, washing aggregates, removing contaminants, removing metals, and sorting aggregates.

The use of mobile equipment on-site and recycling at the site depends on additional restrictions, such as current legislation. In France, with a change in legislation induced by the Syndicat des Enterprises de Déconstruction, Dépollution et Recyclage (SNED) it became possible to install mobile equipment on construction sites (Mignon & Bréquel, 2017).

Another problem that can make CDW reuse difficult is the manual work required to separate the products that will be recycled. Incentives can be provided for the development of new technologies that automate the management of CDW. An interesting example is the Rebrick project (http://www.gamlemursten.eu/) in Denmark, which aims at the automated cleaning of clay bricks through a patented vibration-based process without water use. As burning new bricks consumes energy and resources, reusing bricks saves the environment 0.5 kg of carbon dioxide (CO_2) per brick (Hobbs & Adams, 2017).

11.4.2.4 Recycling centers

According to Mongeard and Dross (2016), the plants of large CDW recycling centers are usually close to urban centers where demolition generates a large amount of waste. Also, there is more significant pressure to search for recycled aggregates in urban centers, as the availability of natural aggregates usually requires long transport distances, which makes them relatively less competitive than secondary sources.

In the economic analysis of the cost of transporting materials, it was found that such analysis should be directed to the local conditions of each use. For example, Braymand et al. (2017a) analyzed several alternatives for transporting natural and recycled aggregates to a concrete production plant in France and determining the cost involved. For the authors, the combination of transport distances in the studied circuit and the replacement of aggregates in concrete generates the need for LCA analysis to verify the best alternative.

For Lauredan et al. (2017), mass recycling or reuse is currently necessary to recover CDW. However, economic criteria such as sales price or manufacturing cost have made it difficult to recover waste. In addition, the issue of quality must be associated so that the reuse of waste allows the production of high-quality materials. In this way, it can have a product that meets economic and technological terms. According to these authors, a recycling plant installed with equipment and quality control can produce materials with a more competitive sale price and very similar in high quality. Continuing, Lauredan et al. (2017) point out that the amount of waste received influences the size and complexity of the processing plant (for example, plants with an elaborate process receive an average of 177 kt per year, while the plant with a more straightforward process receives an average of 35 kt per year). For the authors, the economic advantage obtained with the commercialization of recycled products can guarantee the necessary investment to implement this type of plant.

For Lotfi et al. (2017), in addition to defining the technology needed, the business model to fully recycle and reuse concrete in a new material should involve a set of sustainable solutions that consider:

- Largely replace primary construction raw materials by recycling EOL concrete;
- Achieve a substantial reduction in road transport of construction materials;
- Create a severe cut in CO2 emissions from cement production.

11.4.3 Organizational context

11.4.3.1 Certifiable environmental standards

To increase sustainability in business processes, one of the strategies most used by companies is the use of certifiable and voluntary standards with high added value for society. In general, these norms seek to indicate patterns of behavior that contributed to applying the three fundamental aspects of sustainability: social, economic, and environmental. Among the practices indicated are recycling and the use of recycled materials. Among the primary standards related to the environment in force and applicable in the civil construction sector, the following can be mentioned:

- ISO 14000: Environmental Management Family: a leading group of standards that harmonize international standards for recognizing sustainability practices by requesting stakeholders (https://www.iso.org/iso-14001-environmental-management.html).
- ISO 14024:2018: Environmental labels and declarations—Type I environmental labeling—Principles and procedures: establishes the principles and procedures for developing environmental labeling programs, including the selection of product categories, product environmental criteria, and product function characteristics, and to assess and demonstrate compliance (https://www.iso.org/standard/72458.html).
- ISO 14025:2006: Environmental labels and declarations—Type III environmental declarations—Principles and procedures: establishes the principles and specifies the procedures for developing environmental declaration programs (https://www.iso.org/standard/38131.html).
- ISO 20887:2020: Sustainability in buildings and civil engineering works—Design for disassembly and adaptability—Principles, requirements, and guidance: provides an overview of design principles and strategies aimed at disassembly and adaptability enterprise (https://www.iso.org/standard/69370.html).
- ISO 21930:2017: Sustainability in buildings and civil engineering works—Core rules for environmental product declarations and construction products and services: establishes a basic set of requirements to be considered CPR to develop an EPD for any construction product or service. Complements ISO 14,025 by providing specific requirements for the EPD of construction products and services (https://www.iso.org/standard/61694.html).

In addition to these standards, voluntary certificates attesting to the sustainability of the enterprise during its life cycle are widely used, such as the Building Research Establishment Environmental Assessment Method (BREEAM) of the Building Research Establishment (BRE), Leadership in Energy and Environmental Design for New Construction and Major Renovation (LEED NC) of the Green Building Council (GBC), Haute Qualité Environnementale (HQE) Association and others.

11.4.3.2 *Organizational partnerships*

Recycling CDW requires technological knowledge and the availability of assets to meet the goals. Thus, there is a grouping of organizations interested in sharing best practices and solutions to meet the needs of specialized service providers.

One example is the Association of Cities and Regions for Sustainable Resource Management (ACR+) (https://www.acrplus.org/en/activities/previous-projects) formed by local authorities and networks of European cities linked with nongovernmental organizations, academics, private partners, and consultancies. The international network of cities and regions aims to promote sustainable resource management, including collaborating to optimize the cycle of primary and secondary raw materials (resources, products, and waste).

Another solution suggested by Debacker et al. (2017) is the contracting of companies specializing in the provision of services of companies for crushing and recycling materials at the construction site itself. This solution was possible due to the Hiser project (http://www.hiserproject.eu/), which indicated the need to develop technological and nontechnological solutions to ensure greater efficiency in the recovery of priority waste streams. Among the proposed solutions, the Tracimat subproject (https://www.tracimat.be/) was conceived, aiming to integrate demolition and dismantling contractors, granulate producers for recycling, construction companies, and consultants. Its main objective is to obtain recyclable and reusable materials from the demolition process of buildings and infrastructure and forward them to the recycling process. The Tracimat seal attests to the selective demolition process, identifies the materials obtained, and tracks the flows of materials released during demolition from the construction site to the waste processor. With the operation in a reference work, there was a decrease in waste treatment costs (operational), that is, lower fees paid by demolition companies to crushers, thanks to a higher value of separate fractions of concrete and masonry on-site and the (local) demand assumed for this (Debacker et al., 2017).

A study under the RecyHouse Project of the Belgian Building Research Institute (Center Scientifique et Technique de la Construction—CSTC, 2002, Vrijders et al., 2017) verified that concrete with recycled aggregates on an industrial scale could meet all technical requirements. However, two critical factors for the successful use of recycled aggregates in concrete must be considered: the recycled aggregate quality and the stable supply of aggregates continuously to maintain the production process (with no shortage of supply). This situation is critical, especially if the concrete producer is a company other than the recycling company.

Thus, Vrijders et al. (2017) propose a possible solution for developing a cooperation model between the different actors in the recycling chain: the demolition contractor, the concrete recycling company (crusher), and the concrete producer. Table 11.4 presents an overview of barriers and benefits that can occur for the three different stakeholders.

For the authors, this chain provides benefits and profits if everyone agrees to adopt a more holistic approach. It would mean that the demolition contractor does some 'selective selection,' identifying potential buildings, structures, and parts of the structure of high-quality concrete that can be selectively demolished. Then the recycling

Table 11.4 Overview of barriers and potential benefits in short-chain management.

Stakeholders	Barriers	Potential benefits
Demolition contractor	Work more selectively costs more time and money	Lower disposal fee at crusher
Recycling company	Unsure sources of concrete, quality control and improvement necessary	Fix supply of high-level concrete debris, fix client for high-level RCA
Concrete company	Uncertain supply (volume and quality)	Replacement of virgin aggregate at lower cost

Reproduced from Vrijders, J., Dooms, B., & Itterbeeck, P. V. (2017). Putting recycled concrete into practice in Belgium: From case studies over chain management to standards and quality assurance. In *International HISER conference on advances in recycling and management of construction and demolition waste 21–23 June 2017* (pp. 286–290), Delft University of Technology, Delft, The Netherlands. http://resolver.tudelft.nl/uuid:94a004aa-d118-46a0-950c-02e4688971b1.

company can start the recycling process and make quality recycled aggregates from them. The origin of recycled concrete aggregates becomes known, increasing confidence in recycled concrete, which the third party in the business can produce.

11.4.3.3 CDW management

In several countries, CDW producers are currently required to develop waste management practices. Zorpas and Lasaridi (2013) mention that, based on environmental impacts and available waste management methods, preventing waste generation is the most efficient option, while disposal in landfills is the option with a minor reduction in environmental impact. Sustainable waste management practices act mainly to recover materials and energy to minimize the environmental impact and promote materials with recyclable potential and renewable energy sources (Umar et al., 2017). Fig. 11.3 presents a scheme of possibilities for managing construction waste, considering a most sustainable solution to a less sustainable one.

Thus, various solutions can be used considering the context in which the enterprise and the organization are inserted. European Commission (2021) presents a simplified relationship between the variables that need to be considered in waste management, integrating demolition and construction activities with the Waste Management Plan (WMP), and outcomes for elements, materials, and waste generated. Fig. 11.4 presents a simplified schematic of the keys elements of the demolition and construction phases and their relationship to WMP and the products to be obtained.

Good CDW management practices identified in the EU Construction and Demolition Waste Management Protocol (European Commission, 2016) can be followed for better waste management. The Protocol consists of five components: (1) Waste identification, source separation and collection; (2) Waste logistics; (3) Waste processing; (4) Quality management; and (5) Policy and framework conditions.

Elizar and Wibowo (2017) identified 20 factors that can improve construction waste management, grouping them into five groups: policy, asset, technology, human

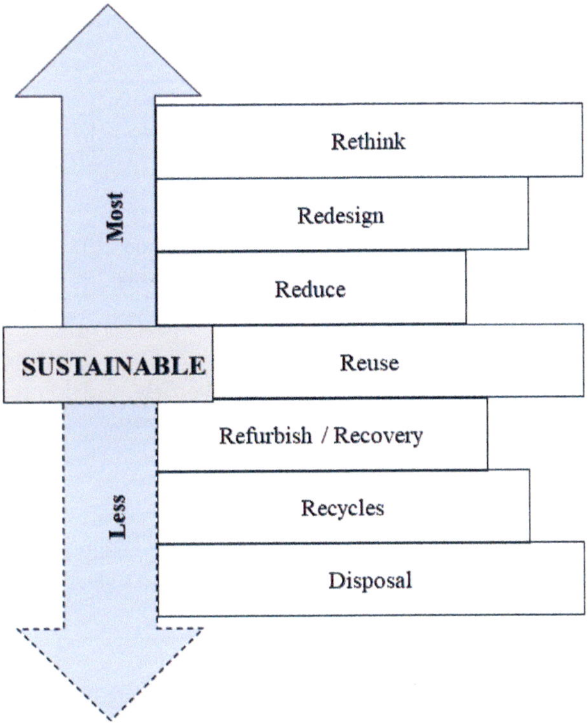

Figure 11.3 Construction waste management solutions.

resources, and knowledge. For the authors, identifying the characteristics of these factors can significantly improve construction waste management. The authors propose a management model based on the following variables and indicators of construction waste management:

- Policy: construction waste management rule, construction waste management certificate, construction waste management incentive, reward, and sanctions;
- Asset: construction waste management operational, human resource asset (creative, initiative, innovative), social asset (care of the environment), Equipment asset (management system);
- Technology: Construction waste management information, Database of waste construction, Low waste of construction method, innovative equipment;
- Human Resource: Degree of study, Skill Certificate, Commitment, Analysis capability; and
- Knowledge; waste construction of knowledge, Construction waste management of knowledge, Reduce, reuse and recycle of knowledge, Improve knowledge skill.

Thus, it can be concluded that the CDW management approach, and the RCA in particular, requires a systemic knowledge of the political, technical/technological, and organizational contexts of the main stakeholders. Investments are still needed to generate knowledge regarding the most sustainable practices of RCA and asset management of recycled concrete.

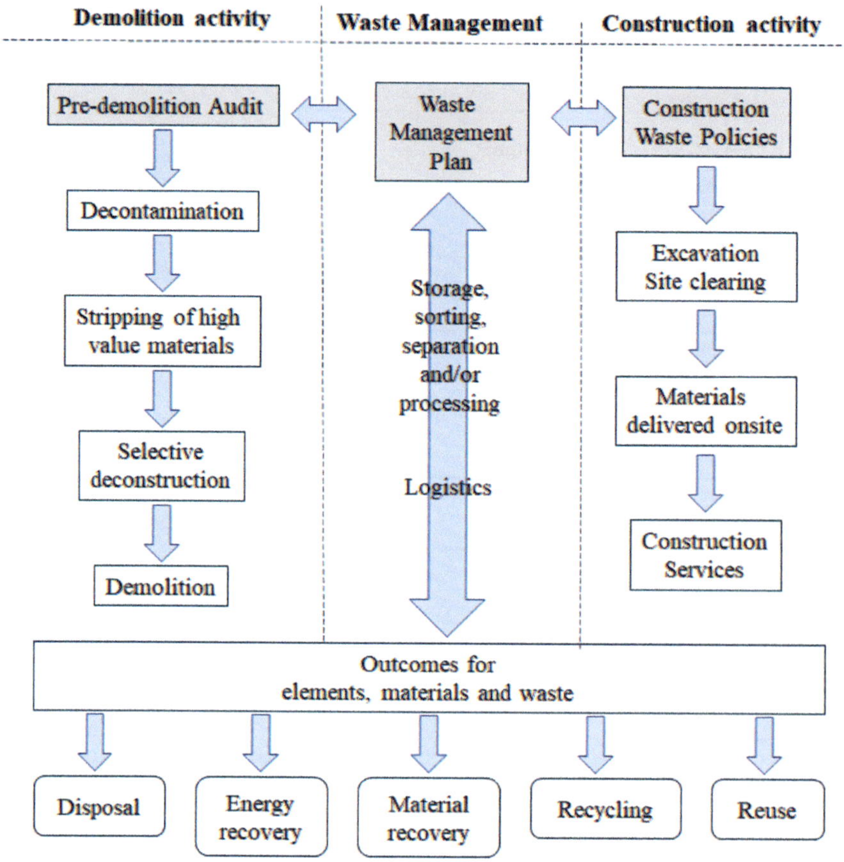

Figure 11.4 Keys elements of the WMP, construction activities and demolition activities Adapted from European Commission, 2016.

11.5 Questions for designing a recycled concrete asset management system

In this section, reflections on issues related to the management of recycled concrete assets will be proposed. These recommendations must include the specific issues of the business that will be undertaken. Thus, they should be considered as insights to improve the success of asset management. In addition, aspects related to the asset management macro-environment, human and technological resources, and production infrastructure will be addressed.

11.5.1 Asset management macro environment

Recycled concrete asset management is strongly influenced by policies and issues external to the organization, strategic internal issues, and stakeholders. About policies and issues external to the organization, what are the government policies for the sector? What is the acceptance level of products that use recycled concrete? Is there an adequate understanding of the sector's dynamics, especially about supply and demand?

Internal strategic questions demand answers to questions such as: is there an adequate understanding of the asset portfolio covered by the scope of the management system? Will the asset management system cover the operational interfaces of external contributors such as suppliers? Is the expected payback period realistic and appropriate to the context of the sector's dynamics? Are the organization's internal policies, plans, processes, procedures (including change control), and knowledge bases on strategic and operational information adequately documented and available to the targeted organizational levels?

Regarding stakeholders, have all the people and organizations that may impact or be impacted by the project been identified? What are the impacts of stakeholder expectations, and how to manage these expectations when conflicting? How to manage stakeholder engagement in business decisions and execution? Is financial and nonfinancial information available and takes into account the decision-making level of stakeholders?

Regulatory and regulatory bodies and public policymakers should be considered essential stakeholders since recycled concrete assets are strongly influenced by technical standards and public policies aimed at using recycled aggregate. In addition, the strong association of assets with environmental sustainability may also make it necessary to consider environmental groups and organizations as potential stakeholders.

11.5.2 Human and technological resources

Regarding human resources, do they have the experience and administrative and technical skills required by asset management? Are there mechanisms for improving skills in line with asset management objectives? The analysis of competencies must consider knowledge, skills, and attitudes, as well as specific situations in the regional and local market, such as, for example, the insufficiency of standards that specify technical requirements for recycled concrete. The new professional may require additional skills. In addition, the scope of the assets may require professionals with specific skills, such as those required by the facilities and utilities of assets that include the production of aggregates.

Another critical point is the team's commitment to the asset management objectives. Are there mechanisms to ensure adequate understanding of the role of each professional for the effectiveness of the management system? The benefits or losses provided by the commitment are identified for the company and the professional?

Increasing the team's commitment may require strategies aimed at behavioral changes, such as those related to aspects of sustainability.

Teams will be formed when appropriate professionals have been appointed based on availability, cost, experience, skills, knowledge, and skills.

About technological resources, they mainly concern management software and information and communication technology (ICT). Therefore, it is essential to determine the degree of contribution to the efficiency and effectiveness of information and communication processes and decisions.

Does the management software have instruments for managing operational actions and maintaining assets, making it possible to increase their useful life? Does it provide tools for monitoring, measuring, analyzing, and evaluating, including through performance indicators? Is it integrable with other company management systems, and does it have the flexibility for updates? Manage interfaces with subcontractors and suppliers?

About information and communication, do ICT enable them to be adequately planned, operationalized, and controlled? Does the information consider all phases of the recycled concrete's life cycle, and is it being collected, produced, distributed, stored, retrieved, and disposed of properly? This includes considering internal and external information, formal and informal, in-person and remotely, through physical and electronic means. It also includes issues related to the urgency, frequency, and format of the information, availability, reliability, accessibility, and suitability for the decisions required by the activities, and the ease of use of information technology.

Information and communication processes must provide answers to questions such as: Which instrument should be used? What is the purpose of communication? Who is responsible for informing and approving? Who should be consulted and informed? How often?

Technical information, including relating to state of the art on recycled concrete, has increased significantly in recent years. This requires information screening system and consequently specialized personnel. This information is strategic for improving operational and management processes and increasing the attractiveness of assets to stakeholders. They can also provide competitive advantages over competitors.

11.5.3 Production infrastructure

Facilities and equipment form the production infrastructure. About facilities, does the location and useable area of recycling plants and production areas enable the best return on investment? Are the work environments safe and healthy? Depending on the type of machinery and the volume of waste received, various facilities require storage areas. In addition, the fact that concrete waste has different dimensions increases the possibility of material slippage, which requires special care about safety.

Is the layout suitable for the process flow, from the truck's arrival with raw material to the dispatch of the waste to the subsequent processing centers and production areas? For example, recycling centers and production lines usually use large machinery. In addition to demanding large spaces, they often require additional spaces, such as those for maintenance stations and cooling systems for equipment.

Does facility design focus on operational efficiency and sustainability? For example, recycled concrete is strongly associated with improving sustainability. Thus, its feasibility depends on how sustainability aspects are considered throughout the entire lifecycle, including the installations. An exciting strategy can be applying the IFD (Industrial, Flexible, Detachable) concept in the design of the facilities and the construction of the facilities with low carbon footprint materials.

Regarding the equipment, are they the most suitable for the waste used? Do production practices respect the characteristics and operating parameters specified for the equipment? The production processes are designed according to the characteristics and specificities of the recycling typology (on the site itself, using mobile equipment, or in permanent installations). Does the planning include activities to prevent and correct equipment operation failures?

11.6 Conclusions

Waste from the manufacturing, transportation, and use of one of the most used materials globally, concrete, has made its recycling increasingly important due to its ability to collaborate to improve sustainability. In addition, recycled concrete has been used for different purposes, such as, for example, as a base for highways and the manufacture of precast concrete, constituting an essential asset for companies that produce construction materials. Both recycled concrete aggregates and derived products can be classified as an organization's "assets" and can enjoy the benefits of asset management.

The asset management of recycled concrete applications is a challenging topic due to the diversity of conceptual approaches related to asset management and the characteristics and specificities of recycled concrete. Therefore, the strategy considered by the authors as the most appropriate to contribute to the theme was to make an expanded approach to the aspects associated with asset management of recycled concrete applications. The proposed chapter considers the conceptualization, systems, standards, and decision-support tools more widespread for asset management. In addition, the political, technological, and organizational contextualization that influence the CDW management are presented, and essential issues for the design of the asset management system.

Finally, the authors hope that the effort and time devoted to this chapter can contribute to their enrichment and professional performance and encourage recycled concrete applications in the built environment.

Acknowledgments

This study was financed in part by the Coordenação de Aperfeiçoamento de Pessoal de Nível Superior—Brasil (CAPES)—Finance Code 001.

References

Alfatih, M. S., Leong, M. S., & Hee, L. M. (2015). Definition of engineering asset management: A review. *Applied Mechanics and Materials, 773–774*, 794–798. https://doi.org/10.4028/www.scientific.net/AMM.773-774.794

Bergmans, J., Dierckx, P., Regel, S. D., Eynde, A. V., Velde, P. V., & Broos, K. (2017). Semi-selective demolition: Current demolition practices in flanders. In *International HISER conference on advances in recycling and management of construction and demolition waste* (pp. 8–11). http://hiserproject.eu/images/mat_na_strone/Proceedings-_HISER_Conference.pdf.

Braymand, S., Fresnet, A. F., Serres, N., & Feugeas, F. (2017). Analyse du cycle de vie de bétons de granulats recyclés: Influence des paramètres de composition et du transport. In *RUGC17 Rencontres Universitaires de Génie Civil de l'AUGC* (pp. 114–117). https://hal.archives-ouvertes.fr/hal-02124845.

Braymand, S., Roux, S., Fares, H., Déodonne, K., & Feugeas, F. (2017). Separation and quantification of attached mortar in recycled concrete aggregates. *Waste and Biomass Valorization, 8*(5), 1393–1407. https://doi.org/10.1007/s12649-016-9771-2

British Precast. (2013). *Precast concrete: A resource efficiency action plan*. https://www.britishprecast.org/Precast/media/BPMediaLibrary/BP%20Sustainability/BP-REAP-October-2013.pdf.

British Standards Institution (BSI). (2008). *PAS 55-1: Specification for the optimized management of physical assets*.

Centre Scientifique et Technique de la Construction (CSTC). (2002). *Recyhouse: Opportunities for using recycled materials in the construction sector* (p. 20p). https://www.cstc.be/homepage/download.cfm?lang=en&dtype=publ&doc=Recyhouse%20xENG.pdf.

Debacker, W., De Regel, S., Amara, O., Lam, W. C., Bergemans, J., Broos, K., & Eynde, A. V. (2017). Assessing potential financial and environmental benefits of a newly developed waste traceability system in Flanders. In *International HISER conference on advances in recycling and management of construction and demolition waste 21-23 June 2017* (pp. 243–246). Delft, The Netherlands: Delft University of Technology. http://resolver.tudelft.nl/uuid:9bbfea03-825b-485c-807a-21de73e4b7c8.

El-Akruti, K. O., Dwight, R., Zhang, T., & Al-Marsumi, M. (2013). The role of life cycle cost in engineering asset management. In *8th world congress on engineering asset management, Hong Kong*. https://plu.mx/plum/a/?repo_url=https://ro.uow.edu.au/eispapers/2491.

Elizar, S., & Wibowo, M. A. (2017). *Model of construction waste management using AMOS-SEM for Indonesian infrastructure projects, 138*. https://doi.org/10.1051/matecconf/201713805005. MATEC Web of Conferences.

European Circular Construction Alliance (ECCA). (2017). *Circular economy: A global challenge for the construction industry*. http://circularconstruction.eu/.

European Commission. (2011). *Construction products regulation (CPR)*. https://ec.europa.eu/growth/sectors/construction/product-regulation_en.

European Commission. (2012). *Strategy for the sustainable competitiveness of the construction sector and its enterprises*. https://eur-lex.europa.eu/legal-content/EN/TXT/?uri=CELEX:52012DC0433&from=EN.

European Commission. (2016). *EU construction & demolition waste management Protocol*. https://ec.europa.eu/docsroom/documents/20509.

European Commission. (2018). *EU construction and demolition waste Protocol and guidelines*. https://ec.europa.eu/growth/content/eu-construction-and-demolition-waste-protocol-0_en.

European Commission. (2021). *Level(s) indicator 2.2: Construction and demolition waste and materials user manual: Introductory briefing, instructions and guidance*. http://susproc.jrc.ec.europa.eu/product-bureau//sites/default/files/2021-01/UM3_Indicator_2.2_v1.1_40pp.pdf.

Federal Highway Administration (FHWA). (2017). *Asset management overview*. https://www.fhwa.dot.gov/asset/if08008/amo_05.cfm.

Global Forum on Maintenance and Asset Management (GFMAM). (2014). *The asset management Landscape* (2nd. Edition). https://gfmam.org/webform/download-the-am-landscape-v2.

Goodship, V. (2010). Management, recycling and reuse of waste composites. In *Woodhead publishing in materials*. Woodhead Publishing Ltd.; CRC Press.

Heinrich, M., & Lang, W. (2019). *Material passports: Best practices*. European Union's Horizon, 2020 https://www.bamb2020.eu/wp-content/uploads/2019/02/BAMB_MaterialsPassports_BestPractice.pdf.

Hobbs, G., & Adams, K. (2017). Reuse of building products and materials: Barriers and opportunities. In *International HISER conference on advances in recycling and management of construction and demolition waste* (pp. 109–113). Delft, The Netherlands: Delft University of Technology. https://www.bamb2020.eu/wp-content/uploads/2017/07/Reuse-of-building-products-and-materials-barriers-and-opportunities.pdf.

International EPD System. (2021). *A short introduction to EPDs*. https://www.environdec.com/all-about-epds/the-epd.

International Organization for Standardization (ISO). (2006). *ISO 14040: Environmental management - Life cycle assessment - principles and framework*. https://www.iso.org/standard/37456.html.

International Organization for Standardization (ISO). (2014). *ISO 55000: Asset management - overview, principles and terminology* (p. 19). International Organization for Standardization. https://www.iso.org/standard/55088.html.

International Union of Laboratories and Experts in Construction Materials, Systems and Structures (RILEM). (1994). Recommendation 121-DRG: Specifications for concrete with recycled aggregates. *Materials and Structures, 27*, 557–559. https://link.springer.com/content/pdf/10.1007/BF02473217.pdf.

Kazaz, A., & Ulubeyli, S. (2016). Current methods for the utilization of the fresh concrete waste returned to batching plants. *Procedia Engineering, 161*, 42–46. https://doi.org/10.1016/j.proeng.2016.08.495

Kelly, E., & McMaster, L. (2017). Building waste: Construction materials from your waste. In *International HISER conference on advances in recycling and management of construction and demolition waste 21-23 June 2017* (pp. 78–81). Delft, The Netherlands: Delft University of Technology. http://resolver.tudelft.nl/uuid:4b727d3f-0a21-477e-bd57-456b26415bba.

Larrard, F. de, & Colina, H. (2019). *Concrete recycling: Research and practice*. CRC Press.

Lauredan, L. G., Ferro, G., & Cazacliu, B. (2017). It is possible to economically upcyle recycled aggregates. In *International HISER conference on advances in recycling and management of construction and demolition waste 21-23 June 2017* (pp. 135–138). Delft, The Netherlands: Delft University of Technology. https://repository.tudelft.nl/islandora/object/uuid%3Ad89170ea-3fd9-4ad6-b9f2-ad56863ecd5e.

Lotfi, S., Rem, P., Deja, J., & Mróz, R. (2017). An experimental study on the relation between input variables and output quality of a new concrete recycling process. *Construction and Building Materials, 137*, 128–140. https://doi.org/10.1016/j.conbuildmat.2017.01.085

Luscuere, L. M. (2017). Materials Passports: Optimising value recovery from materials. *Proceedings of the Institution of Civil Engineers - Waste and Resource Management, 170*(1), 25–28. https://doi.org/10.1680/jwarm.16.00016

Luscuere, L. M., Zanatta, R., Mulhall, D., Hus, S., Bostrom, J., & Elfstrom, L. (2019). *Building as materials Banks (BAMB): Operational material passaports*. https://www.bamb2020.eu/wp-content/uploads/2019/02/D7-Operational-materials-passports.pdf.

Makul, N., Fediuk, R., Amran, M., Zeyad, A. M., de Azevedo, A. R., Klyuev, S., Vatin, N., & Karelina, M. (2021). Capacity to develop recycled aggregate concrete in south East asia. *In Buildings, 11*(6). https://doi.org/10.3390/buildings11060234

Mignon, B., & Bréquel, H. (2017). Comparative case study of the construction and demolition waste practical treatment induced by different legislations in France and Belgium (Wallonia and Flanders). In *International HISER conference on advances in recycling and management of construction and demolition waste 21-23 June 2017* (pp. 319–322). Delft, The Netherlands: Delft University of Technology. https://repository.tudelft.nl/islandora/object/uuid%3A9bd4382e-1908-4919-bc33-0fb3f20bce13.

Mongeard, L., & Dross, A. (2016). La ressource en matériaux inertes recyclables dans le béton en France. Rapport de recherche. *Projet national Recybéton*, 72p. https://www.irex.asso.fr/wp-content/uploads/2019/04/R16RECY032-Theme-3-Ressources-en-inertes-recyclables-dans-les-betons-Lafarge-LC15RECY65.pdf.

Sandrolini, F., & Franzoni, E. (2001). Waste wash water recycling in ready-mixed concrete plants. *Cement and Concrete Research, 31*(3), 485–489. https://doi.org/10.1016/S0008-8846(00)00468-3

Tam, V. W. Y., & Tam, C. M. (2007). Economic comparison of recycling over-ordered fresh concrete: A case study approach. *Resources, Conservation and Recycling, 52*(2), 208–218. https://doi.org/10.1016/j.resconrec.2006.12.005

UK Government. (2013). *Construction 2025: Industrial strategy for construction - government and industry in partnership* (p. 78p). https://assets.publishing.service.gov.uk/government/uploads/system/uploads/attachment_data/file/210099/bis-13-955-construction-2025-industrial-strategy.pdf.

Umar, U. A., Shafiq, N., Malakahmad, A., Nuruddin, M. F., & Khamidi, M. F. (2017). A review on adoption of novel techniques in construction waste management and policy. *Journal of Material Cycles and Waste Management, 19*(4), 1361–1373. https://doi.org/10.1007/s10163-016-0534-8

Union Européenne des Producteurs de Granulats (UEPG). (2020). *European aggregates association: Annual review 2019-2020*. https://uepg.eu/mediatheque/media/UEPG-AR2019 2020_V13_(03082020)_spreads.pdf.

Vanier, D. J. (2001). Why industry needs asset management tools. *Journal of Computing in Civil Engineering, 15*(1). https://doi.org/10.1061/(ASCE)0887-3801 (2001)15:1(35).

Vefago, L. H. M., & Avellaneda, J. (2010). The environmental impact of recycled concrete. In *18th CIB world build congr W116 - smart sustain built environ* (pp. 181–192). https://www.irbnet.de/daten/iconda/CIB_DC24667.pdf.

Vrijders, J., Dooms, B., & Itterbeeck, P. V. (2017). Putting recycled concrete into practice in Belgium: From case studies over chain management to standards and quality assurance. In *International HISER conference on advances in recycling and management of construction and demolition waste 21-23 June 2017* (pp. 286–290). Delft, The Netherlands: Delft University of Technology. http://resolver.tudelft.nl/uuid:94a004aa-d118-46a0-950c-02e4688971b1.

Whittaker, M., Grigoriadis, K., Soutsos, M., Sha, W., Klinge, A., Paganoni, S., Casado, M., Brander, L., Scullin, M., Correia, R., Zerbi, T., Staiano, G., Merli, I., Ingrosso, I.,

Attanasio, A., & Largo, A. (2019). Reuse and recycling of CDW materials and structures in energy efficient prefabricated elements for building refurbishment and construction - RE4. In *Proceedings of the fifth international conference on sustainable construction materials and technologies (SCMT5)* (pp. 158–165). https://doi.org/10.5281/zenodo.3741344

Yuan, H. (2013). A SWOT analysis of successful construction waste management. *Journal of Cleaner Production, 39*, 1–8. https://doi.org/10.1016/j.jclepro.2012.08.016

Zorpas, A. A., & Lasaridi, K. (2013). Measuring waste prevention. *Waste Management, 33*(5), 1047–1056. https://doi.org/10.1016/j.wasman.2012.12.017

Index

Note: 'Page numbers followed by "*f*" indicate figures those followed by "*t*" indicate tables.'

CPI Antony Rowe
Eastbourne, UK
December 02, 2022